Natural and Social Sciences of Patagonia

Despite being an underpopulated region, Patagonia has attracted the attention of scientists since the very beginning of its settlement. From classical explorers such as Darwin or D´Orbigny, to modern science including nuclear and satellite developments, several disciplines have focused their efforts on unraveling Patagonia's natural and social history. Today, scientific and technological research is shifting from being shaped by northern agendas, towards more locally oriented objectives, such as the management of natural resources, the modernization of energy production and distribution, and the coexistence of rural and cosmopolitan social lifestyles. At the intersection of all these topics, new conflicts concerning the economy, human development, population, and the proper and long-standing planification and management of the landscape and its natural resources have emerged. These conflicts, of course, have also caught the attention of many interdisciplinary research groups.

This series is aimed at describing and discussing various aspects of this complex reality, but also at bridging the gaps between the scientific community and governments, policymakers, and society in general. The respective volumes will analyze and synthesize our knowledge of Patagonian biodiversity at different scales, from alleles, genes and species, to ecosystems and the biosphere, including its multilevel interactions. As humans cannot be viewed as being separate from biodiversity, the series' volumes will also share anthropological, archaeological, sociological and historical views of humanity, and highlight the wide range of benefits that ecosystems provide to humanity including provisioning, regulating and cultural services.

Andrea Raya Rey • Ulises Balza
Editors

Biodiversity and Conservation of Isla de los Estados (Chuani-sin)

A XXI Century Terra Incognita

Springer

Editors

Andrea Raya Rey
Centro Austral de Investigaciones Científicas (CADIC-CONICET), Ushuaia, Tierra del Fuego, Argentina

Instituto de Ciencias Polares Ambiente y Recursos Naturales
Universidad Nacional de Tierra del Fuego
Ushuaia, Tierra del Fuego, Argentina

Wildlife Conservation Society, Representación Argentina, Buenos Aires, Argentina

Ulises Balza
Centro Austral de Investigaciones Científicas (CADIC-CONICET)
Ushuaia, Tierra del Fuego, Argentina

ISSN 2662-3463 ISSN 2662-3471 (electronic)
Natural and Social Sciences of Patagonia
ISBN 978-3-032-13629-9 ISBN 978-3-032-13630-5 (eBook)
https://doi.org/10.1007/978-3-032-13630-5

This Springer imprint is published by the registered company Springer Nature Switzerland AG
The registered company address is: Gewerbestrasse 11, 6330 Cham, Switzerland

Preface

This book was first conceived under the skies of Isla de los Estados, during many long nights of fieldwork on this wild and extraordinary island. Nearly 30 years have passed since my first visit, yet the memory of that journey remains vivid. In January 1997, still an undergraduate student in Buenos Aires, I was invited to take part in a survey of pinnipeds along the island's rugged coast. Without hesitation, I said yes, knowing little about the place, but curious and eager.

My grandmother, a passionate reader and plant lover, had warned me of the island's dangers. She spoke of shipwrecks and storms, but also of lush vegetation and giant ferns (Fig. 1). I was intrigued.

Fig. 1 Myself on Roca Bay area next to one of the giant ferns that my grandmother told me about

We set sail from Ushuaia early in the new year. The 200-km journey took us through the Beagle Channel under a bright summer sun, with just enough wind to fill the sails. I remember feeling the thrill of watching seabirds and marine mammals while learning the basics of sailboat navigation, guided by Julio Brunet, the captain.

But adventure doesn't always come gently. Once we left the shelter of the channel and passed Isla Nueva, the sun dipped below the horizon, and black clouds quickly gathered. A fierce storm took hold. The boat shook violently. Inside, it was hard to move; outside, it was too dangerous for a young, inexperienced biologist. I lay in my bunk, listening to the roar of the wind and the shouts from the cockpit. I wasn't afraid, not even reminded of my grandmother's fearful tales, I was simply waiting for the storm to pass. Eventually, I fell asleep.

When I woke, the world had changed. Everything was silent. The boat no longer moved. Curious, I stepped outside. The scene that greeted me took my breath away. We were anchored in a calm, protected bay, surrounded by ancient forests that rolled down to the shoreline. An otter swam playfully nearby, and the cheerful calls of small birds filled the air. It was my first dawn on Isla de los Estados, and it was then that I fell in love with "The Island," as we came to call it.

Years have passed, and I have returned to Isla de los Estados countless times to study its seabirds and marine life. The silence of that first morning has given way to the vibrant sounds of thriving colonies, but the magic remains. Every visit reveals something new. The island is not easy, it is remote, wild, and shaped by relentless weather. Thick peat bogs challenge every step, the rain rarely stops, and the wind can be deafening. After 5 days of southern storms, someone once told me he could no longer hear his own thoughts. That is the kind of place this is, you either love it or you don't. But it never leaves you indifferent.

Although many books have been written about Isla de los Estados, a comprehensive volume dedicated to its unique wildlife was still missing. This book seeks to fill that gap. It is the result of years of dedicated research, brought to life by a community of students, volunteers, and fellow researchers, many of whom, like me, fell under the island's spell. Their contributions enrich every chapter.

We love this island deeply, and through these pages, we hope to share its wonder with others. Whether you are a scientist, a nature lover, or an explorer at heart, may you discover something in these chapters that connects you to the beauty, the resilience, and the wild spirit of Isla de los Estados.

Last but not least, I would like to thank a very special person: Dr. Adrián Schiavini. He invited me on that first journey to "The Island," shared with me his deep knowledge and passion for Tierra del Fuego's wildlife and conservation, and helped shape the researcher I am today. For that, I am forever grateful.

Ushuaia, Argentina Andrea Raya Rey

Isla de los Estados is a place where you feel transported into one of those nineteenth-century adventure stories we have all read, and it never loses that magic. Even after 13 field trips, each visit still feels like the first, overwhelming, full of discovery, impossible to fully grasp.

Its remoteness means you cannot just stop for a day or two, which turns it into a rare refuge for endangered aspects of ecological work, like serendipity and quiet contemplation. I would not be surprised if much of my understanding of the natural world (including my views on politics and conservation ethics) was shaped by my time on the island.

This book started as an inspiration and continued as a proposal. An inspiration from the field experience, and a proposal to all people who were privileged enough to work on the island to share their knowledge with us. All authors have contributed to fulfil our personal curiosity, and I am extremely grateful to all of them. My hope with this book is that it will help others to have an informed view of what the island actually looks like, beyond the breathtaking first impression.

This journey included experiences and lessons that go far beyond the scope of this book, but they all share a common thread: they were made possible thanks to Argentina's public institutions. I used specimens collected by Alberto Castellanos, 53 years before I was born, as models to test the size of hoods designed to reduce stress when handling birds of prey (Fig. 2). I doubt Castellanos ever imagined that his museum specimens would serve such a purpose, but those house at the Museo Argentino de Ciencias Naturales (i.e., the Natural History Museum of Buenos

Fig. 2 Adult Striated Caracara (*Phalcoboenus australis*) collected on Isla de los Estados by Alberto Castellanos in 1935, used as a model for testing hood sizes in the Museo Argentino de Ciencias Naturales in Buenos Aires, Argentina. This specimen of a poorly known taxa, guarded by a public institution for almost a century, will serve later to diminish the stress during manipulation of some of its descendants

Aires) were my very first point of contact with a species I spend years studying in the wild. Today, as public institutions in Argentina face increasing threats, it is more important than ever to recognize that the spirit of adventure, reflection, and first-hand learning from nature can form part of a country's scientific tradition, if that tradition is publicly supported. The hoods I used during all these years now decorate my home office, a quiet reminder of projects paused due to lack of funding. My detachment from publicly funded science is, I hope, only temporary. I still dream that Isla de los Estados will become a hub of science for everyone.

When I first began working on the island in 2013, thanks to my background with the raptor rehabilitation and rescue program of Fundación Caburé-í, I had enough experience handling birds of prey to feel confident that my colleague Nicolás Lois and I could catch the species I was after: the Striated Caracara. With some luck, and after a 3000-km road trip, we succeeded. But it wasn't just luck. I am deeply grateful to Andrea Raya Rey, who placed her trust in our (initially theoretical) abilities and supported our studies on the island year after year.

Ushuaia, Argentina Ulises Balza

The original version of the book has been revised. A correction to this book can be found at https://doi.org/10.1007/978-3-032-13630-5_16

Contents

Editors and Contributors

About the Editors

Andrea Raya Rey is a seabird ecologist with 30+ years dedicated to preserving Southern Ocean seabirds and their habitats. Passionate about exploration, she discovered Magellanic penguin and sooty shearwater colonies on Isla de los Estados. Active in national and international conservation groups, she collaborates with academia, NGOs, government, and private sectors to integrate human and natural dimensions in research. A member of the IUCN Penguin Specialist Group and Onashaga Commitment, she promotes conservation awareness to protect endangered species.

Ulises Balza is a Fuegian biologist committed to preserving endangered wildlife since 2007. He explored the threatened Striated Caracara on Isla de los Estados for both graduate and PhD theses, fostering collaborative research, especially with seabirds. He aims to engage communities in understanding evolving predator-prey relationships and resilience in wildlife populations through the conservation of wild landscapes. He is excited about sharing insights and promoting conservation awareness to inspire collective efforts for the protection of our unique and endangered species.

Contributors

Cecilia Paula Alonso ICPA-UNTDF, Ushuaia, Tierra del Fuego, Argentina

Carina I. Argañaraz Center of Excellence in Telesensing of Environment and Model Prediction of Severe events (CETEMPS), University of L'Aquila, L'Aquila, Italy

Sebastián Ávila Instituto de Arqueología, Facultad de Filosofía y Letras, Universidad de Buenos Aires/Consejo Nacional de Investigaciones Científicas y Técnicas (CONICET), Buenos Aires, Argentina

María Bagur CADIC-CONICET, Ushuaia, Tierra del Fuego, Argentina

Ulises Balza Centro Austral de Investigaciones Científicas (CADIC-CONICET), Ushuaia, Tierra del Fuego, Argentina

Instituto de Ciencias Polares, Ambiente y Recursos Naturales, Universidad Nacional de Tierra del Fuego, Ushuaia, Tierra del Fuego, Argentina

Fernando Biganzoli Facultad de Agronomía, Universidad de Buenos Aires, Buenos Aires, Argentina

Gonzalo Bravo IBIOMAR-CONICET, Puerto Madryn, Chubut, Argentina

Daniel O. Bruno Centro Austral de Investigaciones Científicas (CADIC-CONICET), Ushuaia, Tierra del Fuego, Argentina

Instituto de Ciencias Polares, Ambiente y Recursos Naturales, Universidad Nacional de Tierra del Fuego, Ushuaia, Tierra del Fuego, Argentina

Joaquín Cano Laboratorio de Ecología y Conservación de Vida Silvestre, Centro Austral de Investigaciones Científicas (CADIC-CONICET), Ushuaia, Tierra del Fuego, Argentina

Tomás Chalde Centro Austral de Investigaciones Científicas (CADIC-CONICET), Ushuaia, Tierra del Fuego, Argentina

Universidad Tecnológica Nacional, Facultad Regional Tierra del Fuego, Ushuaia, Tierra del Fuego, Argentina

Ignacio Luis Chiesa Centro Austral de Investigaciones Científicas (CADIC-CONICET), Ushuaia, Tierra del Fuego, Argentina

Leonardo Collado Dirección General de Ordenamiento Territorial y Gestión de Ambientes Forestales, Secretaría de Ambiente de la Provincia de Tierra del Fuego, Antártida e Islas del Atlántico Sur, Ushuaia, Tierra del Fuego, Argentina

Sergio M. Delpiani Centro Austral de Investigaciones Científicas (CADIC-CONICET), Ushuaia, Tierra del Fuego, Argentina

Instituto de Ciencias Polares, Ambiente y Recursos Naturales, Universidad Nacional de Tierra del Fuego, Antártida e Islas del Atlántico Sur (ICPA–UNTDF), Ushuaia, Tierra del Fuego, Argentina

Samanta Dodino Centro Austral de Investigaciones Científicas (CADIC-CONICET), Ushuaia, Tierra del Fuego, Argentina

Instituto de Desarrollo Económico e Innovación, Universidad Nacional de Tierra del Fuego, Ushuaia, Tierra del Fuego, Argentina

Instituto de Ciencias Polares, Ambiente y Recursos Naturales, Universidad Nacional de Tierra del Fuego, Ushuaia, Tierra del Fuego, Argentina

Brenda Lía Doti Universidad de Buenos Aires, Facultad de Ciencias Exactas y Naturales, Departamento de Biodiversidad y Biología Experimental (DBBE), Buenos Aires, Argentina

CONICET – Universidad de Buenos Aires, Instituto de Biodiversidad y Biología Experimental y Aplicada (IBBEA), Buenos Aires, Argentina

Marilén Fernández Centro Austral de Investigaciones Científicas (CADIC-CONICET), Ushuaia, Tierra del Fuego, Argentina

Cintia P. Fraysse Laboratorio de Ecología, Fisiología y Evolución de Organismos Acuáticos, Centro Austral de Investigaciones Científicas (CADIC), Consejo Nacional de Investigaciones Científicas y Técnicas (CONICET), Ushuaia, Tierra del Fuego, Argentina

Marina Güller Universidad de Buenos Aires, Facultad de Ciencias Exactas y Naturales, Departamento de Biodiversidad y Biología Experimental (DBBE), Buenos Aires, Argentina

Consejo Nacional de Investigaciones Científicas y Técnicas (CONICET), Buenos Aires, Argentina

Silvina Gutiérrez Programa Pampa Azul Bonaerense, Ministerio de Ambiente de la Provincia de Buenos Aires, La Plata, Buenos Aires, Argentina

Julieta Kaminsky CADIC-CONICET, Ushuaia, Tierra del Fuego, Argentina

Carlos Landa Instituto de Arqueología, Facultad de Filosofía y Letras, Universidad de Buenos Aires/Consejo Nacional de Investigaciones Científicas y Técnicas (CONICET), Buenos Aires, Argentina

Facundo Llompart Centro Austral de Investigaciones Científicas (CADIC-CONICET), Ushuaia, Tierra del Fuego, Argentina

Instituto de Ciencias Polares, Ambiente y Recursos Naturales, Universidad Nacional de Tierra del Fuego, Antártida e Islas del Atlántico Sur (ICPA–UNTDF), Ushuaia, Tierra del Fuego, Argentina

Nicolás A. Lois Instituto de Ecología Genética y Evolución de Buenos Aires (IEGEBA-CONICET), Buenos Aires, Argentina

Departamento de Ciencias de la Atmosfera y de los Océanos, Facultad de Ciencias Exactas y Naturales, Universidad de Buenos Aires (DCAO-FCEyN-UBA), Buenos Aires, Argentina

María Eugenia Lopez Consejo Nacional de Investigaciones Científicas y Técnicas (CONICET), Ushuaia, Tierra del Fuego, Argentina

Dirección Regional Patagonia Austral, Administración de Parques Nacionales (APN), Ushuaia, Tierra del Fuego, Argentina

Universidad Nacional de Tierra del Fuego (UNTDF), Ushuaia, Tierra del Fuego, Argentina

Centro Austral de Investigaciones Científicas (CADIC), Ushuaia, Tierra del Fuego, Argentina

Sofía Calderón López Universidad de Buenos Aires, Facultad de Ciencias Exactas y Naturales, Departamento de Biodiversidad y Biología Experimental (DBBE), Buenos Aires, Argentina

CONICET – Universidad de Buenos Aires, Instituto de Biodiversidad y Biología Experimental y Aplicada (IBBEA), Buenos Aires, Argentina

Paula Romina Mansilla Instituto de Ciencias Polares, Ambiente y Recursos Naturales, Universidad Nacional de Tierra del Fuego, Antártida e Islas del Atlántico Sur (UNTDF), Ushuaia, Tierra del Fuego, Argentina

María Paz Martinoli CADIC-CONICET, Ushuaia, Tierra del Fuego, Argentina

Luciana Mestre Dirección General de Biodiversidad y Conservación, Secretaría de Ambiente de la Provincia de Tierra del Fuego, Antártida e Islas del Atlántico Sur, Ushuaia, Tierra del Fuego, Argentina

Emanuel Pereira Universidad de Buenos Aires, Facultad de Ciencias Exactas y Naturales, Departamento de Biodiversidad y Biología Experimental (DBBE), Buenos Aires, Argentina

CONICET – Universidad de Buenos Aires, Instituto de Biodiversidad y Biología Experimental y Aplicada (IBBEA), Buenos Aires, Argentina

Analía F. Pérez Laboratorio de Invertebrados Marinos, Centro de Ciencias Naturales, Ambientales y Antropológicas (CCNAA), Universidad Maimónides, Consejo Nacional de Investigaciones Científicas y Técnicas (CONICET), Ciudad Autónoma de Buenos Aires, Argentina

Juan Federico Ponce Centro Austral de Investigaciones Científicas (CADIC-CONICET), Ushuaia, Tierra del Fuego, Argentina

Universidad Nacional de Tierra del Fuego, Antártida e Islas del Atlántico Sur, Instituto de Ciencias Polares, Ambiente y Recursos Naturales, Ushuaia, Tierra del Fuego, Argentina

Klemens Pütz Antarctic Research Trust, Bremervörde, Germany

Alejandra Raies Instituto de Arqueología, Facultad de Filosofía y Letras, Universidad de Buenos Aires/Consejo Nacional de Investigaciones Científicas y Técnicas (CONICET), Buenos Aires, Argentina

Andrea Raya Rey Centro Austral de Investigaciones Científicas (CADIC-CONICET), Ushuaia, Tierra del Fuego, Argentina

Instituto de Ciencias Polares, Ambiente y Recursos Naturales, Universidad Nacional de Tierra del Fuego, Ushuaia, Tierra del Fuego, Argentina

Wildlife Conservation Society, Representación Argentina, Buenos Aires, Argentina

Maríano Rodríguez ICPA-UNTDF, Ushuaia, Tierra del Fuego, Argentina

Javier H. Rojo Centro Austral de Investigaciones Científicas (CADIC-CONICET), Ushuaia, Tierra del Fuego, Argentina

Amira Salom Consejo Nacional de Investigaciones Científicas y Técnicas (CONICET), Ushuaia, Tierra del Fuego, Argentina

Laboratorio de Ecología y Conservación de Vida Silvestre, Centro Austral de Investigaciones Científicas (CADIC-CONICET), Ushuaia, Tierra del Fuego, Argentina

Julián Santiago Universidad de Buenos Aires, Facultad de Ciencias Exactas y Naturales, Departamento de Biodiversidad y Biología Experimental (DBBE), Buenos Aires, Argentina

Adolfina Savoretti CIT-Tierra del Fuego-CONICET, Río Grande, Tierra del Fuego, Argentina

ICPA-UNTDF, Ushuaia, Tierra del Fuego, Argentina

Adrián Schiavini Laboratorio de Ecología y Conservación de Vida Silvestre, Centro Austral de Investigaciones Científicas (CADIC-CONICET), Ushuaia, Tierra del Fuego, Argentina

Universidad Nacional De Tierra Del Fuego, AeIAS (UNTDF), Ushuaia, Tierra del Fuego, Argentina

Wildlife Conservation Society Representación Argentina, Buenos Aires, Argentina

Francisco J. Sola Consejo Nacional de Investigaciones Científicas y Técnicas (CONICET), Ushuaia, Tierra del Fuego, Argentina

Universidad Nacional de Tierra del Fuego (UNTDF), Ushuaia, Tierra del Fuego, Argentina

Daniel Testoni Parque Nacional Los Glaciares, Administración de Parques Nacionales (PNLG-APN), El Calafate, Santa Cruz, Argentina

Angélica Montserrat Tivoli Centro Austral de Investigaciones Científicas (CADIC-CONICET), Ushuaia, Tierra del Fuego, Argentina

Alejandro Eduardo Jorge Valenzuela Instituto de Ciencias Polares, Ambiente y Recursos Naturales, Universidad Nacional de Tierra del Fuego, Ushuaia, Tierra del Fuego, Argentina

Consejo Nacional de Investigaciones Científicas y Técnicas (CONICET), Buenos Aires, Argentina

Otter Specialist Group, Species Survival Commission, International Union for Conservation of Nature, Gland, Switzerland

Atilio Francisco J. Zangrando Centro Austral de Investigaciones Científicas (CADIC-CONICET), Ushuaia, Tierra del Fuego, Argentina

Diego G. Zelaya Universidad de Buenos Aires, Facultad de Ciencias Exactas y Naturales, Departamento de Biodiversidad y Biología Experimental (DBBE), Buenos Aires, Argentina

Consejo Nacional de Investigaciones Científicas y Técnicas (CONICET), Buenos Aires, Argentina

Chapter 1
At the Edge of the Known World: Rediscovering Isla de los Estados (Chuani-sin)

Andrea Raya Rey and Ulises Balza

Abstract The wind strikes first. Cold, relentless, and full of salt, it arrives long before land is visible, whispering across the deck like a warning or a welcome. Then, through the mist, Isla de los Estados rises jagged peaks carved by ice and time, forests clinging to cliffs, seabirds wheeling above crashing surf. At 54 degrees south, east of Tierra del Fuego, the island feels more like a secret than a destination, forgotten by many, truly known by few.

Keywords Subantarctic ecosystem · Biodiversity and conservation · Invasive species · Climate change and resilience

> December 17th, 1832-A little after noon we doubled Cape St. Diego, and entered the famous strait of Le Maire. We kept close to the Fuegian shore, but the outline of the rugged, inhospitable Statenland was visible amidst the clouds. Charles Darwin. The voyage of the Beagle. Chapter X

The wind strikes first. Cold, relentless, and full of salt, it arrives long before land is visible, whispering across the deck like a warning or a welcome. Then, through the mist, Isla de los Estados rises jagged peaks carved by ice and time, forests clinging to cliffs, seabirds wheeling above crashing surf. At 54 degrees south, east of Tierra

A. Raya Rey (✉)
Centro Austral de Investigaciones Científicas (CADIC-CONICET), Ushuaia, Tierra del Fuego, Argentina

Instituto de Ciencias Polares, Ambiente y Recursos Naturales, Universidad Nacional de Tierra del Fuego, Ushuaia, Tierra del Fuego, Argentina

Wildlife Conservation Society, Representación Argentina, Buenos Aires, Argentina
e-mail: arayarey@cadic-conicet.gob.ar

U. Balza
Centro Austral de Investigaciones Científicas (CADIC-CONICET), Ushuaia, Tierra del Fuego, Argentina

A. Raya Rey, U. Balza (eds.), *Biodiversity and Conservation of Isla de los Estados (Chuani-sin)*, Natural and Social Sciences of Patagonia,
https://doi.org/10.1007/978-3-032-13630-5_1

del Fuego, the island feels more like a secret than a destination, forgotten by many, truly known by few.

For centuries, cartographers marked this land with uncertainty. Even today, it remains largely absent from the maps of global biodiversity, hidden by distance, weather, and the ghosts of its past. Known to the Yaghan people as *Chuani-sin* "Region of Abundance" and to sailors as a place of wrecks and refuge, Isla de los Estados has always stood at a crossroads between myth and reality. It is a place where history, isolation, and wildness converge.

Yet in the twenty-first century, as ecosystems collapse and the planet warms, this so-called *terra incognita* has never mattered more. Although the island may seem inhospitable to humans, the human footprint is clearly visible to anyone who visits. Beneath its mists lie fragile breeding grounds for many seabirds, windswept peatlands storing ancient carbon, and kelp forests teeming with life, but also invasive species, and, above all, our plain ignorance. This sentinel of the South Atlantic, this natural laboratory for understanding resilience, adaptation, and the consequences of neglect, must be urgently and fully dedicated to science and conservation.

This book is an invitation to rediscover Isla de los Estados, not only through the lens of science, but also through memory, connection, and care. From its barren ridges to its underwater forests, from the footsteps of indigenous peoples to the flight paths of seabirds, this island reminds us that the edge of the known world is not an end, but a beginning.

1.1 Island at the Edge: The Geographic and Ecological Significance of Isla de los Estados

Isla de los Estados stands as a sentinel at the southeastern tip of South America, only ~1000 kilometers from Antarctica. Often referred to as the "island at the edge," it marks the last outpost of the Andes, the easternmost extent of the Subantarctic Evergreen Forest, and a unique convergence zone of terrestrial and oceanic systems (Ponce and Fernández, Chap. 2, this volume). Its extraordinary position, beneath the Subantarctic Front, at the southern extreme of the Patagonian Shelf, and just 12 km from the shelf break, makes it a globally significant area for understanding climate change, ecological resilience, and biogeographic transitions (Orsi et al. 1995; Piola and Rivas 1997; Fig. 1.1).

The island's striking landscape is the result of over 150 million years of geological evolution, including tectonic uplift, volcanic activity, the formation and closure of an ancient sea, and repeated glaciations. Today, Isla de los Estados exhibits an alpine glacial terrain, with sharp peaks, broad valleys, knife-edged ridges, and a series of deep fjords and bays that support diverse ecosystems, including evergreen forests, grasslands, shrublands, and peat bogs (Ponce and Fernández, Chap. 2, this volume, Salom et al., Chap. 4, this volume). These ecosystems provide refuge to several native and endangered species, some of which reach their distributional

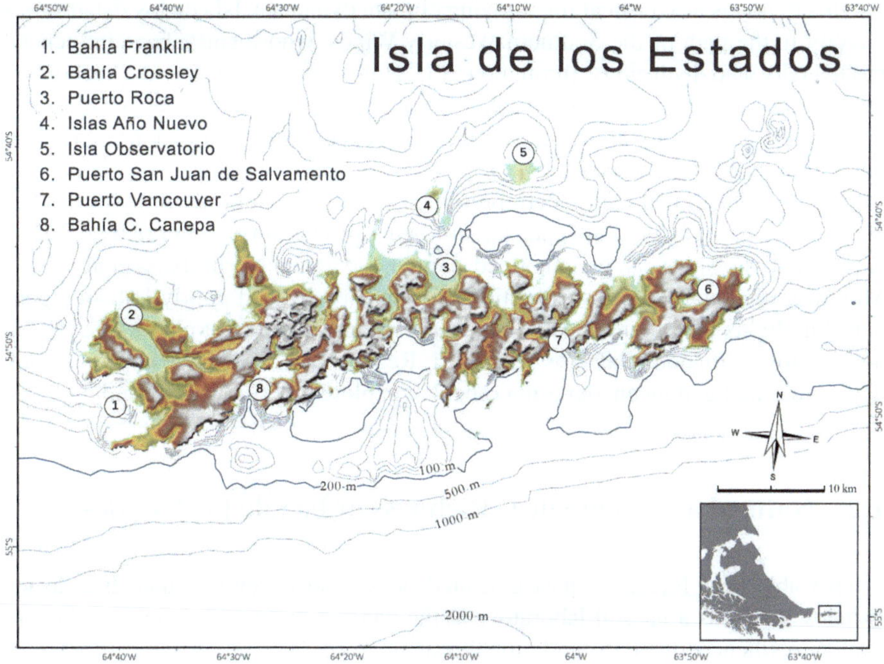

Fig. 1.1 Map of Isla de los Estados highlighting main toponyms and key oceanographic features. (© Valeria Falabella, Wildlife Conservation Society, Argentina Program)

limits on the island (Balza et al., Chap. 12, this volume, Schiavini and Valenzuela, Chap. 13, this volume). However, much remains to be studied. For instance, despite its close proximity to the mainland, the flora of Isla de los Estados may resemble that of a true oceanic island. Nevertheless, the island's updated floristic data, hopefully until now (Salom et al., Chap. 4, this volume), has not been easily accessible for global analyses (Guerrero et al. 2025; Roeble et al. 2024).

The island's waters are equally remarkable. Located at the intersection of the Antarctic Circumpolar Current and the Patagonian Shelf, oceanography in the area includes strong tidal currents (up to 4 knots) and frontal systems such as the Malvinas Current and shelf-break fronts, zones of high biological productivity (Sánchez and Ciechomski 1995). The Le Maire Strait, which separates Isla de los Estados from Isla Grande de Tierra del Fuego, is known for its dynamic hydrodynamics, with complex interactions between tidal flows and prevailing westerlies creating tide rips and whirlpools (Parker et al. 1997).

Such productivity underpins the island's unique marine and coastal ecosystems. For instance, kelp forests along the coast, among the southernmost in the world, have shown remarkable stability over time. Using Landsat imagery and historical field data from 1973, Friedlander et al. (2020) found little change in the abundance of kelp, sea urchins, and sea stars over 45 years, highlighting the resilience of these habitats despite broader environmental changes.

This region is also critical for studying climate dynamics. Isla de los Estados lies directly in the path of the Southern Westerly Winds, whose shifts have influenced biodiversity and hydrology throughout the Holocene (Sánchez and Ciechomski 1995; Ponce and Fernández, Chap. 2, this volume). In recent decades, increases in temperature and precipitation have been observed, consistent with intensification of the Southern Westerly Winds.

In this understudied and underrepresented study area, not only new species but also new biogeographical insights are still being proposed (e.g., Porta et al. 2024). Biodiversity hotspots can go unnoticed, even in the global North (Noss et al. 2015). Moreover, the ongoing mass extinction of island species and poorly-known organisms highlights the urgent need to better understand and protect places like Isla de los Estados (Fernández-Palacios et al. 2021; Régnier et al. 2015). We hope this book helps close that gap by encouraging others to explore and study the island.

1.2 Natural and Ecological Richness of Isla de los Estados

From stable kelp forests to past glaciated peaks and endemic fauna, Isla de los Estados stands as a natural laboratory at the edge of the world, a unique place to explore resilience, diversity, and global change.

1.2.1 Terrestrial Flora and Plant Communities

The flora of Isla de los Estados is shaped by its Subantarctic climate and isolation. Based on historical records and recent surveys, 173 vascular plant species have been documented, representing 50 families, most prominently Poaceae and Asteraceae (Salom et al., Chap. 4, this volume). Although some introduced species are present, native plants dominate, including regionally endemic and biogeographically important species. These plants are distributed across varied ecosystems such as evergreen forests, moorlands, and maritime grasslands. A new vegetation map, developed using drone imagery and high-resolution satellite data, provides the most detailed and up-to-date depiction of plant communities, reinforcing the island's importance as a botanical reserve and as a model for studying vegetation dynamics under changing conditions. Further, Savoretti and Martinoli (Chap. 6, this volume) presents the first preliminary checklist of mosses from Isla de los Estados, compiling 765 records and identifying 66 species. It lays the groundwork for future research on the island's bryophyte diversity.

1.2.2 Underwater Forests and Marine Richness

Beneath the island's turbulent seas lie equally rich and less explored ecosystems. Kelp forests, formed primarily by large brown macroalgae, and animal forests made up of sessile invertebrates, thrive in the cold, nutrient-rich waters (Kaminsky et al., Chap. 5, this volume). A diving expedition revealed over 170 species of macroalgae, mobile invertebrates, and fishes, including several new records for the area. These organisms act as ecosystem engineers, increasing habitat complexity and supporting diverse marine life. The underwater forests of Isla de los Estados are among the southernmost in the world, and their long-term stability, documented over decades, suggests notable ecological resilience.

1.2.3 Marine Invertebrate Diversity

Despite logistical challenges, surveys have begun to uncover the invertebrate biodiversity of Isla de los Estados. Chiesa et al. (Chap. 7, this volume) recorded 29 species of isopods and 22 species of amphipods from 18 sampling stations, most showing a Magellanic distribution. While the reported diversity is relatively low compared to other parts of the Magellan region, the underrepresentation of species, especially in deeper waters, suggests that many more species await discovery.

Starfish diversity also reflects this rich but poorly known marine life. Fraysse and Pérez (Chap. 8, this volume) documented 29 species of Asteroidea from five taxonomic orders, with Forcipulatida being the most common. Remarkably, *Cryptasterias brachiata* was reported as a new species record for the island.

Zelaya and Güller (Chap. 9, this volume) added to the picture with a comprehensive review of mollusks, reporting 166 species: 96 gastropods, 48 bivalves, and several others including cephalopods. More than 90% of these species are shared with Patagonia and/or the Malvinas Islands, placing the island firmly within the Magellan biogeographic province. However, with over half of the species represented by one or two records, it's clear that our knowledge remains far from complete.

1.2.4 Aquatic and Terrestrial Invertebrates

Macroinvertebrates on Isla de los Estados remain poorly known, despite their key ecological roles. Recent surveys have begun to uncover the island's rich but understudied invertebrate fauna, including endemic and Gondwanan-relict species. López et al. (Chap. 10, this volume) compiles current knowledge and new data on both aquatic and terrestrial macroinvertebrates, offering a first comprehensive overview of these critical communities.

1.2.5 *Freshwater and Coastal Fish*

Rojo et al. (Chap. 11, this volume) noted the stark contrast between the island's freshwater and marine fish diversity. Only one native freshwater species has been confirmed, while at least 16 coastal marine species and an additional 25 species from offshore waters have been recorded, some from specimens collected over a century ago. There is an urgent need for updated inventories to support conservation and ecosystem management in this data-deficient region, especially in the context of a regional-scale invasion by introduced salmonids (Pascual and Ciancio 2007).

1.2.6 *Terrestrial and Seabird Diversity*

Birdlife in Isla de los Estados is equally impressive. Approximately 79 species have been recorded, with about 40 breeding regularly (Balza et al., Chap. 12, this volume). The island supports globally important populations of penguins, petrels, and raptors, among others, and it also serves as a vital feeding area for pelagic seabirds such as albatrosses and shearwaters, making it a key area for avian conservation in the South Atlantic.

1.2.7 *Marine Mammals and Other Vertebrates*

Pinniped populations around the island have recovered remarkably in recent decades from past exploitation (Milano et al. 2020a, b). South American Fur Seals now exceed 9500 individuals, while South American Sea Lions and Southern Elephant Seals are also increasing (Schiavini and Valenzuela, Chap. 13, this volume). The Southern River Otter is confirmed to be breeding on the island, but its habitat is threatened by introduced deer and goats. Similarly, native rodents face competition and predation from invasive rats.

1.3 Isla de los Estados: A Land of Myths, Exploration, and Serendipitous Discovery

As mentioned above, Isla de los Estados remains one of the most mysterious and least explored regions of southern South America. Isolated by fierce seas, buffeted by relentless westerly winds, and cloaked in fog and rain for much of the year, its harsh beauty and remote geography have both protected and obscured its extraordinary natural and cultural richness. Few people have seen the island firsthand, and even fewer have studied it in depth. Yet those who have arrived, often by chance or necessity rather than proper design, have encountered a place where myths, history, and biodiversity intertwine.

Long before European explorers named and mapped the island, Indigenous groups from Tierra del Fuego included it in their oral traditions. Though the island lacks evidence of permanent settlement, archaeological remains show that hunter-gatherer-fisher peoples visited its shores seasonally, exploiting its abundant seabirds and marine mammals (Zangrando et al., Chap. 3, this volume). Each culture attributed different symbolic meanings to Isla de los Estados, shaping a tapestry of mythologies around a land they perceived as sacred, dangerous, or powerful.

The arrival of European expeditions in the seventeenth century brought the island into global awareness, often through a lens of peril and wonder. Captain James Cook, during his 1774 voyage, marveled at the dense wildlife of Observatorio Island, describing the penguins, shags, and seals that blanketed its rocky shores. Later, the island's aura of isolation and survival inspired Jules Verne to write *Le Phare du bout du monde* (1901), a novel set on the island's easternmost point, Cape San Juan. Verne's imagination, fueled by reports from early explorers and whalers, projected the island into literary legend as the "Lighthouse at the End of the World."

Yet despite its poetic and historical significance, scientific exploration of Isla de los Estados has remained sporadic and largely opportunistic. Much of what we know about its flora and fauna comes not from targeted campaigns, but from chance encounters: a stranded scientist waiting for the weather to clear, a diver joining a sailing expedition, or an ornithologist detouring from another mission. As this book reveals, many new records and even previously undescribed species and breeding grounds were found unexpectedly, highlighting how little systematic research has been carried out on the island.

The chapters included in this volume are the product of both perseverance and serendipity. From the underwater kelp forests to the remote seabird colonies surveyed by drone and on foot, discoveries often came during brief windows of opportunity. Despite the logistical challenges, the researchers were driven by a shared spirit of exploration and curiosity. What unites them all, biologists, archaeologists, oceanographers, is a deep sense of awe and a desire to reveal the hidden wonders of this forgotten land.

1.4 From Local Protection to Global Significance: Safeguarding Isla de los Estados

Recognized for its outstanding natural, historical, and scientific value, Isla de los Estados has been under provincial protection since 1991 as a *Reserva Provincial Ecológica, Histórica y Turística*, a designation that reflects its exceptional importance within Argentina. However, the ecological and conservation value of this remote island goes far beyond national borders. As one of the southernmost landmasses in South America, and lying beneath the Subantarctic Front, Isla de los

Estados is a critical site for understanding biodiversity, ecosystem functioning, and global environmental change.

In 2022, protection efforts expanded with the creation of the nearby *Área Natural Protegida Península Mitre* (Ley Provincial N°1461), which includes the marine and terrestrial areas surrounding Isla de los Estados and neighboring islets. This milestone represents an important step toward integrated conservation across land and sea in the sub-Antarctic region. Additionally, the nearby Argentine *Yaganes Marine Protected Area* and Chile's *Diego Ramírez–Paso Drake Marine Park* protect vast offshore ecosystems (Massardo 2020). These waters are known to be used by Southern Rockhopper Penguins from Isla de los Estados (Dodino et al. 2024; Rosciano et al. 2022), demonstrating the ecological connectivity between terrestrial colonies and oceanic feeding areas.

The regional, coordinated management of this area is crucial for conserving mobile species, from fish larvae to seabirds. However, legal protection alone is not enough. Many of these protected areas still lack concrete management plans, in-site staff, or adequate resources for effective implementation. Without targeted actions, the integrity of these unique ecosystems remains at risk.

Although Isla de los Estados is currently uninhabited and remote, it has a deep history of human disturbance. One of the most persistent threats comes from invasive mammals introduced by sealers, whalers, and sailors during the 18th and 19th centuries. Originally home to only two native terrestrial mammals, the island now hosts red deer, goats, rats, and other exotic species that alter plant communities, trample vegetation, and prey on native fauna. These invaders have reshaped entire ecosystems. Invasive species interact across trophic levels and ecosystems, compounding their ecological impacts. Long-term fieldwork, including vegetation surveys, diet studies, and camera traps, confirms the urgent need for invasive species control and habitat restoration.

Fisheries pose an additional, less visible but equally important threat. The productive waters surrounding Isla de los Estados lie at the intersection of major ocean currents and support abundant marine life. Yet the impact of fisheries, through bycatch, competition, or noise disturbance, on marine mammals, seabirds, and benthic communities remains largely unknown (Dellabianca et al. 2023). Addressing these gaps through dedicated studies is vital to ensure the conservation of both marine and coastal ecosystems.

Despite these pressures and complexities, Isla de los Estados remains one of the last relatively intact ecological systems in the region. Its unique location, where the Atlantic and Pacific Oceans meet, makes it a natural observatory for climate change, oceanographic shifts, and species' adaptations to extreme environments. The island provides habitat for rare and globally threatened species. Its forests, peatlands, fjords, and underwater kelp forests support complex ecological communities that are highly vulnerable.

This volume brings together decades of dispersed and often serendipitous research into a unified vision for Isla de los Estados. Long overlooked, this remote corner of the world holds lessons that reach far beyond its shores. In compiling these chapters, we attempt to rediscover the island *amidst the clouds*, lifting the veil on a place where nature endures and reveals its quiet resilience.

Safeguarding this singular island is no longer just a local or national challenge; it is part of our shared global responsibility to preserve the wild edges of the known world.

References

Dellabianca NA, Torres MA, Ordoñez C, Fioramonti N, Raya Rey A (2023) Marine protected areas in the southern south-west Atlantic: insights from marine top predator communities. Aquat Conserv Mar Freshwat Ecosyst 33(5):472–487. https://doi.org/10.1002/aqc.3935

Dodino S, Balza U, Riccialdelli L, Polito MJ, Pütz K, Raya Rey A (2024) Pre-molt dispersal and use of marine protected areas by Southern Rockhopper Penguins (Eudyptes chrysocome) at the southernmost oceanic regions of South America. Prog Oceanogr 229:103369. https://doi.org/10.1016/j.pocean.2024.103369

Fernández-Palacios JM, Kreft H, Irl SDH, Norder S, Ah-Peng C, Borges PAV, Burns KC, De Nascimento L, Meyer J-Y, Montes E, Drake DR (2021) Scientists' warning—the outstanding biodiversity of islands is in peril. Glob Ecol Conserv 31:e01847. https://doi.org/10.1016/j.gecco.2021.e01847

Friedlander AM, Ballesteros E, Bell TW, Caselle JE, Campagna C, Goodell W, Hüne M, Muñoz A, Salinas-de-León P, Sala E, Dayton PK (2020) Kelp forests at the end of the earth: 45 years later. PLoS One 15(3):e0229259. https://doi.org/10.1371/journal.pone.0229259

Guerrero PC, Contador T, Díaz A, Escobar C, Orlando J, Marín C, Medina P (2025) Southern Islands Vascular Flora (SIVFLORA) dataset: a global plant database from Southern Ocean islands. Sci Data 12(1):397. https://doi.org/10.1038/s41597-025-04702-9

Massardo F (2020) Reserva de la Biosfera Cabo de Hornos y Parque Marino Islas Diego Ramírez-Paso Drake. An Inst Patagon 48(3):39–44. https://doi.org/10.4067/S0718-686X2020000300039

Milano VN, Grandi MF, Schiavini ACM, Crespo EA (2020a) Recovery of south American fur seals from Fuegian Archipelago (Argentina). Mar Mamm Sci 36(3):1022–1032. https://doi.org/10.1111/mms.12686

Milano VN, Grandi MF, Schiavini ACM, Crespo EA (2020b) Sea lions (Otaria flavescens) from the end of the world: insights of a recovery. Polar Biol 43(6):695–706. https://doi.org/10.1007/s00300-020-02672-9

Noss RF, Platt WJ, Sorrie BA, Weakley AS, Means DB, Costanza J, Peet RK (2015) How global biodiversity hotspots may go unrecognized: lessons from the North American Coastal Plain. Divers Distrib 21(2):236–244. https://doi.org/10.1111/ddi.12278

Orsi AH, Whitworth T, Nowlin WD (1995) On the meridional extent and fronts of the Antarctic Circumpolar Current. Deep-Sea Res I Oceanogr Res Pap 42(5):641–673. https://doi.org/10.1016/0967-0637(95)00021-W

Parker G, Paterlini MC, Violante R (1997) El fondo marino. El Mar Argentino y Sus Recursos Pesqueros 1:65–87

Pascual MA, Ciancio JE (2007) Introduced anadromous salmonids in Patagonia: risks, uses, and a conservation paradox. In: Bert TM (ed) Ecological and genetic implications of aquaculture activities, vol 6. Springer, Dordrecht, pp 333–353. https://doi.org/10.1007/978-1-4020-6148-6_18

Piola AR, Rivas AL (1997) Corrientes en la plataforma continental. El Mar Argentino y Sus Recursos Pesqueros 1:119–132

Porta AO, Parra-Gómez A, Poy D, Kreps G, Mackenzie R, Martínez Pastur G, Fernández LD (2024) The southernmost record for a symphylan: Hanseniella guerreroi sp. nov. (Myriapoda, Scutigerellidae), an inhabitant of the Tierra del Fuego archipelago. Zoosystematics Evol 100(4):1569–1584. https://doi.org/10.3897/zse.100.133632

Régnier C, Achaz G, Lambert A, Cowie RH, Bouchet P, Fontaine B (2015) Mass extinction in poorly known taxa. Proc Natl Acad Sci 112(25):7761–7766. https://doi.org/10.1073/pnas.1502350112

Roeble L, Van Benthem KJ, Weigelt P, Kreft H, Knope ML, Mandel JR, Vargas P, Etienne RS, Valente L (2024) Island biogeography of the megadiverse plant family Asteraceae. Nat Commun 15(1):7276. https://doi.org/10.1038/s41467-024-51556-7

Rosciano NG, Pütz K, Polito MJ, Raya Rey A (2022) Where's the best supermarket deal? Female Southern Rockhopper Penguins (*Eudyptes chrysocome*) show variable foraging areas during the guard stage at Isla de los Estados, Argentina. Can J Zool 100(1):46–55. https://doi.org/10.1139/cjz-2021-0020

Sánchez RP, Ciechomski D (1995) Spawning and nursery grounds of pelagic fish species in the sea-shelf off Argentina and adjacent areas. Sci Mar (España) 59:3–4

Chapter 2
Geological and Paleoenvironmental Evolution of Isla de los Estados

Juan Federico Ponce and Marilén Fernández

Abstract Isla de los Estados exhibits an alpine-type glacial landscape characterized by sharp mountain peaks and knife-edged ridges, gently undulating lowlands, broad valleys, bays, and fjords. Its current appearance is the result of a long and complex geological history that began approximately 150 million years ago, including volcanic activity, the formation and closure of an ancient sea, mountain uplift, sea-level fluctuations, and the development of extensive glaciers during glaciations. Several global climatic events affected the island after the Last Glaciation, influencing the development of postglacial vegetation and the environment as a whole. These changes included variations in precipitation and temperature, mainly driven by shifts in the position and intensity of the Southern Westerly Winds (SSW). In the last decade, an increase in both temperature and precipitation has been observed, associated with a strengthening of the SWW.

Keywords Geodiversity · Tectonic evolution · Glacial dynamics · Sedimentology · Paleoclimate

J. F. Ponce (✉)
Centro Austral de Investigaciones Científicas (CADIC – CONICET), Ushuaia, Tierra del Fuego, Argentina

Universidad Nacional de Tierra del Fuego, Antártida e Islas del Atlántico Sur, Instituto de Ciencias Polares, Ambiente y Recursos Naturales, Ushuaia, Tierra del Fuego, Argentina
e-mail: jfponce@conicet.gov.ar

M. Fernández
Centro Austral de Investigaciones Científicas (CADIC – CONICET), Ushuaia, Tierra del Fuego, Argentina

A. Raya Rey, U. Balza (eds.), *Biodiversity and Conservation of Isla de los Estados (Chuani-sin)*, Natural and Social Sciences of Patagonia,
https://doi.org/10.1007/978-3-032-13630-5_2

2.1 Introduction

Isla de los Estados is located at the southernmost tip of South America, southeast of Isla Grande de Tierra del Fuego, Argentina, from which it is separated by the 30 km-wide Le Maire Strait. The island covers an area of 496 km^2, with a maximum length of 62 km in an east–west direction and an average width of 6 km, ranging from a minimum of 550 m (between Cook and Vancouver fjords) to a maximum of 17 km (between Capes San Antonio and Kempe) (Fig. 2.1). From a physiographic and geological standpoint, Isla de los Estados represents the southeasternmost extension of the Andes Cordillera.

The landscape of Isla de los Estados has been strongly shaped by past glacial activity (Ponce and Rabassa 2012). The island features a highly dissected coastline with numerous inlets, bays, and fjords (Fig. 2.1). A longitudinal mountain chain approximately 50 km in length defines the main axis of the island, with peaks reaching elevations between 400 and 800 m a.s.l. (Ponce and Fernández 2014).

Geotectonically, Isla de los Estados lies at the northernmost end of the Scotia tectonic plate, approximately 30 km south of the Magallanes–Fagnano fault system, which marks the boundary between the South American plate to the north and the Scotia plate to the south (Fig. 2.2). The island forms part of the Argentine Continental Shelf, which connects it with both the Malvinas Islands and Isla Grande de Tierra del Fuego. To the south, the island borders the continental slope, which reaches depths of nearly 1800 m only 20 km offshore.

Geological studies of Isla de los Estados are relatively limited. The first one was conducted by Harrington (1943), who identified the two main lithostratigraphic units exposed on the island. Dalziel et al. (1974a) revised some of Harrington's stratigraphic interpretations and proposed a different structural framework. Later, Caminos (1976) conducted research on the dynamometamorphism affecting the Jurassic volcanic series, including tuffs and lavas. Caminos and Nullo (1979) subsequently presented a complete stratigraphic framework for the island. Additional

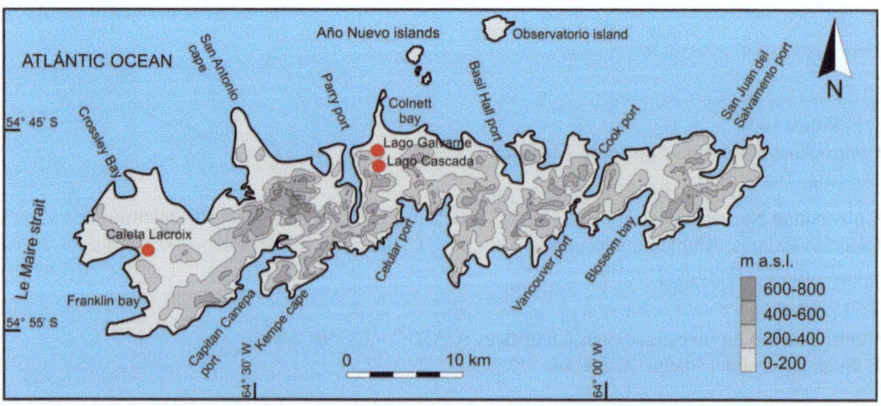

Fig. 2.1 Topographic map of Isla de los Estados

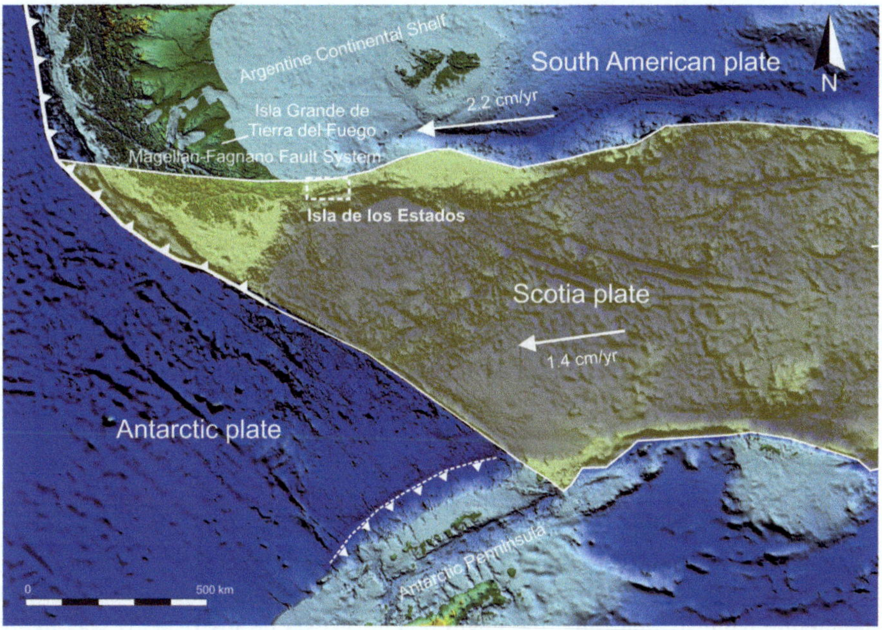

Fig. 2.2 Tectonic setting of Isla de los Estados

works (Dalziel and Elliot 1973; Dalziel et al. 1974b, 1975) have addressed the geotectonic position of Isla de los Estados within the Scotia Arc. Several studies have also focused on the geomorphology and paleoenvironmental evolution of the island. Ponce et al. (2010) conducted an inventory and classification of the fjords. Ponce and Rabassa (2012) analyzed its glacial geomorphology, presenting a map of the main glacial erosional landforms and a non-chronological model of ice cover. Ponce and Fernández (2014) outlined the paleoenvironmental history of the island, including information on geology, geomorphology, and Quaternary paleoenvironmental conditions. Additional paleoenvironmental studies have been conducted using pollen (Ponce et al. 2011a, 2017) and diatom analyses (Unkel et al. 2010; Fernández et al. 2013; Fernández 2013).

2.2 Pre-Quaternary Geological History

The oldest geological event recorded on Isla de los Estados was volcanic in nature and began approximately 150 million years ago. This event involved the eruption of large volumes of lava and ash from volcanic island arcs, resulting from a tectonic plate separation process. The corresponding geological unit is the Lemaire Formation, a stratified volcanic complex composed of tuffs and acidic lavas interbedded with tuffaceous breccias, tuffaceous sandstones, silt–claystones, black shales, and fine conglomerates dating from the Middle to Upper Jurassic (Caminos and Nullo 1979). Outcrops of this formation form the island's mountainous ridge and much of its southern coast (Figs. 2.3 and 2.5d).

The tectonic plate separation process also promoted the formation of an inland sea. As the sea expanded, substantial thicknesses of sediments and products of volcanic activity accumulated on its floor. These deposits correspond to the Beauvoir Formation, which consists of shale and slatey siltstones, fine grauvacas, massive quartz sandstones, calcareous siltstones, micritic limestones, and shales, dated from the Upper Jurassic to the Lower Cretaceous (Dalziel et al. 1974a). This formation exhibits pronounced diagenesis and incipient dynamic metamorphism, with well-developed foliation in certain areas (Caminos and Nullo 1979). Rocks of the Beauvoir Formation outcrop in the northwestern sector of Isla de los Estados and on the Año Nuevo Islands (Figs. 2.3 and 2.5c).

A significant change in tectonic activity occurred around 100 million years ago, when the regime shifted from plate separation to plate collision and compression. This transition led to the closure of the inland sea and initiation of the uplift of the Fuegian Andes, including the mountains ranges of Isla de los Estados. This period was marked by intense deformation of both rocks and sediments.

About 40 million years later, the Fuegian Andes had already emerged as a defined mountain chain. At that time, the environment on Isla de los Estados was likely continental, dominated by rivers and torrential streams flowing at the base of the rising mountains. Geological evidence of this environment is preserved in limited areas along the coast of Bahía Crossley, at the northwestern extremity of the island (Fig. 2.3). These deposits consist of conglomerates and sandstones arranged in a

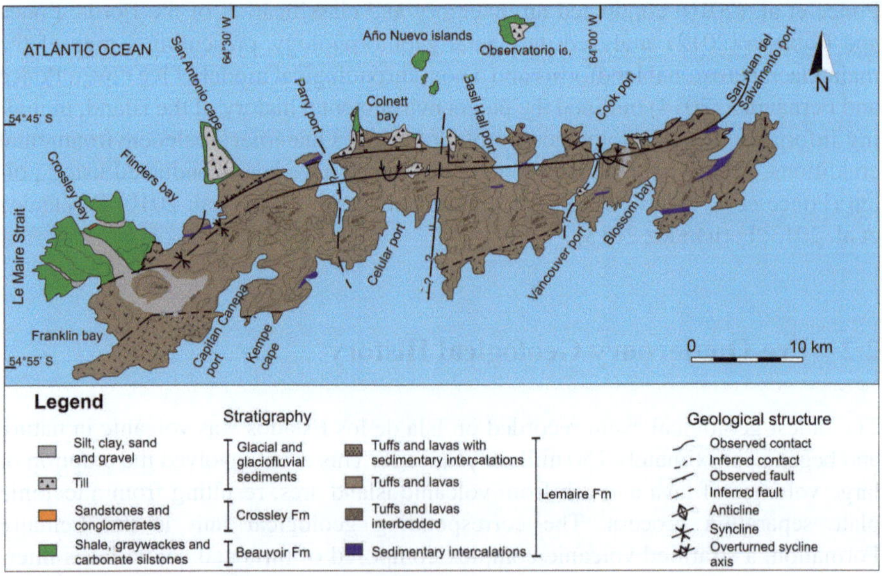

Fig. 2.3 Geological map of Isla de los Estados. (Modified from Ponce and Fernández (2014))

horizontal to subhorizontal layers, overlying the folded strata of the Beauvoir Formation (Ponce and Martínez 2007).

The compression, deformation, and uplift processes that shaped the mountains of the Isla de los Estados concluded around 23 million years ago, as rest of the Fuegian Andes. Since then, the island's geological evolution has been dominated by erosion driven by rivers, streams, and glaciers.

2.3 Structural Control on Topography

The compressional deformation resulting from tectonic plate convergence was the most significant event in the island's geological history, as it gave rise to the mountainous range. The island's current morphology is closely associated with the primary geological structure: a large syncline whose axis describes a broad S-shape with an overall east–west orientation, formed by the folding of the Lemaire and Beauvoir formations (Dalziel et al. 1974a, b). The intensity of this folding increases from west to east, strongly influencing the characteristics of the mountainous relief (Figs. 2.3 and 2.5a).

In the eastern sector of the island, the fold is tight and asymmetric, overturned toward the south, such that the strata of the southeastern limb are found in an inverted position. In the central portion, the fold becomes more open and gentler, though still somewhat asymmetric and steeply inclined to the south, with decreasing intensity toward the west (Ponce and Rabassa 2012). At the island's westernmost end, folding is absent and the strata lie in a subhorizontal position (Figs. 2.3 and 2.5a). Numerous secondary folds are developed within the broader synclinal structure.

The orientation of the central mountain range, as well as the elongation of the island itself, follows the strike of the folded structure. The folding event is assigned to the Middle Cretaceous (Caminos and Nullo 1979).

2.4 The Quaternary Glaciations

During the Quaternary, the Earth experienced several periods of global cooling that led to glacial periods, or "glaciations." These glaciations were separated by relatively warm intervals, known as "interglacial", characterized by climatic conditions more similar to those of the present. During the glaciations, large portions of the continental landmasses in both hemispheres, especially in high-latitude regions, were covered by extensive ice sheets.

The most recent glaciation to affect the planet reached its maximum extent in southern Patagonia approximately between 30,000 and 20,000 years ago (Davies et al. 2020). This event is known as the Last Glacial Maximum (LGM). During the

LGM, global sea level dropped between 120 and 140 m below its present level, exposing vast areas of continental shelves worldwide.

Throughout this major cold event, much of the Fuegian Archipelago was covered by glacial ice. On Isla de los Estados, the ice cover was likely nearly complete, with only the lowest parts of the currently emerged landscape and much of the submarine platform remaining ice-free (Ponce and Rabassa 2012). The glaciers were likely of local origin, nourished by the numerous cirques present on the island and possibly by a small, local ice cap that was physically disconnected from the Fuegian Andes ice field (Fig. 2.4). None of these cirques are currently occupied by ice, indicating that the present snowline lies above 800 m a.s.l., the island's highest elevation. However, it may lie only slightly higher, given the frequent occurrence of substantial perennial snow patches. The glaciers developed on Isla de los Estados are estimated to have had an average thickness of approximately 300 m along their central flow axes (Ponce and Rabassa 2012).

During the LGM, significant differences likely existed between the types of glaciers that developed on the northern and southern coasts of Isla de los Estados. The southern coast lies no more than 4 km from the −140 m isobath (the estimated minimum sea level during the LGM) and is very close to the continental slope. This suggests that the southern shoreline did not shift significantly from its present position. Consequently, glaciers along this coast were strongly influenced by marine conditions and likely by the Antarctic Circumpolar Current, which would have contributed high levels moisture, frequent precipitation, and persistent cold winds from the Antarctic sector. Under these conditions, glaciers in this region were probably more extensive than those in the north and were subjected to calving, a process by which ice blocks (icebergs) breaks off from glacier fronts into the sea.

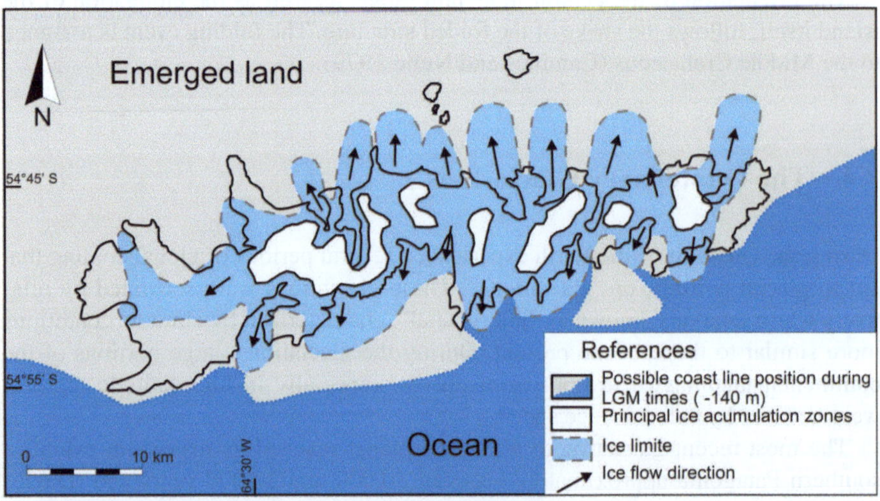

Fig. 2.4 Isla de los Estados during the Last Glacial Maximum times

In contrast, the geological context of the northern coast (Fig. 2.5e) is markedly different, as it borders the broad Argentine Continental Shelf. During the LGM, the northern shoreline was located approximately 100 km north of its present-day position. A large portion of the submarine platform would have been exposed, resulting in the formation of an expansive lowland plain that extended along the eastern coast of the Pampas and Patagonia. This ancient plain was continuous with Isla de los Estados to the south, creating a distinctively different paleogeographic setting. As a result, the northern sector of the island experienced more continental climatic conditions during the LGM. In the drier conditions, glaciers along the northern coast were likely less extensive than those on the southern side and terminated on the surface of this vast plain, where they gradually melted through surface ablation.

Although this model of ice coverage is presented in the context of the LGM, given the magnitude of the quaternary glaciations in Patagonia, it likely applies to each of the major glacial events that occurred over the past 1 million years.

Fig. 2.5 (**a**) Aerial photograph taken from west to east over Isla de los Estados, showing the low and gently undulating relief of the western sector, and the mountainous terrain of the central and eastern sectors of the island. (**b**) San Juan del Salvamento Fjord. (**c**) Shales of the Beauvoir Formation in Crossley Bay. (**d**) Volcanic rocks of the Lemaire Formation in the San Juan del Salvamento Fjord. (**e**) Cliffs in glacigenic sediments at Colnett Bay, northern coast. (**f**) Cirques in the central sector of the island

2.5 The Glacial Landscape

The surface of Isla de los Estados was primarily shaped by glacial activity, which carved out mountains and valleys during each of the glaciations. Over time, the mountains acquired increasingly sharp and jagged forms, resulting in classic alpine glacial landforms such as arêtes, pinnacles, and horns. As a result, an alpine-type glacial landscape was established across much of the island.

Glacial valleys are the most widespread landforms, covering approximately 42% of the island's surface area (Ponce and Rabassa 2012). These valleys are characterized by steep walls, relatively straight courses, and distinctive U-shaped cross-sections. Larger U-shaped valleys are found along the northern coast, likely due to the gentler relief in this sector compared to the steeper terrain of the southern part of the island, where glacial valleys are generally shorter. Today, many of these valley floors are occupied by lakes, ponds, and peat bogs.

Another abundant landform is the cirque (Fig. 2.5f), amphitheater-shaped, steep-walled depressions formed by glacial erosion at the heads of glacial valleys. Cirques cover about 7% of the island's surface, a notably high proportion for an island of intermediate size (Ponce and Rabassa 2012). Many cirques contain small lakes or ponds known as tarns. The elevation range of the cirque basins on Isla de los Estados is broad, extending from sea level up to approximately 460 m a.s.l. These areas are currently ice-free and lack perennial snowfields, and in some cases, they contain small lakes or ponds known as tarns.

The postglacial sea-level rise that followed the LGM led to the flooding of the island's lower glacial valleys, resulting in the formation of the fjords and bays that now characterize its coastline. Isla de los Estados contains a total of 18 fjords of varying dimensions, with an average length of 3.10 km, a maximum of 7.10 km and a minimum of 0.75 km (Fig. 2.5b). Their depths range from 11 to 190 m (Ponce et al. 2010). The fjords are bordered by steep lateral walls rising between 150 and 400 m. The longest fjords are found along the northern coast of the island, likely because the broad continental shelf exposed during the last glaciation, with its flat and extensive topography, favored greater longitudinal development of glacial valleys (Ponce et al. 2010).

2.6 Postglacial Sea-Level Rise and the Paleographic Evolution

The postglacial sea-level rise brought significant changes to the geographic configuration of the southern Fuegian Archipelago. A rapid reduction of the coastal plain located north of Isla de los Estados occurred between 14,000 and 11,500 cal. yr BP (Fig. 2.6a, b).

The disconnection of Isla de los Estados from the mainland and the formation of the Le Maire Strait took place when sea level rose above −84 m. Based on the data

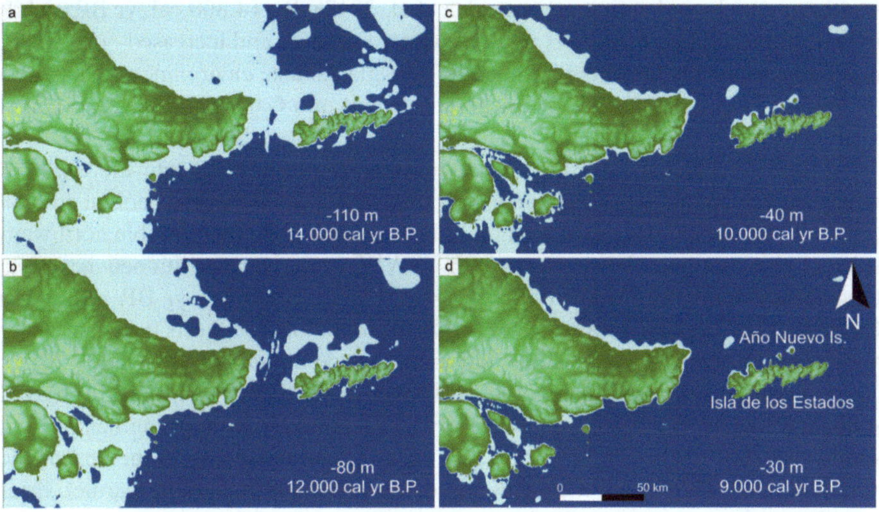

Fig. 2.6 Paleogeographic evolution of Isla de los Estados during the Late Glacial and Early Holocene times. (**a**) Configuration with a coastal plain present north of the island. (**b**) Formation of the Le Maire Strait. (**c**) Stage prior to the separation of the Año Nuevo Islands. (**d**) Separation of the Año Nuevo Islands and formation of the main fjords

presented by Guilderson et al. (2000), this likely occurred around 12,500 cal. yr BP (Fig. 2.6b). On the other hand, the formation of the Año Nuevo islands likely occurred around 10,000 cal. yr BP, when sea level exceeded elevations between −40 and −30 m below present sea level (Fig. 2.6c, d).

By approximately 9000 cal. yr BP, the geographic configuration of Isla de los Estados was likely very similar to its present-day form, including the presence of most of the fjords that currently characterize its coastline (Fig. 2.6d).

2.7 Paleoenvironmental Evolution during Late Glacial and Holocene Times

The paleoenvironmental reconstruction of Isla de los Estados is based on the analysis of several proxies such as pollen, diatoms, geochemical composition, and XRF data, obtained from peat profiles and a lacustrine sequence recovered through sediment cores collected at various sites across the island (Unkel et al. 2008, 2010; Ponce 2009; Fernández et al. 2013; Fernández 2013; Ponce et al. 2017).

Deglaciation on Isla de los Estados began shortly after 16,000 cal. yr BP, followed by rapid glacier retreat during a warming period marked by alternating dry and wet phases until around 12,800 cal. yr BP (Unkel et al. 2008). Between 16,000 and 15,200 cal. yr BP, at Lago Cascada (Fig. 2.1), the dominance of fragilaroid diatoms indicates cold, seasonally ice-covered conditions with strong physical weathering and wind activity (Fernández et al. 2013). By 14,800 cal. yr BP, a shift in diatom assemblages suggests reduced winter ice cover and increased open water, marking a transition to a more productive, post-glacial lake environment. This interpretation is supported by geochemical data from Lago Galvarne (Fig. 2.1) (Unkel et al. 2008).

Around 14,500 cal. yr BP, paleoenvironmental proxies indicate the onset of windy and/or arid to semi-arid conditions (Fig. 2.7), which favored aeolian erosion (Unkel et al. 2008; Ponce et al. 2011a, b; Björck et al. 2012). A possible northward shift of the Southern Westerlies Winds (SWW) may have positioned near Isla Grande de Tierra del Fuego. Between 14,400 and 12,800 cal. yr BP, at Laguna Cascada (Fig. 2.1), the dominance of *Aulacoseira* diatoms suggests turbulent water conditions driven by strong winds and increased precipitation (Fernández 2013). These findings align with palynological (Ponce 2009; Ponce et al. 2011a, b) and sedimentological (Unkel et al. 2008) evidence pointing to a strengthening and southward shift of the SWW during this interval, corresponding to the end of the Antarctic Cold Reversal (ACR) climatic event and the transition to warmer, drier and glacier-free conditions with increased aquatic productivity.

In Bahía Franklin (see Fig. 2.1), the pollen record indicates that between approximately 12,700 and 10,300 cal. yr BP, vegetation was dominated by herbaceous communities, with an absence of trees and only sparse shrubs (Ponce 2009; Ponce et al. 2011a, b) (Fig. 2.7). These assemblages likely developed in locally wet, low-slope areas with slow surface runoff and were influenced by meltwater from nearby glaciers, under moderately cold climatic conditions (Ponce 2009; Ponce et al. 2011a, b). Between 9200 and 5000 cal. yr BP, multiple lines of evidence, including stratigraphic data, geochemical analyses and diatom studies, suggest strong wind conditions, increased precipitation, and an intensification of surface runoff (Fig. 2.7) (Unkel et al. 2010).

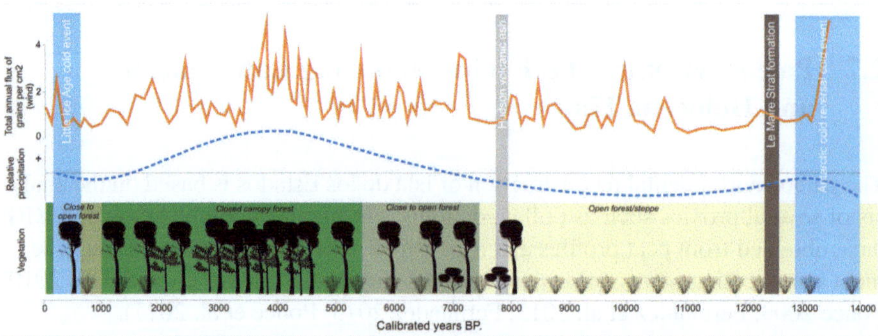

Fig. 2.7 Paleoenvironmental and paleoclimatic reconstruction of Isla de los Estados over the last 14,000 years. Top: curve of mineral grain fluxes per year, related to wind intensity, modified from Björck et al. (2012). Middle: curve of relative precipitation based on the interpretation of diatom and palynological studies in lakes and peat bogs. Bottom: vegetation evolution over time based on palynological data interpretation

The low concentration of *Nothofagus* pollen during this period indicates that forest development had not yet been established (Ponce 2009; Ponce et al. 2011a, b). However, by approximately 8400 cal. yr BP, pollen data from Caleta Lacroix reflect the expansion of closed-canopy *Nothofagus* forests, likely including *N. betuloides*, alongside shrubs and herbaceous vegetation. The expansion of *Nothofagus* from multiple ice-free refuges was not synchronous and possibly was conditioned by the altitudinal gradient, the topographic characteristics, and the soil type, as well as by the W–E and N–S precipitation gradients. The gradual expansion of forest communities in the south of Tierra del Fuego and the development of minerotrophic peat bogs seem to coincide with the gradual decrease in temperature in the region (Coronato et al. 2022), after the Holocene Thermal Maximum, which started around 11,500 cal. yr BP (Mulvaney et al. 2012) and with the low intensity of the SWW in these latitudes according to Björck et al. (2012).

Diatom assemblages from Laguna Cascada support this interpretation, indicating high lake levels and elevated humidity. Additionally, around 8700 cal. yr BP, a thick volcanic ash layer was deposited in the Laguna Cascada sediment core, likely corresponding to the first major eruption of the Hudson volcano (southern Patagonia), providing a key chronological marker within the regional paleoenvironmental record (Unkel et al. 2010).

Between approximately 5500 and 994 cal. yr BP, the pollen record from Bahía Franklin shows a significant shift in vegetation, with the dominance of the Subantarctic Evergreen Forest, primarily composed of *N. betuloides* and *Drimys winteri* (Fig. 2.7). This change is thought to reflect a shift toward more severe, very cold, hyper-humid and stormy conditions (Ponce 2009). Diatom data indicate that between 4900 cal. yr BP and 1885 cal. yr BP, very humid conditions prevailed, with increase vegetation development along the margins of Laguna Cascada. The growing abundance of diatom species associated with aquatic macrophytes and the onset of peat- formation processes suggest a gradual hydrological transition toward peatland environments. Around 4000 cal. yr BP records from Isla de los Estados show high tree concentration, indicating increased precipitation and a likely southward shift in the position of the SWW to around 55° S (Ponce 2009; Ponce et al. 2011a, b). Between 500 and 50 cal. yr BP, a decline in forest density on the island may reflect the coldest and driest conditions associated with the Little Ice Age (LIA) (Ponce et al. 2017).

2.8 Recent Climate Trends (1950–2024)

Meteorological information derived from ERA5 Land monthly reanalysis data, with a spatial resolution of 0.1 × 0.1 degrees (Hersbach et al. 2020), was used to analyze climatic trends over the past 75 years (1950–2024) in the central-western sector of Isla de los Estados (Fig. 2.8). The annual mean temperature data shows a marked increase between 1950 and 1980, followed by a period of relative stabilization from 1980 to 2005. This is succeeded by a decline between 2005 and 2015, and finally, a

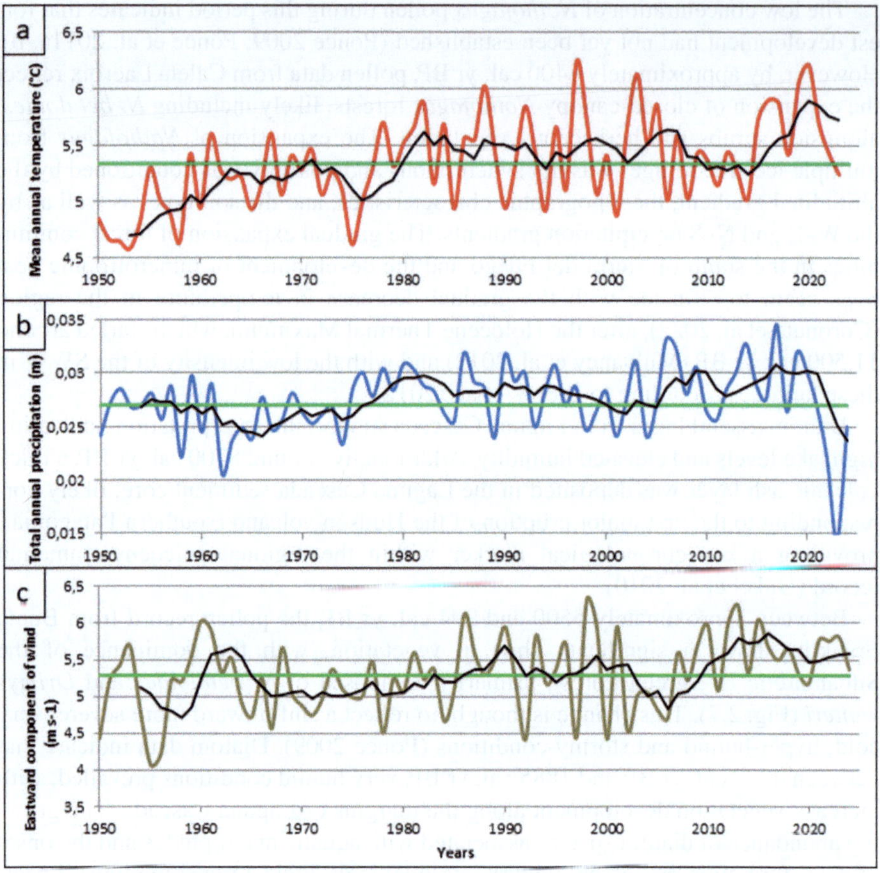

Fig. 2.8 Meteorological information of Isla de los Estados from 1950 to 2024. (**a**) ERA5-Land 2 m mean annual temperature curve (°C). (**b**) ERA5-Land total annual precipitation curve (m). (**c**) Eastward component of the 10 m wind (m s^{-1}). Black lines represent the 5 years moving mean and green lines represent the average values for the entire series

sharp increase in recent years. Total precipitation records indicate lower values between 1960 and 1975, higher values between 1980 and 1995, and again from 2006 to 2020, followed by a pronounced decrease until 2024. The westerly wind component does not exhibit a clear long-term trend since 1950; although, relatively low values are observed between 1950 and 1985, and higher values from 2010 onward.

Overall, the climatic data reveal a direct relationship between the intensity of the westerly winds and trends in both total precipitation and mean temperature. Periods of stronger westerly winds are associated with increases in both precipitation and temperature, whereas weaker winds correspond to reductions in these variables.

2.9 Conclusions

The present-day configuration of Isla de los Estados is the outcome of a complex geological history that began approximately 150 million years ago. Key events in this evolution include compressional deformation driven by tectonic plate interactions and the uplift of the mountain range. The island's current morphology reflects the structure of a large, east–west-trending fold with a gentle sigmoidal shape, generated during these tectonic processes.

Glacial activity during successive Quaternary glaciations played a central role in the final landscape modeling. Glaciers intensely eroded mountain ridges and valley floors, progressively sharpening peaks, deepening and widening valleys, and generating numerous closed depressions. Many of these depressions are now occupied by lakes, ponds, and peat bogs. Following the Last Glacial Maximum, postglacial sea-level rise led to the inundation of the lowest valley sections, forming the fjords and bays that characterize the island's modern coastline.

The establishment of *Nothofagus* forests on Isla de los Estados began approximately 8000 years ago, reaching a distribution similar to the present landscape around 5500 years ago. This vegetation expansion coincided with the onset of a cold and humid climatic regime, likely associated with a southward displacement of the Southern Westerlies to latitudes near the island.

Over the past few decades, the island has experienced marked climatic changes, including significant increases in temperature and precipitation. These trends may be linked to a strengthening or latitudinal shift of the westerly wind belt, consistent with broader patterns of regional climate change.

Future paleoenvironmental studies should prioritize high-resolution reconstructions of climate variability over the last millennium. Such research will be essential to improving our understanding of how global-scale climatic phenomena—such as the Medieval Climate Anomaly, the Little Ice Age, and the recent anthropogenic warming—have impacted the environments of Isla de los Estados and the climate of Fuegian region.

References

Björck S, Rundgren M, Ljung K, Unkel I, Wallin A (2012) Multy-proxy analysis of a peat bog on Isla de los Estados, easternmost Tierra del Fuego: a unique record of the variable Southern Hemisphere Westerlies since the last deglaciation. Quat Sci Rev 42:1–14

Caminos R (1976) Tobas y pórfidos dinamometamorfizados de la Isla de los Estados, Tierra del Fuego. Actas VI Congreso Geológico Argentino. II

Caminos R, Nullo F (1979) Descripción Geológica de la Hoja 67 e Isla de los Estados, Territorio Nacional de Tierra del Fuego, Antártida e Islas del Atlántico Sur. Servicio Geológico Nacional. Boletín 175:52

Coronato A, Borromei A, Ponce JF, Candel S, Musotto L, Fernández M, Laprida C, Mehl A, Montes A, San Martin C, Savoretti A, Cusminsky G, Gordillo S, Orgeira MJ (2022) Holocene environmental changes in the fuegian forest and steppe, Argentina. J S Am Earth Sci 119:103952

Dalziel IWD, Elliot DH (1973) Scotia arc and Antarctic margin. In: Nair AEM, Stehil FG (eds) Ocean basins and margins, vol 1. Plenum Press, New York, pp 171–245
Dalziel IWD, Caminos R, Palmer KF, Nullo FE, Casanova R (1974a) South extremity of Andes: geology of Isla de los Estados, Argentina, Tierra del Fuego. AAPG Bull 58:2502–2512
Dalziel IWD, de Wit MJ, Palmer KF (1974b) Fossil marginal basin in southern Andes. Nature 250:291–294
Dalziel IWD, Dott RH, Winn RD, Bruhn RL (1975) Tectonic relations of South Georgia Island to the southernmost Andes. Geol Soc Am Bull 86:1034–1040
Davies BJ, Darvill CM, Lovell H, Bendle JM, Dowdeswell JA, Fabel D, García JL, Geiger A, Glasserg NF, Gheorghiu DM, Harrison S, Hein A, Kaplan K, Martin J, Mendelov M, Palmer A, Pelto M, Rodés A, Sagredo E, Smedley R, Smellie J, Thorndycraft V (2020) The evolution of the Patagonian Ice Sheet from 35 ka to the present day (PATICE). Earth Sci Rev 204(2020):103152
Fernández M (2013) Los paleoambientes de Patagonia meridional, Tierra del Fuego e Isla de los Estados en los tiempos de las primeras ocupaciones humanas. Estudio basado en el análisis de diatomeas. Tesis doctoral inédita. Facultad de Cs. Naturales y Museo. Universidad Nacional de La Plata, La Plata, p 228
Fernández M, Björck S, Wolfarth B, Maidana NI, Unkel I, Van der Putter N (2013) Diatom assemblage changes in lacustrine sediments from Isla de los Estados, southernmost South America, in response to shifts in the southwesterly wind belt during the last deglaciation. J Paleolimnol 50(4):433–446
Guilderson TP, Burkle L, Hemming S, Peltier WR (2000) Late Pleistocene sea level variations derived from the Argentine shelf. G-Cubed 1(12). https://doi.org/10.1029/2000GC000098
Harrington HJ (1943) Observaciones geológicas en la Isla de los Estados. Anales Museo Argentino de Ciencias Naturales. Geologia 29:29–52
Hersbach H, Bell B, Berrisford P et al (2020) The ERA5 global reanalysis. Q J R Meteorol Soc 146:1999–2049. https://doi.org/10.1002/qj.3803
Mulvaney R, Abram NJ, Hindmarsh RCA, Arrowsmith C, Fleet L, Triest J, Sime LC, Alemany O, Foord S (2012) Recent Antarctic peninsula warming relative to Holocene climate and ice-shelf history. Nature 148:141–144
Ponce F (2009) Palinología y geomorfología del Cenozoico tardío de la Isla de los Estados. Tesis Doctoral Inédita. Universidad Nacional del Sur, Bahía Blanca
Ponce JF, Fernández M (2014) Climatic and environmental history of Isla de los Estados, Argentina. Editorial Springer, p 128
Ponce JF, Martínez O (2007) Hallazgo de depósitos sedimentarios postcretácicos en Bahía Crossley, Isla de los Estados, Tierra del Fuego. Rev Asoc Geol Argent 62(3):467–470
Ponce JF, Rabassa J (2012) Geomorfología glacial de la Isla de los Estados, Tierra del Fuego, Argentina. Rev Soc Geol Esp 25(1–2):67–84
Ponce JF, Rabassa JO, Martínez O (2010) Fiordos en Isla de los Estados: Descripción morfométrica y génesis de los únicos fiordos en la Patagonia Argentina. Rev Asoc Geol Argent 65(4):638–647
Ponce JF, Borromei AM, Rabassa JO, Martinez O (2011a) Late quaternary palaeoenvironmental change in western Staaten Island (54°.5° S; 64° W), Fueguian Archipelago. Quat Int 233:89–100
Ponce JF, Rabassa J, Coronato A, Borromei AM (2011b) Paleogeographic evolution of the Atlantic coast of Pampa and Patagonia since the last glacial maximum to the middle Holocene. Biol J Linn Soc 103:363–379
Ponce JF, Borromei AM, Menounos B, Rabassa J (2017) Late-Holocene and little ice age palaeoenvironmental change inferred from pollen analysis, Isla de los Estados, Argentina. Quat Int 442:26–34
Unkel I, Björck S, Wohlfarth B (2008) Deglacial environmental changes on Isla de los Estados (54.4 S), southeastern Tierra del Fuego. Quat Sci Rev 27:1541–1554
Unkel I, Fernandez M, Björck S, Kjung K, Wolfarth B (2010) Records of environmental changes during the Holocene from Isla de los Estados (54.4° S), southeastern Tierra del Fuego. Glob Planet Chang 74:99–113

Chapter 3
Human Impact on the Isla de los Estados Environment

Atilio Francisco J. Zangrando, Sebastián Ávila, Silvina Gutiérrez, Carlos Landa, Alejandra Raies, and Angélica Montserrat Tivoli

Abstract Archaeological and historical evidence reveals that human activity on Isla de los Estados has induced ecological disturbances at different scales. The island was intermittently occupied by hunter-gatherer-fisher groups between approximately 2800 and 1375 cal. yr BP possibly through seasonal movements for hunting of seabirds and marine mammals. Following the arrival of European exploration vessels in the seventeenth century, Western interventions on the island primarily involved timber exploitation, oil extraction from marine fauna, the construction of shelters, and the introduction of non-native species. This chapter examines the anthropogenic influences and factors that could have shaped the environment of Isla de los Estados along history.

Keywords Hunter-gatherer-fisher groups · Historical exploitation · Anthropogenic disturbance · Invasive species · Conservation management

A. F. J. Zangrando (✉) · A. M. Tivoli
Centro Austral de Investigaciones Científicas (CADIC – CONICET), Ushuaia, Tierra del Fuego, Argentina
e-mail: fzangrando@conicet.gov.ar; tivoli.angelica@conicet.gov.ar

S. Ávila · C. Landa · A. Raies
Instituto de Arqueología, Facultad de Filosofía y Letras, Universidad de Buenos Aires/Consejo Nacional de Investigaciones Científicas y Técnicas (CONICET), Buenos Aires, Argentina
e-mail: savila@unaj.edu.ar; carloslanda@conicet.gov.ar

S. Gutiérrez
Programa Pampa Azul Bonaerense, Ministerio de Ambiente de la Provincia de Buenos Aires, La Plata, Buenos Aires, Argentina
e-mail: silvina.gutierrez@hotmail.com

A. Raya Rey, U. Balza (eds.), *Biodiversity and Conservation of Isla de los Estados (Chuani-sin)*, Natural and Social Sciences of Patagonia,
https://doi.org/10.1007/978-3-032-13630-5_3

3.1 Introduction

Individuals, communities, and entire societies continuously make land-use decisions that can profoundly impact the environment (Redman 1999). These human influences span a broad spectrum of societies, from hunter–gatherer–fisher groups to globalized colonial systems (Butzer 1982). Numerous cases documented around the world suggest that the arrival of humans on non-colonized islands is often linked to significant environmental changes (Armstrong 1994; Erlandson and Rick 2008).

In the Fuegian archipelago, the geographic separation between the mainland, and the islands, as well as among the islands themselves, fostered the development of distinctive historical ecologies for many plant and animal species. As hunter-gatherer-fishers developed seafaring capabilities—beginning at least 7300 cal. yr BP (Orquera and Piana 2009)—these once-pristine island environments came increasingly under human influence. This impact intensified with the arrival of European state-sponsored explorations in the southern South Atlantic from the seventeenth century onward.

Isla de los Estados lies at the easternmost edge of the Fuegian Archipelago, approximately 24 km east of Isla Grande de Tierra del Fuego. The island covers an area of 520 km^2 and stretches predominantly west to east over a maximum length of 75 km. Its average width is 6 km, varying from a minimum of 550 m between Puerto Cook and Puerto Vancouver to a maximum of 17 km between Capes San Antonio and Kempe. Isla de los Estados represents the south easternmost extension of the Andes mountain range and is marked by a 50-km-long chain of hills rising between 400 and 800 m a.s.l. The terrain is highly dissected, with many lower valleys submerged by the sea, forming fjords (Ponce et al. 2011).

Archaeological and historical evidence indicates that Isla de los Estados was explored and occupied by humans during two distinct periods. The first occurred between 2800 and 1375 cal. yr BP and was associated with a hunter-gatherer-fisher economy. The second spanned from the seventeenth to the twentieth centuries and was linked to capitalist activities. The environmental impact of these historical processes on Isla de los Estados varied significantly in both magnitude and character.

The objective of this chapter is to examine the long-term implications of human-induced environmental changes on Isla de los Estados. Specifically, we aim to address the following question: What types of impacts did past human populations have on the island, and how did these impacts vary over time and across different historical contexts?

3.2 Discontinuous Forager Occupations on Isla de los Estados

Although the interior channels of the Fuegian archipelago were inhabited by hunter-gatherer and fisher groups as early as 7300 cal. yr BP (Orquera and Piana 2009), the islands located at the rim of the continental shelf were colonized at a later stage

(Legoupil 1993–94). These outer islands presented contrasting conditions for such populations: while their isolation—due to hazardous navigation environments—posed significant challenges, they also offered notable subsistence advantages, particularly the abundance of seabird and marine mammal colonies. Ethnographic (Fitz Roy 2016 [1839]; Martial 1888) and archaeological evidence (Lefevre 1993–94; Legoupil 1993–94, 1995) suggests that these islands were likely used seasonally, accessed from population nodes within the interior channels. However, Isla de los Estados exhibits a markedly lower intensity of occupation compared to other outer sectors of the Fuegian archipelago (Zangrando et al. 2024). The forager occupations on the island occurred over a relatively brief period, between 2800 and 1375 cal. yr BP, indicating an intermittent human presence (Horwitz 1990, 1993).

Following that period, evidence suggests a phase of abandonment of Isla de los Estados. In this context, the island became part of a different cultural and symbolic system, reflected in the mythological worldviews of both the Selk'nam and Yagan peoples during ethnographic times (Chapman 1987; Horwitz and Weissel 2011). Navigators, explorers, and cartographers who sailed around Isla de los Estados from the early seventeenth century made no mention of native inhabitants. However, ethnographic records from the late nineteenth and early twentieth centuries highlight the island's significance through the oral traditions of Indigenous groups. In Selk'nam and Haush mythology, Isla de los Estados was known as *Jáius*—the roots, or the place of origin—where powerful shamans and elemental forces engaged in colossal battles. These land-dwelling peoples could see the island's peaks from the easternmost shores of Tierra del Fuego, referring to it as the "Root Mountain Range" or *Kéoin Harri*. The Yagan, seafaring people (Fig. 3.1), called the island *Chuani-sin*, meaning "Region of Abundance".

The archaeological record of the Late Holocene is limited to a few bays located at the western end of Isla de los Estados. Reflecting the intermittent nature of human

Fig. 3.1 Yagans sailing in bark canoes. (Photo taken during the French scientific expedition to Cape Horn (1882–1883))

occupations, the exploitation of faunal resources by hunter-gatherer-fisher groups can be examined through zooarchaeological evidence from the Bahía Crossley I and Bahía Franklin 3 sites. At Bahía Crossley I, the assemblages include remains of pinnipeds, cetaceans, various seabird species, and fish, with a clear predominance of bird bones. At Bahía Franklin 3, only avian remains have been identified, with penguin bones being the most abundant (Zangrando et al. 2024).

The seasonal use of Isla de los Estados, as proposed by several authors, has been linked to the exploitation of seabird and marine mammal colonies (Horwitz 1990; Lefevre 1993–94; Horwitz and Weissel 2011). While some researchers have not confirmed the seasonality of these occupations, others—based on the identification of neonate pinniped remains—have suggested that the Bahía Crossley I sequence reflects occupations during the mid to late summer months (Lanata et al. 1992). A more recent study analyzing mortality profiles, based on sex and age determinations of pinniped remains from the same site, suggests that different hunting strategies may have been employed (Martinoli 2019). Although individuals under 1 year old are present, the predominance of juvenile and subadult remains indicates the targeting of isolated animals and/or groups located in non-reproductive areas at various times of the year. This information challenges a strictly seasonal interpretation of both resource exploitation and site occupation (Martinoli 2019). As previously noted, seabird remains are the most abundant faunal component. Several authors have emphasized that the presence of numerous articulated portions of penguin carcasses suggests minimal processing. This pattern may reflect the proximity of breeding colonies and the ease with which these resources could be acquired (Caviglia 1987; Horwitz and Weissel 2011: 154–155). While a detailed taphonomic analysis is essential to better understand the formation of these deposits (Cruz 2007; Massigoge et al. 2015)—particularly given the low frequency of cut marks observed on the bones (Lanata et al. 1992)—the incomplete exploitation of penguin carcasses, as seen in the Bahía Crossley I assemblage, may indicate a subsistence strategy aimed on obtaining immediate returns with minimal effort, made possible by the accessibility of nearby colonies. Given the migratory behavior of penguins on Isla de los Estados (Boersma et al. 2013; Pütz et al. 2013), the resource availability likely occurred between October and March. In this context, the overwhelming predominance of penguin throughout nearly all stratigraphic levels at Bahía Crossley I may point to a seasonal site use, a pattern that may also be inferred at Franklin Bay 3 (see discussion in Zangrando et al. (2024)).

Although the occupation of Isla de los Estados represents the most extensive geographic expansion of hunter-gatherer-fisher populations within the Fuegian archipelago, it also demonstrates that bird colonies and pinniped rookeries were accessible to humans during the Late Holocene. However, there is no indication in the archaeological record suggesting that these human activities had a significant ecological impact. While the exploitation of breeding colonies of pinnipeds cannot be definitively confirmed at this stage, the temporally and spatially limited nature of the occupations suggests that, even if such exploitation occurred, it was neither intensive nor sustained over time.

Previous studies have suggested that canoeing populations introduced terrestrial species into island environments in the South Atlantic region. For example, one hypothesis proposes that the extinct fox-like canid (*Dusicyon australis*) was transported to the Malvinas Islands by indigenous groups traveling by canoe from Tierra del Fuego, possibly as a domesticated hunting companion (Buckland and Edwards 1998; Hamley et al. 2022). More recently, Franklin (2022: 27) concludes: "*I propose as the only large and medium-sized, native terrestrial mammal species on Tierra del Fuego, guanacos and culpeo foxes (reverted from Fuegian dogs) were selectively recolonised by the canoe culture using bark canoes from the mainland to Tierra del Fuego during the waves of new settlers some 1000–3000 years after the 7750 YBP Hudson eruption (Volcano-barrier Straits-canoe hypothesis 3 above)*". Isla de los Estados offers a good geographical context to evaluate these hypotheses, however archaeological evidence provides no indication that such processes occurred on the island prior to the arrival of Europeans, as no canid or guanaco remains have been recorded to date in Holocene cultural or natural deposits. This absence supports the conclusions of other studies, based on archaeological, genetic, and stable isotope data which question the occurrence of species introductions by canoeing populations (Clark et al. 2022; Leggieri et al. 2024; Zangrando and Borrero 2022). A different scenario begins with the arrival of European navigators and, later, the establishment and eventual abandonment by the Argentine Navy penal colony on Isla de los Estados—an episode examined in the following sections.

3.3 European Expeditions to the End of the World

The emergence of capitalism at the end of the fifteenth century was marked by the process of primitive accumulation of capital (Marx 1867), which laid the basis for the rapid commercial and industrial expansion that followed in subsequent centuries (Wallerstein 1979). Although maritime trade already existed, the colonization of the Americas and the rise of transoceanic commerce during the seventeenth and eighteenth centuries expanded it to a global scale. Africa, the Americas, and Europe became part of the system that enabled the development of mercantile capitalism. Powers such as Portugal, Spain, the Netherlands, and France sought new markets, raw materials, and sources of enslaved or cheap labor—culminating in Great Britain's emergence as the leading force of industrial growth in the latter half of the eighteenth century. This process consolidated a new global order and redirected the ambitions of imperial powers toward the exploration and domination of strategically valuable territories.

In this context, maritime transport played a central role, enabling expeditions for a variety of purposes—including military, commercial, geographical, colonizing, and scientific endeavors—particularly in the race to control interoceanic passages. In 1520, amid the rivalry between the Spanish and Portuguese crowns over dominance in the Americas, Ferdinand Magellan and Juan Sebastián Elcano discovered the Atlantic–Pacific interoceanic passage (Pigafetta 2001 [1522]). This event marked a turning point in navigation toward the southern seas and the Atlantic-facing coast of Patagonia.

Within this framework of explorations in search of new inter-oceanic passages, and given the exclusive use of the Strait of Magellan by the recently founded Dutch East India Company (Vereenigde Oostindische Compagnie), the Dutch navigators Jacob Le Maire and Willem Schouten discovered a strait between Tierra del Fuego and a new island in 1616, which they named Stantelandt (Isla de los Estados): "*in honor of our States [referring to the States General, the parliament of the United Provinces of the Netherlands]*" (Le Maire 1619: 33). The strait was later known as the Strait of Le Maire.

This expedition was followed by many others. In 1619, the Spanish Nodal brothers confirmed the insularity of Isla de los Estados, and in 1643, the yacht Dolphin, under the command of Captain Brouwer, completed a full circumnavigation of the island. During the eighteenth and nineteenth centuries, various explorers and expeditions sighted, landed on, and charted the islands. Among them was Elizalde y Ustariz, Spanish governor of the Malvinas Islands, who, during his 1765 patrols for maritime control of the region, encountered Englishmen hunting pinnipeds along the coast. Other notable figures such as James Cook (Banks 1896), Robert Fitz Roy (2016 [1839]) and Charles Darwin (2019 [1839]) also visited or mentioned from the distance the island.

Successive visits to Isla de los Estados not only expanded knowledge of its geography, navigational hazards, and cartography but also led to the unintentional introduction of non-native species, such as the black rat (*Rattus rattus*) and the brown rat (*Rattus norvegicus*) (Salom and Schiavini, Chap. 14, this volume). These expeditions also yielded critical information about the island's natural resources, a key factor in enabling future ventures in the region. Furthermore, during this period and continuing into more recent times the island became the site of numerous shipwrecks. The remains of these wrecks, including wood, iron, and other materials, were scattered along its beaches and shores. From the late nineteenth century to the present, such remnants have been observed, described, and documented by a wide range of visitors, from sailors to scientific expeditions (Bove [1883] 2005; De Rosa et al. 2010; Santos and Elkin 2015; Zangrando and Tessone 2018, among others).

3.4 Sawmills and Seal Hunting Stations: The Experience of Luis Vernet and the Government of the United Provinces (1828–1830)

Following the dissolution of the Viceroyalty of the Río de la Plata and the subsequent declaration of independence, the government of the United Provinces of the Río de la Plata granted Soledad Island (Malvinas Islands) and Isla de los Estados to the Hamburg businessman Luis Vernet in 1828.[1] The concession aimed to establish a permanent colony within 3 years, with the dual purpose of asserting sovereignty over the territories inherited from Spain and promoting Vernet's economic ventures, which focused on the exploitation of feral cattle and fishing.

In this context, and with the goal of consolidating Argentine governmental presence in the region, the Political and Military Command of the Malvinas Islands and adjacent areas around Cape Horn in the South Atlantic was established in 1829, under the leadership of Luis Vernet. Headquartered on Soledad Island, the command exercised authority over surrounding islands, including Isla de los Estados. Due to the scarcity of timber in the colony's main settlement, Puerto Luis (Malvinas Islands), Isla de los Estados—about 550 km away—became a vital source of wood. To meet this need, Vernet signed a contract in 1829 with businessman Samuel Brooks for the installation of a sawmill to supply six shipments of timber. The agreement also authorized Brooks and his crew to hunt fur seals and sea lions, provided that such activities did not interfere with logging operations.

In June 1829, the brig Betsy departed from Buenos Aires with the mission of transporting Luis Vernet, his family, and a group of settlers to the Malvinas Islands, in order to secure the establishment of the Puerto Luis colony. At the same time, the vessel carried four families and six single British men tasked with developing a sawmill on Isla de los Estados (Parish to Aberdeen, April 25, 1829, in FO 78/1/16–18). During the voyage, the ship stopped at Puerto Luis, where it loaded 14 steers, two cows, and two bulls—marking the first recorded introduction of live cattle to Isla de los Estados. Additionally, three workers involved in the exploitation of feral cattle in the Malvinas Islands joined the expedition. They departed for the island on August 16, 1829. One of them, Loreto Sáez, Vernet's brother-in-law, left a written account describing the existing infrastructure: "*...there we found the lobería that your uncle had established the previous year next to some wooden settlements, and which for several years afterwards we always referred to as the Antigua Lobería*" (Loreto Sáez in AGN, LVE01-S7-131-224).

In the maps created by Vernet (AGN, LVE01-S7-134-028-002), these sites are marked at various locations on the island, along with the corresponding years to their periods of exploitation (Fig. 3.2). Possible worker dwellings, labeled as "*habite maderas*" (wooden shelters), are also indicated on the maps.

Almost 2 months later, the Betsy returned from Isla de los Estados to Malvinas/Flakland Islands with the first shipment of timber, which was used to construct fences and corrals essential for the exploitation of feral cattle (Vernet 1829 in Gutiérrez 2022: 119). The ship also brought back "*...two hundred feet of trees of three distinct varieties; the bark of one of them stings like chili pepper, another has a yellow color that many believe to be a dyeing stick, and the bark of the third is*

[1] Luis Vernet first arrived in the Malvinas Islands in 1826, accompanied by his brother Emilio Vernet and his brother-in-law Loreto Sáez (the brother of María Sáez, Vernet's wife). During this expedition, they not only conducted surveys of the Malvinas Islands but also explored Isla de los Estados, including its interior, harbors, and coastlines (AGN, LVE01-S7-134-028-002).

Fig. 3.2 Map created by Vernet depicting the *loberías* and occupied areas. (Source: AGN, LVE01-S7-134-028-002)

fragrant, like the quinaquina that the Indians take to Buenos Aires" (Sáez de Vernet 1829 in AGN, LVE01-S7-141-005-0030).

Less than a month later, the brig set sail once more to Isla de los Estados, this time carrying an additional 16 steers. It returned 3 weeks later with another cargo of timber. On this occasion, they also collected "*a bush full of Statenland fruits, very similar to blackberries, though not as sweet, with leaves resembling those of myrtle*" (Sáez de Vernet [1829] 2020: 59). Notably, this shipment marked the first time that timber from the island was exported to Montevideo (AGN, LVE01-S7-131-224).

This monthly rotation between Isla de los Estados and the mainland not only facilitated the extraction of timber but also ensured a steady resupply of provisions for the workers and their families living on the island, mainly through the transport of cattle for food. However, the vessel designated for this task—the *Belleville*—was shipwrecked off the coast of Río Negro, leaving the workers stranded without supplies for over 6 months. It was not until June 1830, with the arrival of the *María Antonia*, that the workers ultimately decided to abandon the island.

Vernet attempted to persuade the workers to continue their activities on Isla de los Estados; however, they refused (AGN, LVE01-S7-141-012). On August 29, 1830, the *Maria Antonia* returned to Puerto Luis carrying 50 tons of lumber. After this point, vessels departing from Puerto Luis in search of timber began sailing directly to the Strait of Magellan. In this region, they were also able to obtain horses through trade with the Tehuelche people.

Following the departure of the sawmill personnel, seal-hunting operations (*loberías*) began to be established on the island, primarily conducted by British crews such as that of the *Superior*. In 1831, Vernet seized the *Superior* in the Malvinas Islands for illegally exploiting fishing resources without official authorization. Seven members of its crew, who had been operating a *lobería* on Isla de los Estados, were left stranded until they were rescued by the *Breakwater*, another schooner previously captured by Vernet, which had managed to escape from the Malvinas Islands.

In summary, Vernet's interventions on Isla de los Estados initiated significant transformations of the island's ecosystem. In addition to the systematic extraction of timber and the construction of dwellings, cattle were introduced for the first time. Although the animals failed to establish due to limited forage availability and the island's rugged topography, this early introduction of an exotic species marked a critical precedent in the anthropogenic modification of the island. Similarly, *loberías* were established near the sawmill and at other locations, resulting in the intensive exploitation of marine resources as well. The remains of habitational structures identified by biologists Manfred Niekisch and Adrián Schiavini (1998) at Puerto Basil Hall may correspond to these settlements.

3.5 Penguin and Sea Lion Oil Factories and Introduced Domestic Species: Luis Piedra Buena at Isla de los Estados (1862–1875)

Luis Piedra Buena was an Argentine sailor and trader whose activities in Patagonia and the South Atlantic played a crucial role in consolidating Argentine sovereignty over these territories. Beginning in 1859, Piedra Buena focused on the exploitation of pinnipeds and the salvage of shipwrecks along the Patagonian coasts and surrounding islands. In 1862, he began establishing huts on Isla de los Estados, with the first built at Puerto Cook. Concurrently, he introduced exotic species, such as goats (*Capra hircus*), to provide a reliable food source for castaways (Fig. 3.3). In 1868, Piedra Buena secured official ownership of Isla de los Estados from the national government and established a penguin oil factory. This venture occurred during the global peak of exploitation of such resources (1850 and 1880), driven by the demand for oils with properties similar to whale oil, including those derived from elephant and sea lion oil (Haller and García 2022).

Piedra Buena established his factory at Puerto Roca, where King penguins—along with other penguin species—were exploited. The colony, estimated by G. H. Gardiner, one of the workers who maintained a daily diary, numbered approximately 20,000 individuals (Haller and García 2022). To facilitate this process, trails and corrals were constructed, as the extraction method involved herding penguins into the corral and then clubbing them to death. The penguins were subsequently skinned, stacked, and their fat extracted and cooked in 50-gallon vats. A press was

Fig. 3.3 Feral goats in Flinders Bay (Isla de los Estados)

employed to extract the remaining fat, known as greaves, which was repurposed as fuel. Finally, the oil was collected in a vat and transferred into *cascos* or wooden barrels crafted from local timber.

In terms of infrastructure, the site included a wooden house with French tiles where the four workers, as well as a wooden shed used for processing activities. To facilitate mobility while the penguins fattened, the workers built a boat, which allowed them to explore the surrounding islands. During these excursions, they discovered evidence of previous occupation in Puerto Cook. These included fishermen's shelters made from stone, as well as boards nailed to trees bearing the names of visiting ships. Additional small structures, built from sticks or stones, were also found in the hills.

When Piedra Buena failed to return to the factory within the agreed timeframe, the workers were left completely isolated for 11 months. During this period, they subsisted by hunting a range of animals for sustenance, including ducks, otters, fur seals, red mice, doves or snowy sheathbills, and geese. They also made use of available wild resources such as strawberries and wild celery.

Notably, the workers established the first known vegetable gardens on the island, successfully planting and harvesting radishes, lettuce, and beets, although attempts to cultivate onions was unsuccessful. Similarly, the introduction of pigs (*Sus domesticus*) and goats to a nearby island yielded poor results: all the pigs died and only three goats survived after several months.

Following this experience, which concluded in 1870, Piedra Buena returned to the island on multiple occasions. Even after being shipwrecked in 1873 aboard his vessel *Espora*, he persisted in his endeavors and returned in 1875 to expand his exploitation of both fur seals and sea lions at Crossley Bay (Bascopé-Julio and

García-Oteiza 2024). Recent archaeological surveys at Crossley Bay have identified a historic settlement that likely correspond to the location of these activities (Zangrando and Tessone 2018). Preliminary analyses of materials recovered from the site (e.g., glass containers, ceramics, and cartridge case) suggest that the settlement dates to the post 1870 period. Additional evidence of resource exploitation, particularly involving bird and marine mammal colonies, has also been recorded at Flinders Bay (Zangrando and Tessone 2018).

Piedra Buena's intervention on the island brought about significant transformations. He not only carried out the systematic exploitation of penguin and sea lion colonies, but also introduced exotic fauna and flora, some of which, such as the goat, successfully adapted and becoming feral in the island. Furthermore, shelters were established in various inland bays, and timber was extracted for the constructions of houses, sheds, corrals and containers. Finally, in light of the territorial dispute with Chile over national boundaries at the time, Piedra Buena's sustained presence and activities on Isla de los Estados and in Patagonia served as a symbolic and strategic assertion of Argentine sovereignty over these contested regions.

3.6 Lighthouse, Sub-prefecture and Prison: The Argentine Navy (1884–1902)

In 1881, the Argentine Geographic Institute organized an Argentine-Italian expedition, led scientifically by Giacomo Bove and militarily commanded by Piedra Buena. The expedition spent 2 months conducting a range of studies at Rockery Penguin Bay, later renamed Puerto Roca. During their stay, they constructed houses and a scientific camp, and carried out hydrographic, zoological, botanical, and geological research. They also ascended and measured several mountains of the island, making their presence with physical indicators. It was Bove (2005 [1883]) who proposed using the island as a penal colony to address the penitentiary challenges facing the young Argentine state, an idea that helped facilitate permanent settlement and the consolidation of Argentine sovereign in the region. The expedition members also reiterated a proposal, originally made by Piedra Buena to construct a lighthouse to prevent future shipwrecks in the area (Bove 2005 [1883]).

The following year, Piedra Buena began planning the South Atlantic Expeditionary Division, though he passed away in 1884 before the mission could be realized. The goal of the expedition was to establish a permanent settlement on the island by creating a sub-prefecture by troops and their families, alongside a small civilian prison. Under the command of Commodore Augusto Lasserre, the mission arrived at San Juan Salvamento in 1884. There, the personnel were tasked with building a rescue station, a lighthouse, a sub-prefecture, a stone pier, and barracks for military prisoners. These developments significantly transformed the island's landscape and environment. Notably, the construction materials used were not native to the island: the wood consisted of pine boards imported from Buenos Aires, an iron was introducc for hinges, roofs, and windows frames, marking a significant shift in the materials employed in the island's historical architecture.

On May 25, 1884,[2] Commodore Lasserre inaugurated the San Juan de Salvamento Lighthouse, the southernmost lighthouse in the world, famously referred as the "lighthouse at the end of the world" in Jules Verne's novel, along with the presidio at Punta Lasserre and the "Subprefectura Marítima". The latter was designed to serve as a port of call for ships in distress and a rescue station for shipwrecked individuals. In a report to authorities in Buenos Aires, Lasserre summarized the infrastructure as follows: "*Two groups of buildings have been erected, Mr. Minister, separated from each other. The first consists of four buildings: one for the Subprefecture office with adjoining rooms for the accommodation of the Subprefect and his assistant, two spacious blocks for the seamen's brigades, and a shed*" (Vairo 1997: 61). The same report notes that access to this cluster of buildings was made possible via a pinewood staircase supported by pillars constructed from island-sourced logs. Lasserre also described the newly constructed lighthouse, which stood adjacent to the telegraph office, an installation that also functioned as a meteorological observatory and the residence of the officer in charge. In the years that followed, a cemetery was established nearby, requiring significant clearing, leveling, and trenching due to the island's steep and rugged terrain. Continuing a pattern set by Vernet and Piedra Buena in earlier years, the expedition introduced exotic fauna. According to the inventory recorded at the time of their departure for Ushuaia in 1884, it included "*Three hundred and ten live animals. Eight cows. Two bulls*" (Vairo 1997: 70).

Initially, the island was inhabited by a garrison of approximately 24 men and 10 civilian prisoners transferred from Buenos Aires. These individuals were the only civilian prisoners ever housed on the island and were specially selected to assist with the construction of essential infrastructure. Subsequent prison populations consisted exclusively of contingent's convicted military personnel. In 1890, the first military prisoners arrived, some accompanied by their wives and children. A few years later, the post office was established. By the end of the nineteenth century, a population census recorded approximately 154 inhabitants on the island (www.biblioteca.indec.gov.ar).

By 1899, the sub-prefecture and military prison were relocated to Puerto Cook, which offered better shelter and facilities for repair. At this new site, a range of structures was constructed, including dungeons for prisoners, housing for guards corrals, a bakery, sheds, a pharmacy, two kitchens, bathrooms, workshops, a meteorological observatory, iron water tanks, a chicken coop, a kennel, and additional support buildings (Fig. 3.4). Housing units were erected on wooden frames

[2] On October 16, 1884, the Argentine government enacted Law No. 1532, known as the Law on the Organization of National Territories, which subdivided the extensive Patagonian region into the Governments of Río Negro, Neuquén, Chubut, Santa Cruz, and Tierra del Fuego. The boundaries of the Government of Tierra del Fuego were defined as follows: Article 6°: "Governorate of Tierra del Fuego, with its natural limits, according to the treaty of July 23, 1881, and including Isla de los Estados."

Fig. 3.4 Puerto Cook military prison. (Image taken from García Basalo (1998))

supported by stones, after removing the bay's characteristic peat soil. This design mitigated humidity by elevating the buildings approximately 40 centimeters above ground level (Vairo 1997). Local stones were used to build the dungeon walls, while wood and zinc sheeting were employed for the residential structures. A cemetery was established, along with scattered ranches that housed the families of the convicts. Noteworthy among these is the prison's chief house constructed using materials salvaged from local shipwrecks. Additionally, ten bathrooms were built using "machine" bricks. Technology advanced for the time, the prison wards were equipped with electric bells, allowing for internal communication between buildings.

In addition to carrying out various prison-related duties, the inmates also crafted "trinkets" from the island's natural resources, which they attempted to sell or barter for alcoholic beverages with visiting ships. The conditions under which they lived were unhealthy and precarious. This is evident in the report submitted by Commodore Howard, who described the facility as "*such an inhuman prison, comparable to the Siberia of autocratic Russia, in a country with such free and benign laws as those of this republic*" (García Basalo 1998).

Ultimately in 1902, the prison was relocated to Ushuaia for humanitarian, security, and disciplinary reasons. At that time, in addition to the 46 Army and Navy personnel assigned as guards, the facility housed 119 convicts, some of whom attempted to escape during the initial transfer. That same year, the San Juan de Salvamento Lighthouse was decommissioned and replaced by the Observatorio Island Lighthouse, situated about 20 km north of its original location.

The presence of the Argentine Navy on the island left a mark on the landscape. Even today, the remnants of its constructions San Juan de Salvamento and Puerto Cook remain clearly visible. In contrast, efforts to introduce species such as cattle and vegetables proved unsuccessful due to the island's harsh environmental conditions.

Since then, there have been several additional attempts to establish settlements for extractive purposes, along with the introduction of more exotic species, such as the red deer (*Cervus elaphus*) (Niekisch and Schiavini 1998). Today, the Argentine Navy maintains a permanent detachment on the island, and Isla de los Estados is also regularly visited by scientists from various disciplines, including biology, archaeology, geology, and others.

3.7 Conclusion

Archaeological and historical evidence indicates that the ecological impact of hunter-gatherer-fisher populations during the Late Holocene was minimal. In contrast, significant environmental changes began with the arrival of European navigators and the subsequent establishment of settlements from the seventeenth century onward. Human influence on the island became substantial, particularly through the exploitation of native species and to the introduction of non-native ones. These introductions, ranging from domestic animals to exotic plants, were intended to support the settlement efforts but had unintended and often detrimental effects on the island's ecosystem. The primary disturbance are outline below:

- Cattle and Livestock: Cattle were introduce to provide a sustainable food source; however, this effort failed due to the island's rugged terrain and limited pasture. The steep, rocky environment hindered effective grazing, and the sparse vegetation could not support livestock. Additionally, the presence of cattle likely intensified environmental stress through overgrazing, and their introduction may have disrupted local flora and fauna by spreading disease or competing for resources.
- Vegetation: Efforts to cultivate vegetables were largely unsuccessful due to harsh climatic conditions, including low temperatures, strong winds, and nutrient-poor soil. These environmental constraints rendered agriculture impractical. Moreover, the introduction of non-native plants may have led to competition with or displaced of native vegetation.
- Invasive Species: Feral goats and rats, both introduced species, caused significant ecological disruption. Feral goats, in particular, are known to degrade native plant communities and promote erosion through trampling (Petracci et al. 2016). Although the introduction of pigs ultimately failed, they may have also contributed to soil disturbance and habitat degradation.
- Human Impact on Native Species: Human activities in the nineteenth century, including the construction of buildings, a prison, and resource extraction, further altered the island's natural landscape. Infrastructure such as the lighthouse, mili-

tary facilities, and associated buildings displaced native habitats and potentially introduced pollutants. Most notably, the exploitation of marine species, including fur seals, sea lions and penguins, led to exterminations of local populations of these species (e.g., Salom and Schiavini, Chap. 14, this volume).

The legacy of these disturbances is still evident in the ecological challenges facing the island today. Preserving its native flora and fauna requires careful management of the long-term consequences of past human activities. Conservation efforts must address these enduring impacts to ensure that future development does not exacerbate existing environmental degradation.

Competing Interests We would like to extend our special thanks to Andrea Raya Rey and Ulizes Balza for inviting us to contribute to this volume and for their valuable comments and suggestions on the chapter's edition. We are also grateful to Laura Smith and Federico Guerrero for their logistical support aboard the Ocean Tramp in 2017, which was instrumental in carrying out the research presented here. This work was supported by grants from the Agencia I + D + i (PICT 3153/2019 and PICT 1230/2017) and CONICET (PIP 0200/2017).

References

Armstrong PH (1994) Human impact on the Falkland Islands environment. Environmentalist 14(3):215–231

Banks J (1896) Journal of the right honourable sir Joseph Banks during captain cook's first voyage in H.M.S. In: Endeavour in 1768–71 to Tierra del Fuego, Otahite, New Zealand, Australia, the Dutch East Indies, etc. Mcmillan and Co., London

Bascopé-Julio J, García-Oteiza S (2024) Diarios del viaje de Piedrabuena en Fuegopatagonia, 1863-1874.Magallania, 52(15):1–57

Boersma D, Garcia-Borboroglu P, Frere E, Kane O, Pozzi L, Pütz K, Raya Rey A, Rebstock GA, Simeone A, Smith J, Yorio P, Van Buren A (2013) Magellanic penguins. In: García-Borboroglu PG, Boersma PD (eds) PENGUINS: natural history and conservation. University of Washington Press, Seattle, pp 233–264

Bove G (2005 [1883]) Expedición a la Patagonia. Un viaje a las tierras y mares australes (1881–1882). Ediciones Continente, Buenos Aires

Buckland PC, Edwards KJ (1998) Palaeoecological evidence for possible pre-European settlement in The Falkland Islands. J Archaeol Sci 25(6):599–602. https://doi.org/10.1006/jasc.1998.0297

Butzer K (1982) Archaeology as human ecology. Cambridge University Press, Cambridge

Caviglia SE (1987) Análisis del material faunístico del sitio I Bahía Crossley. In: Chapman A (ed) La Isla de los Estados en la prehistoria. Primeros datos arqueológicos. EUDEBA, Buenos Aires, pp 95–112

Chapman A (1987) La Isla de los Estados en la prehistoria. Primeros datos arqueológicos. EUDEBA, Buenos Aires

Clark TJ, Newton J, Wakefield ED (2022) Comment on "Evidence of prehistoric human activity in the Falkland Islands". Sci Adv 8:eabo0928

Cruz I (2007) Avian taphonomy: observations at two Magellanic penguin (Spheniscus magellanicus) breeding colonies and their implications for the fossil record. J Archaeol Sci 34(8):1252–1261. https://doi.org/10.1016/j.jas.2006.10.016

Darwin C (2019 [1839]) Los viajes del Beagle. Diario y observaciones (1832–1836). In: Colección Reservada del Museo del Fin del Mundo. Eudeba, Buenos Aires

De Rosa H, Ciarlo N, Svoboda H (2010) Estudio de clavos de aleación de cobre asociados a los restos de una embarcación hallada en la Isla de los Estados. XI Congreso Iberoamericano de Metalurgia y Materiales (IBEROMET) y X Congreso Internacional de Metalurgia y Materiales (CONAMET/SAM). Viña del Mar, Chile, (2 al 5 de noviembre)

Erlandson JM, Rick TC (2008) Archaeology, marine ecology and human impacts on marine environments. In: Rick TC, Erlandson JM (eds) Human impacts on ancient marine ecosystems. A global perspective. University of California Press, Berkeley, Los Angeles, and London, pp 1–19

Fitz Roy R (2016 [1839]) Los viajes del Beagle. Informes de la segunda expedición 1831–1836. EUDEBA, Buenos Aires

Franklin W (2022) Guanaco colonisation of Tierra del Fuego Island from mainland Patagonia: Walked, S\van1, or by canoe? Geo Geogr Environ 9:1–35. https://doi.org/10.1002/geo2.110

García Basalo JC (1998) Historia del presidio militar San Juan del Salvamento, Isla de los Estados, “La prisión del Diablo”; y el sangriento motín de 1902. Revista Todo es Historia 366

Gutiérrez S (2022) Así se construyó Malvinas. El diario de Emilio Vernet (Puerto de la Soledad, 1828–1831). Editorial AGN. Serie Histórica 01. Buenos Aires

Haller S, García S (2022) La producción de aceite de pingüino en Patagonia y Malvinas (1850–1880). Nuevo Mundo Mundos Nuevos 10:1–21

Hamley KM, Gill JL, Krasinski KE, Groff DV, Hall BL, Sandweiss DH, Southon JR, Brickle P, Lowell TV (2022) Response to comment on “evidence of prehistoric human activity in The Falkland Islands”. Sci Adv 8(17):eabo6765. https://doi.org/10.1126/sciadv.abo6765

Horwitz V (1990) Maritime settlement patterns in Southeastern Tierra del Fuego (Argentina). PhD thesis, University of Kentucky

Horwitz V (1993) Maritime settlement patterns: the case from Isla de los Estados (Staten Island). Arqueología Contemporánea 4:163–176

Horwitz V, Weissel M (2011) Arqueología de Isla de los Estados: La frontera de la abundancia. In: Zangrando AF, Vázquez M, Tessone A (eds) Los cazadores recolectores del extremo oriental fueguino. Arqueología de Península Mitre e Isla de los Estados. Sociedad Argentina de Antropología, Buenos Aires, pp 143–170

Lanata J, Weissel M, Caracotche M, Belardi J, Muñoz S, Savanti F (1992) Dos mil huesos de viaje submarino: análisis faunístico del sitio Bahía Crossley 1, Isla de los Estados. Palimpsesto Revista de Arqueología 1:9–24

Le Maire J (1619) Relación diaria del viaje de Jacobo Le Maire y Guillermo Schouten, en que se descubrieron el nuevo Estrecho y pasaje del mar del norte a la mar del sur, a la parte austral del Estrecho de Magallanes/Jacobo Le Maire. Biblioteca Virtual Miguel de Cervantes. https://www.cervantesvirtual.com/nd/ark:/59851/bmchm558

Lefevre C (1993–94) Las aves en los yacimientos del Cabo de Hornos y del Seno Grandi. Anal Inst Patagon (Serie Ciencias Humanas) 22:123–136

Leggieri LR, Anello M, Peralta D, Túnez JI, Di Rocco F, Poljak S, Flores C, Alunni D, Belardi JB, Gutiérrez MA, Álvarez MC, Massigoge A, Kaufmann C, Borrero L, Borghi C, Demartini J, Petracci P, Martín S, Marbán L, Kathiravan P, Rudolf P, Marín JC, Cárcamo JG, Caramachai P (2024) Delineating genomic features for wild guanaco conservation. Biol J Linn Soc 143(3). https://doi.org/10.1093/biolinnean/blae087

Legoupil D (1993–94) El archipiélago del cabo de Hornos y la costa sur de la isla Navarino: poblamiento y modelos económicos. Anal Inst Patagon (Serie Ciencias Humanas) 22:101–121

Legoupil D (1995) Des indigènes au Cap Horn: conquête d’un territoire et modèle de peuplement aux confins du continentsud-américain. J Soc Am 81:9–45. https://doi.org/10.3406/jsa.1995.1582

Martial LF (1888) Historie du voyage. En Mission Scientifique du Cap Horn (1882–1883), vol I. Gauthier-Villars, París

Martinoli M (2019) La explotación de pinnípedos en Isla de los Estados: el caso de las ocupaciones de Bahía Crossley I. Rev Mus Antropol 12(3):43–60. https://doi.org/10.31048/1852.4826.v12.n3.24514

Marx K (1867) El capital: Crítica de la economía política, Tomo I, 5a edn. Siglo XXI Editores, México

Massigoge A, Rafuse DJ, Álvarez MC, González ME, Gutiérrez MA, Kaufmann CA, Scheifler NA (2015) Beached penguins on the Atlantic Coast in the pampas region of Argentina: Taphonomic analysis and implications for the archaeological record. Palaeogeogr Palaeoclimatol Palaeoecol 436:85–95. https://doi.org/10.1016/j.palaeo.2015.06.045

Niekisch M, Schiavini A (1998) Desarrollo y Conservación de la Isla de los Estados (Tierra del Fuego, Argentina), Informe preparado para el Gobierno de tierra del Fuego, Antártida e Islas del Atlántico Sur: Deutsche Gesellschaft für Technische Zusammenarbeit (GTZ) GmbH de la República de Alemania PN 95.3582.4–008.00

Orquera LA, Piana EL (2009) Sea nomads of the Beagle Channel in Southernmost South America: over six thousand years of coastal adaptation and stability. J Isl Coast Archaeol 4(1):61–81. https://doi.org/10.1080/15564890902789882

Petracci P, Sáenz Samaniego RA, Raya Rey A (2016) Aspectos reproductivos y uso de hábitat del cauquén común (*Chloephaga picta*) y el cauquén real (*Chloephaga poliocephala*) en Isla de los Estados, Argentina. Hornero 31(2):73–81

Pigafetta A (2001[1522]) Primer viaje alrededor del mundo. Elefante Blanco, Buenos Aires

Ponce JF, Borromei AM, Rabassa J (2011) Evolución del paisaje y de la vegetación durante el Cenozoico tardío en el extremo sudeste del Archipiélago Fueguino y Canal Beagle. In: Zangrando AF, Vázquez M, Tessone A (eds) Los cazadores recolectores del extremo oriental fueguino. Arqueología de Península Mitre e Isla de los Estados. Sociedad Argentina de Antropología, Buenos Aires, pp 31–64

Pütz K, Raya Rey A, Otley H (2013) Southern rockhopper penguin. In: García-Borboroglu PG, Boersma PD (eds) PENGUINS: natural history and conservation. University of Washington Press, Seattle, pp 113–130

Redman C (1999) Human impact on ancient environments. The University of Arizona Press, Tucson

Santos H, Elkin D (2015) Patrimonio cultural, histórico y arqueológico: Naufragios en la costa Atlántica de Tierra del Fuego. La Lupa. Colección Fueguina De divulgación científica 8:10–12

Vairo C (1997) La Isla de los Estados y el Faro del Fin del Mundo. Zagier Urruty, Ushuaia

Vernet M (2020) Malvinas, Mi Casa. Vísperas, Diario de María Sáez de Vernet y Apostillas. Editorial EME, Buenos Aires

Wallerstein I (1979) El moderno sistema mundial. Tomo I. Siglo XXI Editores, México

Zangrando AF, Borrero LA (2022) A pre-European archaeology in Malvinas/Falkland Islands? A review. J Isl Coast Archaeol 19(2):433–447. https://doi.org/10.1080/15564894.2022.2077484

Zangrando AF, Tessone A (2018) Prospección arqueológica del litoral norte de la Isla de los Estados. Informe presentado a la Secretaría de Ambiente, Desarrollo Sostenible y Cambio Climático mediante (Resolución 647/2017), Provincia de Tierra del Fuego A.e.I.A.S

Zangrando AF, Martinoli MP, Tessone A, Vázquez MM, Alunni DV, Franch Bach A, Hernández Herrero O, Fernández Ropero MC, Pinto Vargas G, Tivoli AM (2024) El uso humano de espacios externos en el archipiélago de Tierra del Fuego durante el Holoceno: el caso de Isla de los Estados. Bol Soc Chil Arqueol 57:85–119. https://doi.org/10.56575/BSCHA.05700240843

Chapter 4
Flora and Vegetation of Isla de los Estados

Amira Salom, Joaquín Cano, Daniel Testoni, Paula Romina Mansilla, Luciana Mestre, Leonardo Collado, and Fernando Biganzoli

Abstract Isla de los Estados hosts a unique assemblage of vascular plant communities shaped by its geographic isolation, harsh maritime climate, and complex topography. This chapter provides a synthesis of the island's flora and vegetation patterns. We compiled a list of 173 vascular plant species based on historical records and recent fieldwork, classifying each as common, rare, or of uncertain occurrence. A total of 50 families were recorded, with *Poaceae* and *Asteraceae* being the most species-rich. Despite the presence of some introduced species, native flora dominates, including several species of biogeographic interest and regional endemics. We describe the major plant communities, including forests, moorlands, and maritime grasslands, across the archipelago; and provide a detailed vegetation map based on field data, drone images, georeferenced photos, and visual interpretation of high-resolution imagery. Special attention is given to vegetation patterns on the

A. Salom (✉) · J. Cano
Laboratorio de Ecología y Conservación de Vida Silvestre, Centro Austral de Investigaciones Científicas (CADIC-CONICET), Ushuaia, Tierra del Fuego, Argentina
e-mail: amirasalom@cadic-conicet.gob.ar

D. Testoni
Parque Nacional Los Glaciares, Administración de Parques Nacionales (PNLG-APN), El Calafate, Santa Cruz, Argentina

P. R. Mansilla
Instituto de Ciencias Polares, Ambiente y Recursos Naturales, Universidad Nacional de Tierra del Fuego, Antártida e Islas del Atlántico Sur (UNTDF), Ushuaia, Tierra del Fuego, Argentina

L. Mestre
Dirección General de Biodiversidad y Conservación, Secretaría de Ambiente de la Provincia de Tierra del Fuego, Antártida e Islas del Atlántico Sur, Ushuaia, Tierra del Fuego, Argentina

L. Collado
Dirección General de Ordenamiento Territorial y Gestión de Ambientes Forestales, Secretaría de Ambiente de la Provincia de Tierra del Fuego, Antártida e Islas del Atlántico Sur, Ushuaia, Tierra del Fuego, Argentina

F. Biganzoli
Facultad de Agronomía, Universidad de Buenos Aires, Buenos Aires, Argentina

A. Raya Rey, U. Balza (eds.), *Biodiversity and Conservation of Isla de los Estados (Chuani-sin)*, Natural and Social Sciences of Patagonia,
https://doi.org/10.1007/978-3-032-13630-5_4

main island and surrounding islets, emphasizing regional contrasts and the influence of topography, wind exposure, and landform. This chapter provides a valuable baseline for future ecological research and conservation efforts in one of the least-known floristic regions of the Sub-Antarctic.

Keywords Vascular plants · Vegetation assemblages mapping · Plant community dynamics · Tussack grasslands · Botanical exploration

4.1 Introduction

Vegetation represents the largest biotic component of natural ecosystems, determining the distribution and abundance of many other organisms. Since Humboldt's early XIX century work, which emphasized plant communities and their environmental relationships, vegetation has been studied not only for its composition but also for its structure, dynamics, and biogeographic patterns (Nicolson 1987; Whittaker 1962). This laid the foundation for modern vegetation science, which now also includes plant population biology, species strategies and vegetation dynamics (van der Maarel 2005).

Knowing vegetation assemblages have become central to biodiversity conservation, ecological monitoring, and land management (Chytrý et al. 2011). This is typically accomplished by analyzing plant communities through the survey of species composition and structure across plots, an approach established over a century ago (Warming 1895; Clements 1905). This is key for phytosociology, which classifies vegetation units into a hierarchical system seeking to summarize spatial and temporal variation, while considering species-environment relationships and interactions with fauna (Kent 2011). Concepts like synusiae (life-form layers) and physiognomy (general appearance), help interpret vertical and horizontal structure (Barkman 1978; Küchler and Zonneveld 2012). When combined with dominant species and growth forms, these concepts aid in mapping, habitat assessment, and conservation (Dengler et al. 2008). Particularly, vegetation maps are key tools for understanding species distribution, environmental relationships, and habitat use. However, despite technological advances, field-based vegetation data, much needed for input, remains limited, particularly in remote or rugged areas (Jelaska 2009).

Isla de los Estados, a Provincial Ecological, Historical, and Tourist Reserve of Tierra del Fuego, Antarctica and the South Atlantic Islands, is shaped by its geological isolation and harsh maritime climate. These conditions support a rich mosaic of plant communities across the island's rugged, mountainous landscape. Since European contact in 1615, its flora has attracted scientific interest, with early classifications placing it within the Sub-Antarctic Dominion (Cabrera 1976). Yet much remains unknown. Field surveys have been patchy, and while recent efforts have improved our knowledge (Collado 2015; Huertas Herrera et al. 2018), they only focused on forest ecosystems. Key open habitats, such as moorlands and maritime tussock grasslands, remain poorly described, despite their importance for seabirds and marine mammals. Likewise, the latest study which combined satellite data with detailed field surveys was limited to the island's western sector, leaving a substantial knowledge gap across much of its territory (Salom et al. 2025a).

In this chapter, we compile and update the current knowledge of the vascular flora and vegetation assemblages of Isla de los Estados, drawing on new herbarium collections and extensive fieldwork conducted since 2017. We describe major plant communities and produce a detailed vegetation map and cover estimates, integrating field and satellite data. Additionally, we examine patterns across the island's distinct ecological regions. Although Isla de los Estados remains one of the most pristine environments on Earth, largely due to the absence of permanent human settlements and minimal anthropogenic disturbance, its ecological integrity is at risk. Growing pressures from introduced mammalian herbivores and rising tourism increases the likelihood of alien plant introductions, which could compromise the island's fragile ecosystems. In this scenario, this chapter provides a vital tool for conservation planning, long-term monitoring, and a deeper understanding of the island's unique botanical heritage.

4.1.1 Brief History of Botanical Exploration

Schouten and Le Maire encountered the island for the first time in 1615 and named it Isla de los Estados in honor of their homeland, the Netherlands. However, the first naturalists to explore the area did not arrive until late 1774, when Captain James Cook anchored off the Año Nuevo islands during his second voyage. During this expedition, botanists collected plants exclusively from this group of islands, located just off the north coast of Isla de los Estados.

Moore (1983) listed 20 additional expeditions that collected plant material on Isla de los Estados between 1774 and 1974, but most were brief, non-comprehensive stopovers. Notable exceptions include surveys by Spegazzini in 1882 (Spegazzini 1896), Castellanos in 1933–1934, and Nicora, Crespo, Ancibor, and Torres expedition in 1967 (Kühnemann 1976). The most substantial botanical work during this period, however, was conducted by Dudley, Goodall, and Crow between 1968 and 1971 (Dudley and Crow 1983; Moore 1983).

Following the period covered by Moore (1983), further collections were made by one of the authors (F.B.) during three campaigns between 1998 and 2001 (Biganzoli 2003). More recently, between 2017 and 2023, several field campaigns were led by A.S., L.M., D.T. and V. Lencinas. The material collected during these expeditions, now deposited at Museo del Fin del Mundo (MFM) and Centro Austral de Investigaciones Científicas (CADIC-CONICET), both in Ushuaia, Tierra del Fuego, forms the basis of the current floristic update.

4.1.2 Revised Inventory of Vascular Plant Diversity

To compile the list of vascular plants for Isla de los Estados, we reviewed the existing literature and our own collections. We included all species with herbarium specimens that have confirmed identifications, as well as a select number of species that, whilst not yet confirmed, have been suggested for the island. Each species was classified as common, rare, or of uncertain occurrence based on the following criteria: Common species are those represented by several collections in a registered herbarium (Thiers 2025), supplemented by frequent field observations in recent surveys, with no doubts about species identity. Rare species are those represented by few or a single collection in a registered herbarium, by old collections only, with few known localities, and/or by recent records that have not yet been deposited in an herbarium, but whose identification is not in doubt. Uncertain species are those recorded or suggested in floras or species lists for the island but lacking herbarium specimens with confirmed identification.

A total of 173 species were recorded on the island, of which 106 were classified as common, 57 as rare, and 10 as of uncertain status (Table 4.1). These species belong to 50 families, with *Poaceae* (29 species) and *Asteraceae* (22 species) being the most species-rich families (Table 4.1). Several native species with restricted distributions are of particular interest. For instance, *Senecio eightsii* (Fig. 4.1a) is a vulnerable species, that together with *Senecio websteri* (Fig. 4.1b), are endemic to Isla de los Estados and the easternmost tip of Isla Grande de Tierra del Fuego. Others are endemic to the South Atlantic islands, such as *P. flabellata* (Fig. 4.1c), a large tussock-forming grass restricted to Tierra del Fuego, the Diego Ramírez Islands, the Malvinas Islands, South Georgia, and Gough Island (Guerrero et al. 2025). This species is especially noteworthy because it dominates coastal grasslands that form critical breeding habitat for numerous bird and marine mammal species (Groff et al. 2020). Likewise, several species are considered endemic to Patagonia and subantarctic islands such as *Azorella ranunculus* (Fig. 4.1e) and *Lagenophora nudicaulis* (Fig. 4.1f). These last genera of petite herbaceous plants are of particular interest to botanists, because they lack a pappus and sticky fruits while showcasing a transpacific distribution, predominantly inhabiting remote islands in the southern hemisphere (Sancho et al. 2023). Recently, we discovered *Calceolaria fothergillii* specimens (Fig. 4.1d) on Isla de los Estados, which was previously considered endemic to the Malvinas Islands (Baranzelli et al. 2018). Similarly, one of the only two native Antarctic species, *Colobanthus quitensis*, has also been recorded on Isla de los Estados (Fig. 4.1g).

Only seven introduced species were documented: *Taraxacum officinale*, *Cerastium fontanum*, *Sagina procumbens*, *Stellaria media*, *Poa annua*, *Poa pratensis*, and *Rumex acetosella*. These are widespread European-Eurasian ruderal and weedy species typical of temperate regions. Luckily, none are currently dominant on the island (Biganzoli 2003). Their presence has also been documented in the Isla Grande of Tierra del Fuego and other nearby Sub-Antarctic islands, including the Malvinas Islands (Moore 1967), South Georgia (Greene 1964, 1969), and remote

Table 4.1 Vascular plant species recorded for Isla de los Estados and Año Nuevo Islands, their status and some illustrative examples of reference specimens from registered herbarium collections

Family	Species	Status	Examples of herbarium collection
Alstroemeriaceae	*Luzuriaga marginata* (Gaertn.) Benth.	Common	Biganzoli 483 (SI), Testoni 1607 (BA)
Apiaceae	*Apium australe* Thouars	Common	Nicora 7234 (SI), Testoni 1615 (BA)
	Azorella filamentosa Lam.	Common	Castellanos s/n (BA 12948), Dudley 1024 (SI), Nicora 7266 (SI)
	Azorella fuegiana Speg.	Rare	Dudley 1509B (NA), Rodriguez 40 (SI)
	Azorella lycopodioides Gaudich.	Common	Nicora 7284 (SI), Dudley 903 (BAB), Biganzoli 717 (SI)
	Azorella ranunculus d'Urv.	Common	Rodriguez s/n (SI 163395), Biganzoli 666 (SI), Testoni 1635 (BA)
	Azorella selago Hook. f.	Common	Dudley 808 (BAB), Biganzoli 750 (SI)
	Azorella trifurcata (Gaertn.) Pers.	Common	Biganzoli 760 (SI), Testoni 1641 (BA)
	Bolax gummifera (Lam.) Spreng.	Common	Nicora 7282 (SI), Dudley 1737 (SI), Biganzoli 758 (SI), Testoni 1656 (BA)
	Lilaeopsis occidentalis J.M. Coult. & Rose	Rare	Dudley 1613 (BAB), Dudley 557 (NA)
Aspleniaceae	*Asplenium dareoides* Desv.	Common	Dudley 796 (SI), Dudley 1516 (SI), Biganzoli 497 (SI), Testoni 1624 (BA), Testoni 1625 (SI)
Asteliaceae	*Astelia pumila* (G. Forst.) Gaudich.	Common	Testoni 1643 (BA), Biganzoli 476 (SI)
Asteraceae	*Abrotanella emarginata* (Gaudich.) Cass.	Common	Rodriguez 11,956 (BAB), Biganzoli 698 (SI), Dudley 1731 (SI), Nicora 7243 (BAB)
	Chiliotrichum diffusum (G. Forst.) Kuntze	Common	Biganzoli 734 (SI), Dudley 829 (SI), Pichi-Sermoli 7449 (SI), Castellanos s/n (BA)
	Gamochaeta affinis Cabrera	Common	Dudley 1789 (NA), Dudley 1266 (NA)
	Gamochaeta nivalis Cabrera	Common	Testoni 1594 (BA), Biganzoli 661 (SI)
	Gamochaeta spiciformis (Sch. Bip.) Cabrera	Common	Nicora 7224 (SI), Dudley 880 (SI)
	Hypochaeris arenaria Gaudich.	Uncertain	Photograph and specimen in registration process at MFM
	Iocenes virens (Phil.) Ciald., M. Muñoz & Morrone	Common	Biganzoli 769 (SI), Nicora 7201 (BAA), Dudley 1298 (SI)

Family	Species	Status	Examples of herbarium collection
	Lagenophora nudicaulis (Comm. ex Lam.) Dusén	Common	Testoni 1634 (BA), Biganzoli 657 (SI), Castellanos s/n (BA)
	Leptinella scariosa Cass.	Common	Testoni 1646 (BA), Biganzoli 711 (SI), Nicora 7232 (BAA), Castellanos s/n (BA)
	Macrachaenium gracile Hook. f. var. *gracile*	Rare	Biganzoli 664 (SI), Dudley 1790 (NA)
	Nassauvia latissima Skottsb.	Rare	Skottsberg 273 (SI), Dudley 679 (NA)
	Nassauvia pygmaea (Cass.) Hook. f. var. *pygmaea*	Common	Biganzoli 719 (SI), Nicora 7362 (SI), Dudley 1330 (SI)
	Perezia lactucoides (Vahl) Less.	Common	Biganzoli 762 (SI), Castellanos s/n (LP)
	Perezia magellanica (L. f.) Less.	Common	Testoni 1593 (BA), Biganzoli 682 (SI), Holmberg s/n (SI), Dudley 1602 (BAB)
	Senecio candidans DC.	Rare	Castellanos s/n (BA), Dudley 1300 (NA)
	Senecio eightsii Hook. & Arn.	Common	Biganzoli 707 (SI), Dudley 827 (SI), Nicora 7216 (SI)
	Senecio smithii DC.	Common	Testoni 1645 (BA), Biganzoli 714 (SI), Dudley 729 (NA)
	Senecio trifurcatus (G. Forst.) Less.	Common	Biganzoli 683 (SI), Dudley 1694 (SI)
	Senecio websteri Hook. f.	Common	Testoni 1648 (BA), Biganzoli 673 (SI), Dudley 518 (SI), Nicora 7253 (BAA)
	Symphyotrichum vahlii (Gaudich.) G.L. Nesom	Rare	Biganzoli 740 (SI), Dudley 1609 (NA)
	Taraxacum gilliesii Hook. & Arn.	Rare	Dudley 1511 (NA)
	Taraxacum officinale F.H. Wigg.*	Rare	Biganzoli 768 (SI), Dudley 1510 (NA)
Berberidaceae	*Berberis empetrifolia* Lam.	Uncertain	Cited by Spegazzini
	Berberis ilicifolia L. f.	Common	Testoni 1655 (BA), Biganzoli 677 (SI), Nicora 7203 (SI)
	Berberis microphylla G. Forst.	Common	Biganzoli 679 (SI), Nicora 7290 (SI), Dudley 1417 (SI), Roig Juñet s/n (MERL)
Blechnaceae	*Austroblechnum penna-marina* (Poir.) Gasper & V.A.O. Dittrich	Common	Nicora 7197 (SI), Dudley 1060 (LP)
	Lomariocycas magellanica (Meyen) Acevedo & Kausel	Common	Biganzoli 468 (SI), Torres 1179 (LP)
Brassicaceae	*Cardamine geraniifolia* (Poir.) DC.	Common	Dudley 1201 (NA), Dudley 1788A (NA)
	Cardamine glacialis (G. Forst.) DC.	Common	Testoni 1626 (BA), Biganzoli 658 (SI), Nicora 7271 (SI), Dudley 1129 (SI)

Family	Species	Status	Examples of herbarium collection
Calceolariaceae	*Calceolaria fothergillii* Sol	Rare	Photographs Fig. 4.1d
Campanulaceae	*Lobelia oligophylla* (Wedd.) Lammers	Common	Biganzoli 732 (SI), Albov 3872 (SI), Testoni 1610 (BA)
Caryophyllaceae	*Cerastium arvense* L.	Common	Biganzoli 665 (SI), Dudley 1293 (SI), Dudley 1369 (SI)
	Cerastium fontanum Baumg.*	Rare	Dudley 783, Photograph and specimen in registration process at MFM
	Colobanthus quitensis (Kunth) Bartl.	Common	Testoni 1627 (BA), Dudley 1753 (SI)
	Colobanthus subulatus (d'Urv.) Hook. f.	Common	Biganzoli 670 (SI), Hicken s/n (SI 066773), Dudley 1094 (SI), Nicora 7248 (SI)
	Sagina procumbens L.*	Rare	Photograph and specimen in registration process at MFM
	Stellaria debilis d'Urv.	Common	Testoni 1628 (BA), Biganzoli 766 (SI)
	Stellaria media (L.) Cirillo*	Rare	Photograph and specimen in registration process at MFM
Crassulaceae	*Crassula moschata* G. Forst.	Common	Dudley 496 (SI), Biganzoli 669 (SI), Biganzoli 509 (SI)
Cyperaceae	*Carex banksii* Boott var. *banksii*	Common	Castellanos 2459 (SI), Dudley 1271 (SI), Dudley 882 (NA)
	Carex camptoglochin V.I. Krecz.	Rare	Castellanos 264 (SI), Biganzoli 744 (SI), specimen in registration process at MFM
	Carex firmula (Kük.) J.R. Starr	Rare	Biganzoli 718 (SI), Dudley 1134 (NA), Dudley 1171 (NA)
	Carex kingii (Boott) Reznicek	Rare	Rodriguez 2303 (SI), Castellanos 212 (SI)
	Carex meridensis (Steyerm.) J.R. Starr	Uncertain	Testoni 1650 (BA)
	Carex turbaria J.R. Starr	Rare	Biganzoli 739 (SI)
	Carpha schoenoides Banks & Sol. ex Hook. f.	Common	Biganzoli 738 (SI), Testoni 1652 (BA)
	Eleocharis pseudoalbibracteata S. González & Guagl.	Rare	Dudley 586 (NA), Dudley 1479 (NA)
	Isolepis cernua (Vahl) Roem. & Schult.	Common	Testoni 1654 (BA), Biganzoli 705 (SI), Dudley 1033 (SI), Dudley 721 (SI)
	Oreobolus obtusangulus Gaudich.	Common	Castellano 133 (SI), Nicora 7279 (SI), Dudley 1441 (SI), Dudley 1528 (SI)
	Schoenus antarcticus (Hook. f.) Dusén	Rare	Dudley 1384D (NA)
Cystopteridaceae	*Cystopteris apiiformis* Gand.	Rare	Testoni 1623 (BA), Dudley 512 (SI)
Droseraceae	*Drosera uniflora* Willd.	Common	Testoni 1614 (BA), Biganzoli 761 (SI), Nicora 7221 (SI), Dudley 1669 (SI)

Family	Species	Status	Examples of herbarium collection
Dryopteridaceae	*Polystichum chilense* (H. Christ) Diels var. *chilense*	Common	Dudley 1202 (LP), Albov s/n (SI 051215), Dudley 1746 (SI), Dudley 1103 (SI), Skottsberg 116 (SI), Spegazzini s/n (SI 05210), Biganzoli 708 (SI)
	Polystichum multifidum (Mett.) T. Moore var. *Multifidum*	Uncertain	cited by Correa (1998)
Ericaceae	*Empetrum rubrum* Vahl ex Willd.	Common	Testoni 1520 (BA), Biganzoli 733 (SI)
	Gaultheria antarctica Hook. f.	Common	Dudley 736 (BAB), Biganzoli 475 (SI), Testoni 1616 (BA)
	Gaultheria mucronata (L. f.) Hook. & Arn.	Common	Testoni 1619 (BA), Bignazoli 725 (SI), Dudley 1762 (BAB), Brickner 4718 (BAB), Moore 2049 (BAB)
	Gaultheria pumila (L. f.) D.J. Middleton var. *Pumila*	Common	Coll 528 (BAB), Biganzoli 749 (SI), Nicora 7356 (SI), Dudley 905 (BAB)
	Lebetanthus myrsinites (Lam.) Dusén	Common	Dudley 771 (BAB), Moore 2059 (BAA), Albov 644 (SI), Nicora 7336 (BAB), Biganzoli 688 (SI)
Escalloniaceae	*Escallonia serrata* Sm.	Common	Nicora 7235 (SI), Biganzoli 487 (SI), Testoni 1621 (BA)
	Tribeles australis Phil.	Common	Biganzoli 500 (SI), Biganzoli 681 (SI), Dudley 481
Euphorbiaceae	*Dysopsis glechomoides* (A. Rich.) Müll. Arg.	Common	Biganzoli 659 (SI), Dudley et al. 722 (NA)
Gleicheniaceae	*Sticherus cryptocarpus* (Hook.) Ching	Rare	Dudley 1170 (SI), Dudley et al. 1328 (NA)
Grossulariaeae	*Ribes magellanicum* Poir. subsp. *Magellanicum*	Common	Dudley 1124 (NA), Dudley 1450 (NA)
Gunneraceae	*Gunnera lobata* Hook. f.	Common	Dudley 1179 (SI), Testoni 1641 (BA), Biganzoli 743 (SI)
	Gunnera magellanica Lam.	Common	Biganzoli 480 (SI), Biganzoli 692 (SI)
Hymenophyllaceae	*Hymenophyllum caespitosum* Gaudich.	Common	Dudley 758 (SI), Dudley 1575 (BAB), Dudley 1180 (SI), Dudley 1019 (SI), Albov 924 (SI), Nicora 7288 (SI)
	Hymenophyllum darwinii Hook. f. ex Bosch	Rare	Dudley 1060 (BAB)
	Hymenophyllum dentatum Cav.	Common	Dudley 853 (NA), Dudley 1319 (NA)
	Hymenophyllum falklandicum Baker	Common	Dudley 1017 (LP), Testoni 1642 (BA), Skootsberg s/n (SI)
	Hymenophyllum ferrugineum Colla subsp. *berteroi* R. Rodr. & Penneck.	Rare	Dudley 1580 (SI 031670)
	Hymenophyllum secundum Hook. & Grev.	Common	Dudley 704 (LP), Albov 918 (SI), Dudley 1740 (SI 022163), Dudley 1095 (BAB), Torres 1116 (SI),Biganzoli 470B (SI)
	Hymenophyllum tortuosum Hook. & Grev.	Common	Dudley 1323 (LP), Albov s/n (SI), Nicora 7312 (SI), Castellanos s/n (LIL), Biganzoli 470A (SI), Dudley 1589 (SI), Dudley 1097 (LP), Torres 1117 (LP)

Family	Species	Status	Examples of herbarium collection
Juncaceae	*Juncus scheuchzerioides* Gaudich.	Common	Testoni 1651 (BA), Testoni 1653 (BA), Biganzoli 712 (SI)
	Luzula alopecurus Desv.	Rare	Biganzoli 651 (SI), Dudley 1367 (NA)
	Luzula antarctica Hook. f.	Rare	Castellanos s.n. (BA 12871; BA 12873), Dudley 1466 (NA)
	Luzula pumila Hook. f.	Uncertain	Spegazzini 215 (LP)
	Marsippospermum grandiflorum (L. f.) Hook. f.	Common	Biganzoli 504 (SI) Dudley 815 (NA)
	Marsippospermum reichei Buchenau	Rare	Biganzoli 724 (SI)
	Rostkovia magellanica (Lam.) Hook. f.	Rare	Biganzoli 695 (SI), Dudley 1459 (NA)
Juncaginaceae	*Tetroncium magellanicum* Willd.	Rare	Biganzoli 699 (SI 024527)
Lentibulariaceae	*Pinguicula antarctica* Vahl	Common	Dudley 1527 (SI), Biganzoli 482 (SI), Biganzoli 691 (SI)
Lycopodiaceae	*Austrolycopodium alboffii* (Rolleri) Holub	Common	Biganzoli 694 (SI), Biganzoli 477 (SI)
	Austrolycopodium confertum (Willd.) Holub	Common	Spegazzini s/n (SI 049964), Dudley 1738 (NA)
	Austrolycopodium magellanicum (P. Beauv.) Holub	Common	Dudley 1720 (SI), Dudley 1326 (NA)
Misodendraceae	*Misodendrum brachystachium* DC.	Common	Testoni 1611 (BA), Biganzoli 676 (SI), Biganzoli 877 (SI)
	Misodendrum punctulatum DC.	Common	Biganzoli 686 (SI), Dudley 575 (NA)
Montiaceae	*Montia fontana* L.	Common	Biganzoli 728 (SI), Dudley 1476 (NA)
Myrtaceae	*Myrteola nummularia* (Poir.) O. Berg	Common	Biganzoli 656 (SI), Biganzoli 478 (SI)
Nothofagaceae	*Nothofagus antarctica* (G. Forst.) Oerst.	Common	Dudley 500 (NA), Dudley 1416 (NA)
	Nothofagus betuloides (Mirb.) Oerst.	Common	Biganzoli 678 (SI), Biganzoli 506 (SI)
	Nothofagus pumilio (Poepp. & Endl.) Krasser	Uncertain	Dudley 1061 (NA) with contradictory determinations
Onagraceae	*Epilobium ciliatum* Raf.	Uncertain	Dudley 1294 (NA) not found
Ophioglossaceae	*Botrychium dusenii* (H. Christ) Alston	Rare	Castellanos 12,790 (BA)
Orchidaceae	*Codonorchis lessonii* (Brongn.) Lindl.	Common	Biganzoli 1241 (SI), Biganzoli 495 (SI), Nicora 7212 (SI)
Oxalidaceae	*Oxalis magellanica* G. Forst.	Rare	Dudley 1281 (NA)

Family	Species	Status	Examples of herbarium collection
Poaceae	*Agrostis meyenii* Trin.	Rare	Testoni 1597 (BA)
	Agrostis uliginosa Phil.	Rare	Castellanos 12,860 (BAA)
	Anthoxanthum moorei (De Paula) N. Villalobos & Finot	Rare	Moore 2064 (BAB)
	Anthoxanthum redolens (Vahl) P. Royen	Common	Dudley 1308 (SI), Testoni 1608 (BA), Dudley 1544 (SI), Biganzoli 493 (SI)
	Cortaderia egmontiana (Roem. & Schult.) M. Lyle ex Connor	Rare	Castellanos s/n (BA), Testoni 1658 (BA)
	Deschampsia kingii (Hook. f.) E. Desv.	Common	Biganzoli 510 (SI), Testoni 1598 (BA), Hicken 15,044 (BAA)
	Deschampsia parvula (Hook. f.) E. Desv.	Rare	Castellanos 1285 (BAA)
	Elymus magellanicus (E. Desv.) A. Löve	Rare	Nicora 12,853 (BAA)
	Festuca cirrosa (Speg.) Parodi	Common	Spegazzini 276 (BAA), Castellanos 12,865 (BAA), Testoni 1601 (BA), Biganzoli 505 (SI)
	Festuca contracta Kirk	Rare	Biganzoli 633 (SI)
	Festuca magellanica Lam.	Rare	Biganzoli 752 (SI)
	Festuca purpurascens Banks & Sol. ex Hook. f.	Rare	Biganzoli 755 (SI)
	Festuca pyrogea Speg.	Common	Testoni 1602 (BA), Biganzoli 759 (SI)
	Koeleria spicata (L.) Barberá, Quintanar, Soreng & P.M. Peterson	Common	Castellanos s/n (BAA), Dudley 646 (SI), Nicora 7354 (BAA), Spegazzini s/n (LP)
	Leymus arenarius (L.) Hochst.	Uncertain	Photographs
	Nicoraepoa robusta (Steud.) Soreng & L.J. Gillespie	Common	Castellanos 12,845 (BAA), Dudley 1028 (SI), Nicora 7330
	Ortachne rariflora (Hook. f.) Hughes	Uncertain	Photograph and specimen in registration process at MFM
	Phleum alpinum L.	Rare	Hicken s/n (SI 13771)
	Poa alopecurus (Gaudich. ex Mirb.) Kunth subsp. *alopecurus*	Common	Nicora 7261 (BAA), Castellanos 12,854 (BAA), Castellanos 11,829 (BAA)

Family	Species	Status	Examples of herbarium collection
	Poa alopecurus (Gaudich. ex Mirb.) Kunth subsp. *fuegiana* (Hook. f.) D.M. Moore & Dogg.	Common	Testoni 1600 (BA), Dudley 887 (NA)
	Poa alopecurus (Gaudich. ex Mirb.) Kunth subsp. *shuka* (Speg.) Giussani & Soreng	Rare	Spegazzini s/n (LF)
	Poa annua L.*	Rare	Nicora 7217 (SI), Photograph and specimen in registration process at MFM
	Poa darwiniana Parodi	Common	Castellanos s/n (BAA), Nicora 7245 (SI; BAA)
	Poa flabellata (Lam.) Raspail	Common	Dudley 382 (SI), Dudley 1144 (SI), Castellanos 12,836 (BAA), Biganzoli 503 (SI), Nicora 7359 (BAA)
	Poa lanuginosa Poir. var. *patagonica* (Phil.) Giussani & Soreng	Rare	Castellanos s/n (BA 12829)
	Poa pratensis L.*	Uncertain	Specimen in registration process at MFM
	Poa yaganica Speg.	Rare	Testoni 1657 (BA), specimen in registration process at MFM
	Polypogon magellanicus (Lam.) Finot	Common	Moore 2053 (SI), Testoni 1595 (BA), Dudley 1005 (SI), Hicken s/n (SI)
	Puccinellia pusilla (Hack.) Parodi	Rare	Skottsberg 115 (SI), Castellanos s/n (BA 12828)
Polygonaceae	*Rumex acetosella* L.*	Rare	Biganzoli 496 (SI) - in Observatorio island, Photographs in Port Parry
Polypodiaceae	*Notogrammitis angustifolia* (Jacq.) Parris	Common	Biganzoli 473 (SI), Biganzoli 687 (SI), Nicora 7181 (SI), Biganzoli 473 (SI), Albov 861 (SI)
Plantaginaceae	*Callitriche antarctica* Engelm. ex Hegelm.	Common	Nicora 7267 (SI), Nicora 7334 (SI), Nicora 7239 (SI)
	Ourisia breviflora Benth. subsp. *Breviflora*	Rare	Dudley 1717 (BAB), Dudley 1339 (NA)
	Ourisia fuegiana Skottsb.	Rare	Biganzoli 479 (SI)
	Ourisia ruelloides (L. f.) Kuntze	Common	Dudley 1500 (BAB), Biganzoli 1255 (SI), Dudley 1268 (BAB)
	Plantago barbata G. Forst. subsp. *Barbata*	Common	Dudley 847 (NA), Dudley 1360 (NA)
	Veronica elliptica G. Forst.	Rare	Biganzoli 1248 (SI), Dudley 1703 (NA)
Plumbaginaceae	*Armeria maritima* (Mill.) Willd.	Common	Testoni 1618 (BA), Biganzoli 722 (SI), Dudley 1102 (BAB)
Primulaceae	*Lysimachia amoena* (Schl. ex Gaertn.) Hassemer, Funez & R. Trevis.	Common	Testoni 1617 (BA), Biganzoli 713 (SI)
	Primula magellanica Lehm.	Common	Biganzoli 484 (SI), Biganzoli 710 (SI)

Family	Species	Status	Examples of herbarium collection
Ranunculaceae	*Caltha appendiculata* Pers.	Rare	Dudley 1411B (NA)
	Caltha dioneifolia Hook. f.	Common	Dudley 735 (SI), Biganzoli 472 (SI), Biganzoli 697 (SI)
	Caltha sagittata Cav.	Rare	Dudley 1197 (NA)
	Hamadryas magellanica Lam. f. *magellanica*	Common	Nicora 7215 (SI), Biganzoli 720 (SI), Biganzoli 715 (SI)
	Hamadryas magellanica Lam. f. *paniculata* (Hook. f.) Lourteig	Common	Dudley 945 (SI), Dudley 1020 (NA)
	Ranunculus biternatus Sm.	Common	Testoni 1638 (BA), Biganzoli 660 (SI)
	Ranunculus fuegianus Speg.	Rare	Testoni 1640 (BA)
Restionaceae	*Gaimardia australis* Gaudich.	Common	Dudley 1148 (SI), Nicora 7258 (SI), Biganzoli 736 (SI)
Rosaceae	*Acaena antarctica* Hook. f.	Common	Biganzoli 753 (SI), Dudley 1425 (NA)
	Acaena magellanica (Lam.) Vahl	Common	Biganzoli 492 (SI), Biganzoli 747 (SI), Dudley 1299 (NA)
	Acaena ovalifolia Ruiz & Pav.	Rare	Dudley 469 (NA)
	Acaena pumila Vahl	Common	Testoni 1631 (BA), Biganzoli 684 (SI), Dudley 1412 (NA)
	Rubus geoides Sm.	Common	Biganzoli 481 (SI), Testoni 1659 (BA), Biganzoli 730 (SI)
Rubiaceae	*Coprosma granadensis* (Mutis ex L.f.) Heads	Rare	Dudley 883 (NA)
	Galium antarcticum Hook. f.	Common	Testoni 1637 (BA), Biganzoli 703 (SI)
	Galium fuegianum Hook. f.	Rare	Dudley 1047 (NA)
Santalaceae	*Nanodea muscosa* Banks ex Gaertn.	Common	Biganzoli 667 (SI), Biganzoli 501 (SI)
Saxifragaceae	*Chrysosplenium macranthum* Hook.	Rare	Biganzoli 754 (SI), Dudley 1272 (NA)
	Saxifraga bicuspidata Hook. f.	Rare	Castellanos 12,983 (BAA)
	Saxifraga magellanica Poir.	Rare	Bignazoli 723 (SI), Dudley 1244 (NA)
Thymelaeaceae	*Drapetes muscosus* Lam.	Common	Testoni 1612 (BA), Biganzoli 693 (SI), Biganzoli 471 (SI)
Violaceae	*Viola commersonii* DC. ex Ging.	Common	Biganzoli 485 (SI), Testoni 1613 (BA), Bignazoli 498 (SI)
	Viola magellanica G. Forst.	Common	Biganzoli 498 (SI), Nicora 7211 (SI 091720), Testoni 1622 (BA), Dudley 1267 (SI)
	Viola tridentata Menzies ex Ging.	Common	Biganzoli 726 (SI), Dudley 1716 (NA)
Winteraceae	*Drimys winteri* J.R. Forst. & G. Forst. var. winteri	Common	Biganzoli 770 (SI), Dudley 725 (NA)

Introduced species are indicated with an asterisk

Fig. 4.1 Some native species from Isla de los Estados of: (**a**) *Senecio eightsii*, (**b**) *Senecio websteri*, (**c**) *Poa flabellata*, (**d**) *Calceolaria fothergillii*, (**e**) *Azorella ranunculus*, (**f**) *Lagenophora nudicaulis*, (**g**) *Colobanthus quitensis*, (**h**) *Drosera uniflora*, (**i**) *Poa yaganica*. Credits: Rodrigo Iturraspe (**c**, **f**), Amira Salom (**a**, **d**, **e**), Daniel Testoni (**b**, **g**, **h**, **i**)

islands such as Crozet, Kerguelen (Frenot et al. 2001), and Marion (Gremmen and Smith 1999). Although the current impact of exotic species on the island's ecosystems appears negligible, it is crucial to monitor their status and prevent new introductions, since under climate change scenarios and increasing ecotourism, invasive species could potentially disrupt native plant communities (Whinam 2009).

4.2 Mapping the Vegetation Mosaic

Within Argentina, Isla de los Estados presents an ecologically distinctive vegetation. As the only region in the country with fjords, its rugged topography and strong oceanic influence shape its diverse plant communities (Ponce and Fernandez 2014). Year-round intense winds and its edaphic diversity, ranging from peatlands to mineral soils and scree fields, further defines the island's vegetation structure and distribution (Dudley and Crow 1983).

The island hosts Argentina's only Magellanic moorland, an ecosystem shaped by persistent winds, low temperatures, and maritime humidity, with affinities to Sub-Antarctic islands further south (Dudley and Crow 1983; Arana et al. 2021). It also marks the southernmost extent of the temperate subantarctic forest present in Argentina and Chile, which spans a narrow but extensive latitudinal range (35–55° S) along both flanks of the Andes mountain range. This distribution gives rise to distinct forest formations shaped by local adaptation to environmental conditions (Marchelli et al. 2021). The formation present on Isla de los Estados is the subantarctic evergreen forests, characterized by the dominance of evergreen *Nothofagus* species (Dudley and Crow 1983). This type of forests descends directly from mountain slopes to the coast and are exclusive from southern Tierra del Fuego and nearby islands (Veblen et al. 1996). Comparable evergreen broad-leaved *Nothofagus* forests are otherwise found only in upland regions of Tasmania and New Zealand (Box 2002). Similarly, coastal habitats, enriched by seabird-derived nutrients, support unique vegetation formations resembling those in southern archipelagos like the Malvinas Islands (Groff et al. 2020).

This environmental diversity makes defining discrete and uniform vegetation units on Isla de los Estados particularly challenging. Steep gradients and rugged terrain produce a mosaic of plant communities with diffuse boundaries. Even within the same nominal type, floristic composition and structure can vary by location, influenced by longitudinal and altitudinal shifts (Dudley and Crow 1983; Huertas Herrera et al. 2018). Here we adopt and revise the classification scheme of Dudley and Crow (1983), supported by field observations and modern GIS tools. Their work aboard the R/V Hero remains the most thorough botanical survey of the island. Few studies since have matched its scope, with the exception maybe of the collective surveys of our own team (Biganzoli 2003; Collado 2015; Balza et al. 2022; Salom et al. 2025a). Based on these efforts, we propose a set of vegetation units that integrate floristic, structural, and ecological traits reflecting the island's environmental complexity. These units are best viewed as practical frameworks for interpreting ecological patterns, rather than fixed categories.

We also present a vegetation map grounded in field expertise and ecological understanding. Aimed at multiple uses, its development was informed by plant community knowledge and the needs of conservation and land-use planning. As a current snapshot of plant community distribution, the map offers a baseline for monitoring and a framework for interpreting vegetation-environment relationships and fauna distribution. In a dynamic landscape where climate change and invasive species may drive change, vegetation maps will be essential for long-term ecological understanding and informed conservation. Below, we describe the methods used to define, classify, and spatially represent the island's vegetation units.

4.2.1 Mapping Methodology

To generate a land cover map, we used three Sentinel-2 images (26 December 2022) covering the entire study area, selected for minimal interference of snow, clouds, and shadows in the classification process (European Space Agency 2018). Then, we generated a single multispectral following the methods described in Balza et al. (2022). Based on field data, drone images, georeferenced photos, and visual interpretation of high-resolution imagery (Google Earth Pro, Bing Satellite), and using various band combinations of the multispectral image, we defined 560 training areas for a semi-automatic classification process using (SCP) (QGIS Development Team 2020; Congedo 2021).

Classification was performed using the Remotior Sensus package in Python v.3.12.8 (Python Software Foundation 2023; Congedo 2023). We performed a random forest classification using 800 trees, with a minimum of two samples to split a node, and the number of variables considered at each split set to the square root of the total number of input variables ($\sqrt{10}$). After classification, the raster map was vectorized and manually corrected for misclassified areas due to terrain shadows, clouds, or overexposure. Corrections were guided by the same ground and satellital reference data.

To further improve the map's interpretability, we reduced classification noise by reclassifying patches <500 m^2, merging them with the largest adjacent contiguous class, using the *v.clean* function in GRASS GIS v.8.4.0 (GRASS Development Team 2022). Vegetation classes were initially defined to minimize within-class multispectral reflectance variability and maximize separability. These were later merged into final vegetation types based on field data, known plant communities, and classification confidence. Where uncertainty existed, similar classes were merged to ensure that the resulting map accurately reflected dominant patterns, thus preventing misinterpretation of fine-scale noise as distinct communities.

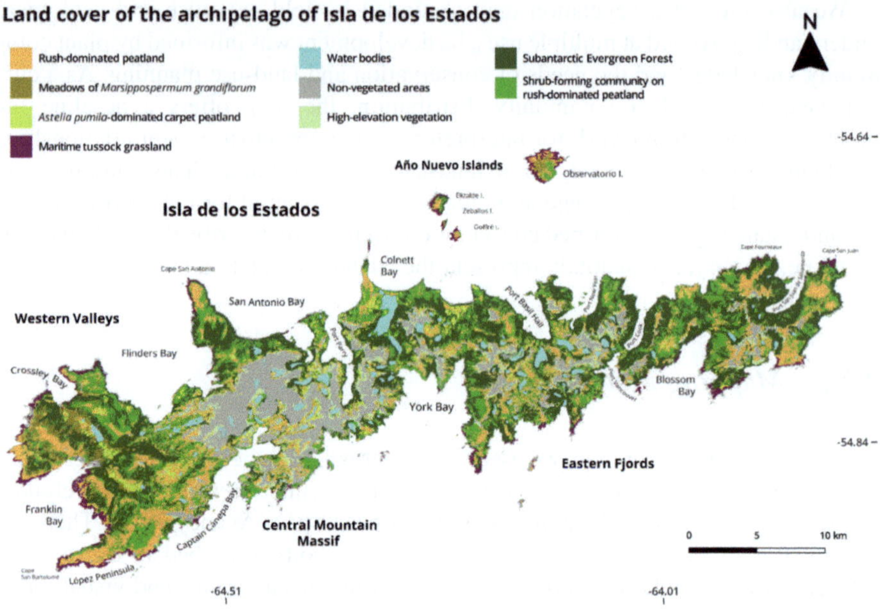

Fig. 4.2 Land cover map of Isla de los Estados, Año Nuevo Islands (Observatorio, Goffré, Zeballos and Elizalde Islands) and adjacent islets. (Shape available at Salom et al. (2025b))

4.2.2 Vegetation Formations

We identified nine major land cover types across Isla de los Estados archipelago, including two non-vegetated classes and seven plant communities with distinct floristic and structural features (Fig. 4.2). These include water bodies, non-vegetated areas, subantarctic evergreen forest, *Astelia pumila*-dominated carpet peatland, rush-dominated peatland, *Marsippospermum grandiflorum* meadows, maritime tussock grassland, shrub-forming community on rush-dominated peatland, and high-elevation vegetation. Some are named after physiognomic dominant species (e.g., *M. grandiflorum*), though true dominance is rare under the island's extreme conditions, except for *Poa flabellata* in some coastal areas.

In terms of area coverage, the archipelago comprises 27.6% of subantarctic evergreen forest, 22.1% of shrub-forming community on rush-dominated peatland, 19.2% of rush-dominated peatland, 17.3% of non-vegetated areas, 3.6% of water bodies, 3.5% of high-elevation vegetation, 2.5% of *Astelia pumila*-dominated carpet peatland, 2.3% of maritime tussock grassland, and 2.0% *Marsippospermum* meadows. These classes represent the archipelago's main vegetation structures and spatial patterns. Below, we present a detailed description of the ecological characteristics and distributional patterns of each class. Additionally, we recognize a littoral formation not mapped here due to its limited extent.

4.2.2.1 Subantarctic Evergreen Forests

Forests cover nearly one-third of Isla de los Estados, over 140 km^2, making them the archipelago's most prominent vegetation type. They are widespread across the main island and the Año Nuevo Islands, with greater structural development in wind-sheltered sites like northern slopes and deep fjords. Two forest types dominate: multi-stratified mixed forests of *Nothofagus betuloides* and *Drimys winteri* (Fig. 4.3b–e), and pure stands of *N. betuloides* (Fig. 4.3a, f).

The multi-stratified mixed forests, found on wet, peaty soils, are the most diverse and representative of the region. These forests typically exhibit a dense structure with full vertical occupation, often making understory access difficult (Fig. 4.3c–e). Up to six vertical strata may develop, including two tree layers, one or two shrub layers, and both herbaceous and subshrub layers, resulting in a structurally complex system (Collado 2015).

Nothofagus betuloides typically form broad crowns in the upper canopy at 15–20 m, rising 5–8 m above the denser lower crowns of *D. winteri* (Veblen et al. 1996). Both species are shade-tolerant and maintain persistent seedling banks, with *D. winteri* seedlings generally being more abundant (Dudley and Crow 1983; Veblen et al. 1996). The shrub layer includes *Lebetanthus myrsinites*, *Gaultheria mucronata*, *Berberis ilicifolia*, and *Chiliotrichum diffusum* (Fig. 4.3c, f).

The herbaceous layer is dense with ferns, mosses, liverworts, and species such as *Luzuriaga marginata*, *Viola magellanica*, *Lagenophora nudicaulis*, *Dysopsis glechomoides*, and *Montia fontana* (Salom et al. 2025a). Common ferns include *Hymenophyllum* spp. and *Asplenium dareoides*, with *Sticherus cryptocarpus*, *Austroblechnum penna-marina*, and *Lomariocycas magellanica* (Fig. 4.3d) more localized (Young 1973; Dudley and Crow 1983; Collado 2015).

With increasing elevation, forest structural complexity (i.e. dominant height, basal area, DBH, and number of strata) declines, accompanied by changes in tree architecture, species richness, and life-form composition (Collado 2015; Huertas Herrera et al. 2018). On Isla de los Estados, only ~20% of forests above 100–200 m a.s.l. retain more than four strata, while above 200 m, stands consist generally of simpler pure *N. betuloides* stands, with closed canopies, sparse shrub layers, and patchy understories of bryophytes, sedges, rushes, and herbs (Fig. 4.3a, b, f). Bare ground and waterlogging vary with canopy openness. Similar elevation-driven patterns in forest structure and composition have been reported for *Nothofagus* forests on the Isla Grande of Tierra del Fuego (Barrera et al. 2000; Mestre et al. 2017), though further studies from high southern latitudes are still needed.

A notable high-elevation feature is the formation of "wind-wave forests" on exposed slopes above 100 m. These show alternating bands of dead trees and age-cohorts of living trees aligned perpendicular to prevailing winds, a structural response to chronic exposure, most common in pure *N. betuloides* stands, though also present in mixed forests.

Fig. 4.3 Internal diversity present in the subantarctic evergreen forest. (Credits: Leonardo Collado (**c–e**); Amira Salom (**a**, **b**, **f**))

4.2.2.2 Shrub-Forming Community on Rush-Dominated Peatland

This community is the second most extensive vegetation type, covering over 112 km^2 (Fig. 4.4). It usually forms transitional zones between forests and peatlands, with diffused boundaries. It corresponds to the scrub formation described by

Fig. 4.4 Internal diversity present in the *shrub-forming community on rush-dominated peatland*. (Credits: Leonardo Collado (**b**, **d**); Amira Salom (**a**, **c**, e, **f**))

Dudley and Crow (1983), comprising the *Nothofagus betuloides-Marsippospermum grandiflorum* and *Nothofagus antarctica* associations. The former is more widespread, with dense, nearly impenetrable *N. betuloides* thickets above a lower layer of *M. grandiflorum* (Fig. 4.4a, d, f). Understory species include *Chiliotrichum diffusum*, *Gaultheria mucronata*, *Berberis ilicifolia*, and *D. winteri*, while *Hymenophyllum tortuosum* occurs in sheltered spots (Fig. 4.4c). In wetter peatlands dominated by *M. grandiflorum* or *Astelia pumila*, *N. betuloides* and *D. winteri* frequently appear stunted and scattered (Fig. 4.4d). This association is widespread but is more frequent on northern and western slopes.

The *N. antarctica* association occupies a narrow elevational band above the evergreen forest and below the alpine zone, where this deciduous species grows as low, wind-sculpted krummholz (1.0–1.5 m), forming dense thickets. These often host *Chiliotrichum diffusum*, *Gaultheria mucronata*, *Berberis ilicifolia*, *Hymenophyllum tortuosum*, the moss *Sphagnum* spp., and occasionally *M. grandiflorum* in wetter microsites (Fig. 4.4b). These formations often grade into Magellanic moorland and *M. grandiflorum* meadows at lower elevations (Fig. 4.4e).

On Observatorio Island, where forests are absent, this vegetation type occupies similar wind-sheltered sites and is dominated by shrubby species such as *Berberis ilicifolia*, *B. microphylla*, *Chiliotrichum diffusum*, *Empetrum rubrum*, and *Gaultheria mucronata*, the latter dominant in the southeast. *Drimys winteri* appears only as scattered individuals up to 4 m tall (Fig. 4.5b). The lower layer is dominated by *Gaultheria pumila*, growing on bryophyte-rich peatlands interspersed with patches of exposed organic soil (Fig. 4.5c).

Fig. 4.5 Vegetation formations on Observatorio Island: (**a**) Aerial view of Observatorio Island, with the slightly higher Año Nuevo Islands visible in the background; (**b**) Shrub-forming community on rush-dominated peatlands in the southeastern sector, where individuals of *D. winteri* up to 4 m tall can be seen; (**c**) Rush-dominated peatland in the foreground, interspersed with the shrub-forming community on rush-dominated peatlands on the left and a *Marsippospermum grandiflorum* meadow in the background; (**d**) Maritime tussock grassland along the beach. (Credits: Joaquín Cano (**b**, **d**), Leonardo Collado (**c**), Rodrigo Iturraspe (**a**))

4.2.2.3 Rush-Dominated Peatland

Rush-dominated peatland forms a mosaic of two plant associations described by Dudley and Crow (1983), both found on peat substrates with *M. grandiflorum*: the *Empetrum rubrum* and the *Caltha* spp. associations. Widespread across the archipelago, it occupies nearly 97 km^2 and occurs on hills, slopes, and exposed plains unsuitable for forests or shrublands.

The *Empetrum rubrum* association is the dominant and most distinctive subunit of the Magellanic moorland (Fig. 4.6a, b), characterized by *E. rubrum*, *Gaultheria mucronata*, *Chiliotrichum diffusum*, *Berberis ilicifolia*, and *M. grandiflorum*. It also includes herbaceous species such as *Gunnera magellanica*, *Austroblechnum penna-marina*, *Luzuriaga marginata*, and scattered, stunted *D. winteri* and *N. betuloides*. Dense *Sphagnum* mounds and other bryophytes often surround the shrub bases, especially *Empetrum*, though *Sphagnum* does not form monospecific peat bogs as on Isla Grande of Tierra del Fuego. Drier sites, particularly promontories, support flat patches of *Rubus geoides* associated with other herbs such as *Lobelia oligophylla*. *Anthoxanthum redolens* appears locally, especially at ecotones between *M. grandiflorum* meadows and maritime tussock grasslands.

Fig. 4.6 Internal diversity of moorland vegetation types including the *Rush-dominated peatlands* (**a–e**) and *Astelia pumila-dominated carpet peatlands* (**f**): (**a, b**) *Empetrum rubrum* association; (**c, d**) *Caltha* spp. association; (**e**) *Bolax gummifera* dominated peatland at Cape San Bartolomé. (Credits: Amira Salom)

The *Caltha* spp. association occurs in localized areas where *E. rubrum* and *M. grandiflorum* are sparse (Fig. 4.6c, d). It is dominated by *Caltha dioneifolia* and *C. appendiculata*, often accompanied with *Astelia pumila*, *Myrteola nummularia*,

Gaultheria antarctica, *G. pumila*, *Gunnera lobata*, *Oreobolus obtusangulus*, *Tribeles australis*, *Bolax gummifera*, *Perezia magellanica*, and *Drapetes muscosus*. High-elevation bogs also feature *Rostkovia magellanica* and *Carpha schoenoides*.

A distinctive expression of this community occurs at Cape San Bartolomé (López Peninsula), the island's driest area, where vegetation reaches a near-climax stage dominated by *Bolax gummifera* (Fig. 4.6e). Here, *N. betuloides* and *D. winteri* are stunted by persistent southwesterly winds and confined to stream banks (Dudley and Crow 1983). On the cape's windswept slopes, *Poa flabellata* and large cushions of *B. gummifera* (1.0–1.5 m wide) dominate extensive *M. grandiflorum* peatlands, with scattered *Gunnera magellanica*, *A. penna-marina*, *Cardamine glacialis*, *Senecio websteri*, and *S. eightsii*.

On the southern slopes, where the substrate is loose rock, *B. gummifera* remains dominant, accompanied by cushion plants like *Colobanthus quitensis* and *Armeria maritima*, and tussock grasses including *Festuca* spp. (Fig. 4.8a). *Abrotanella emarginata*, typically alpine, grows only within *B. gummifera* cushions.

4.2.2.4 Astelia Pumila-Dominated Carpet Peatland

This distinct but relatively minor community is part of the Magellanic moorland and corresponds to the *Astelia pumila* association defined by Dudley and Crow (1983). Characterized by its rigid, carpet-forming growth (Fig. 4.6f), it occurs throughout the archipelago but fully develops on steep upper slopes and low valleys in the western and northern sectors of the main island. In these areas, it forms flat, firm patches composed almost entirely of *A. pumilia* within broader rush-dominated peatlands. It often intermingles with the *Caltha* spp. association, forming large, uniform flat mats.

Associated species may include *Gunnera lobata*, *Gaultheria pumila*, *Drosera uniflora*, *Oreobolus obtusangulus*, *Caltha dioneifolia*, *C. appendiculata*, *Abrotanella emarginata*, *Azorella lycopodioides*, and *Bolax gummifera*. The insectivorous *D. uniflora* often forms dense, purple-hued patches in wet areas such as snowmelt zones, adding striking visual contrast to the landscape (Fig. 4.1h).

Myrteola nummularia, *Perezia magellanica*, *Acaena pumila*, *Galium antarcticum*, and *Nanodea muscosa* are frequently found alongside these secondary species but are absent from pure *Astelia pumilia* stands. *Marsippospermum grandiflorum* may also be present but at very low cover. The rare fern *Sticherus cryptocarpus* is found almost exclusively within this community.

4.2.2.5 *Marsippospermum grandiflorum* Meadows

This community, composed almost entirely of *M. grandiflorum*, is the least extensive vegetation type in the archipelago, covering only 10 km^2. These true rush meadows occur almost exclusively in valleys and flat areas in the western region of the main island and on Observatorio Island. It can also extend into subalpine zones and summit ridges, particularly in shallow, moist, and sheltered depressions.

With over 80% ground cover, these meadows grow on fern and bryophyte-rich substrates. The ferns *Hymenophyllum* spp. and *Asplenium dareoides* are commonly present, while the moss *Sphagnum sp.* is often found nestled between and atop the rush stems. Shrubs such as *Chiliotrichum diffusum*, *Gaultheria mucronata*, *Berberis microphylla*, and *B. ilicifolia* may occur scattered across the landscape (Fig. 4.7b), along with stunted or shrub-like individuals of *N. betuloides* and *D. winteri*. In wetter and more open patches, *Gunnera magellanica* and *Ranunculus biternatus* contribute herbaceous diversity to this otherwise rush-dominated system.

4.2.2.6 Maritime Tussock Grassland

Though covering just 12 km^2, about 2% of the archipelago, Maritime Tussock Grasslands are notable for their proximity to the coast and importance as seabird nesting sites (Fig. 4.7c). They are most extensive on the exposed slopes and sea cliffs of the main island's western side (Fig. 4.7e) and are especially well-developed on Año Nuevo and Observatorio Islands (Fig. 4.5d), which have a more oceanic character.

This community is dominated by *Poa flabellata*, a perennial tussock-forming grass endemic to the South Atlantic. Seedlings of *P. flabellata* grow atop pedestals of decaying organic matter (Smith and Clymo 1984), and mature tussocks may reach up to 4 m in height, with crowns rising 2 m above ground level (Strange et al. 1988). These dense canopies often suppress the growth of other plants beneath them. However, browsing by introduced ungulates in the main island frequently limits *P. flabellata* size and vigor (Salom and Schiavini, Chap. 14, this volume).

In such disturbed areas, secondary grasses such as *Anthoxanthum redolens*, *Polypogon magellanicus*, *Poa yaganica* (Fig. 4.1i), and the exotic *Poa annua* are common, the latter often forming dense turf-like patches (Fig. 4.7d). Open areas between tussocks may support scattered herbs like *Montia fontana*, *Leptinella scariosa*, *Lobelia oligophylla*, *Viola magellanica*, *Cardamine glacialis*, *Apium australe*, *Austroblechnum penna-marina*, and *Dysopsis glechomoides*, along with *Asplenium dareoides* and bryophytes such as *Marchantia* spp. *Senecio smithii* is often abundant in lower, wetter areas, while scattered shrubs of *Gaultheria mucronata*, *Berberis ilicifolia*, *D. winteri*, and *Chiliotrichum diffusum* may occur throughout.

The best-preserved coastal grasslands are found on Alférez Goffre Island (Año Nuevo group) and Observatorio Island, both free of ungulates. Here, *Poa flabellata* and *M. grandiflorum* tussocks serve as nesting sites for Magellanic penguins (*Spheniscus magellanicus*). On Isla de los Estados, Southern rockhopper penguins (*Eudyptes chrysocome*) and Imperial shags (*Leucocarbo atriceps*) are typically associated with this grassland, while Magellanic penguins nest more often in moorland areas. In sites heavily used by seabirds, marine mammals, and ungulates, exotic species such as *Stellaria media*, *Sagina procumbens*, and *Poa annua* are commonly found (Fig. 4.7c). Similarly, in abandoned seabird colonies, natural vegetation recovery begins with rapid colonization by mosses and liverworts that stabilize the soil. These are followed by dense patches of *Colobanthus quitensis* (Fig. 4.1g), and

Fig. 4.7 *Meadows of Marsippospermum grandiflorum* (**a**, **b**) and *maritime tussock grasslands* (**c**–**e**). (Credits: Leonardo Collado (**e**), Amira Salom (**a**–**d**))

later by small grasses like *Agrostis meyenii*, along with *P. annua* and *S. media*. However, regeneration is often impeded by browsing and trampling from introduced ungulates.

4.2.2.7 High-Elevation Vegetation

High-elevation vegetation develops around 450 m but may occur at lower altitudes depending on exposure and soil conditions. Wind-swept peaks and ridgelines above the treeline support sparse, scattered vegetation, mostly confined to sheltered hollows and peat pockets among rocks in typically open alpine moorland (Fig. 4.8c). This community covers about 18 km^2, similar in extent to water bodies on the island, and is especially prominent around the peaks above Port San Juan del Salvamento. It forms a transition zone between non-vegetated rocky summits and rush-dominated peatlands (Fig. 4.8d). Vegetation here is dominated by cushion-forming and prostrate species (Fig. 4.8a). *Nothofagus antarctica* and *Empetrum rubrum* are common in subalpine zones but appear dwarfed and close to the ground (Fig. 4.8b). Typical alpine cushion species include *Azorella selago*, *A. lycopodioides*, *Bolax gummifera*, *Abrotanella emarginata*, *Caltha dioneifolia*, *Oreobolus obtusangulus*, *Gaimardia australis*, and *Drapetes muscosus*. *Gaultheria pumila*, *G. antarctica*, and the tussock-forming grass *Poa darwiniana* are also widespread. Ferns such as *Hymenophyllum falklandicum* and *H. caespitosum* grow in crevices between rocks, while *Senecio eightsii* is locally common in protected hollows but more typical at lower elevations. In contrast, *Austrolycopodium confertum* and *A. magellanicum* appear restricted to the highest areas and adjacent to Magellanic moorlands.

4.2.2.8 Non-vegetated Areas

The archipelago is characterized by a high proportion of non-vegetated areas, occupying almost 88 km^2. These are most extensive in the central sector and on the southern side of the main island, where harsh climatic conditions prevail. In the highest, exposed or coastal heights, it is characterized by rocky outcrops, while sand dunes and beaches dominate the western and northern maritime coasts (Fig. 4.8e). Bare ground is also found in areas disturbed by seabird colonies.

4.2.2.9 Water Bodies

Despite its steep and rugged topography, Isla de los Estados has a remarkably high proportion of freshwater bodies, covering nearly 18 km^2. The main island alone hosts almost 300 lakes, ponds, and smaller water bodies, with the highest concentration located in its central-eastern sector (Fig. 4.8f).

4.2.2.10 Littoral Formation

This plant community develops along rocky shores and cliffs, occupying small peatland areas just above the high-tide line and crevices within the intertidal zone. Dudley and Crow (1983) noted that it typically includes cushion or mat-forming

Fig. 4.8 Internal diversity of the *high-elevation vegetation* (**a–d**) and examples of beaches in the *non-vegetated areas* (**e**) and *water-bodies* (**f**). (Credits: Leonardo Collado (**b**, **c**, **f**), Amira Salom (**a**, **d**, **e**))

species such as *Colobanthus subulatus*, *C. quitensis*, *Crassula moschata*, *Armeria maritima*, *Poa darwiniana*, *Plantago barbata*, and *Azorella filamentosa*, although the latter two were not directly recorded by our team.

Above the rocky coastline, common species include *Leptinella scariosa*, *Ranunculus biternatus*, *Gunnera magellanica*, *Apium australe*, *Cardamine glacialis*, *Anthoxanthum redolens*, *Senecio eightsii*, and *S. websteri*. Bryophytes are also abundant in these moist, shaded microsites.

On sandy beaches, typical species include *Acaena magellanica*, *Apium australe*, *Anthoxanthum redolens*, *Juncus scheuchzerioides*, *Cardamine glacialis*, *Puccinellia pusilla*, *Deschampsia* sp., various *Festuca* species, and occasionally *Poa flabellata*. In sheltered bays, a narrow, dense shrub fringe often develops between the high-tide line and the forest. This fringe is typically dominated by *Escallonia serrata*, and occasionally includes *Veronica elliptica*, *Berberis microphylla*, and *Ribes magellanicum*. Similar shrub associations are frequently observed along stream and river margins, especially in areas inaccessible to introduced herbivores, where *E. serrata*, *R. magellanicum* and *Chiliotrichum diffusum* tend to persist.

4.3 Ecological Gradients Across the Archipelago of Isla de los Estados

For environmental analysis, the archipelago was divided into six sectors, following the framework proposed by Collado (2015). Isla de los Estados was subdivided into three major regions: the Western Valleys, the Central Mountain Massif, and the Eastern Fjords. Observatorio Island and the Año Nuevo group were treated as distinct sectors due to their contrasting topographical and ecological features. A sixth sector includes adjacent rocky islets located more than 500 m from the main island (Table 4.2).

Although not all sectors represent discrete ecological units, patterns in vegetation cover reveal key environmental differences among them. These ecological gradients are shaped primarily by topography, elevation, exposure, and local microclimates. Among these factors, persistent and often intense wind exposure plays a central role in determining vegetation structure and species composition throughout the archipelago.

4.3.1 Western Valleys

This region spans approximately 12 km in both east-west and north-south directions, covering around 120 km^2. It includes key geographic features such as Franklin Bay to the west, Crossley Bay to the north, and the western edge of Flinders Bay to the northeast (Fig. 4.2).

This is the island's least dissected coastal area, characterized by broad interior valleys and rounded, fully vegetated hills reaching just over 400 m (Fig. 4.9a). Unlike other regions, it lacks rocky peaks and has the lowest proportion of non-vegetated areas, which are mostly limited to rocky shores, cliffs, and active dunes along the coast, landforms shaped by persistent westerly winds and oceanic swell.

Table 4.2 Regional division of Isla de los Estados and Año Nuevo Islands

Natural regions	Area (km^2)	Percentage (%)
Western Valleys	119.8	23.6
Central Mountain Massif	126	24.8
Eastern Fjords	252.4	49.7
Observatorio Island	4.7	0.9
Año Nuevo Islands	2.5	0.5
Isolated rocky islets	2.1	0.4
Total	507.5	100

Fig. 4.9 Ecological regions of Isla de los Estados: (**a**) Western Valleys Region; (**b**) Central Mountain Massif Region; (**c**) Eastern Fjords Region. (Credits: Amira Salom (**a**, **b**), Leonardo Collado (**c**))

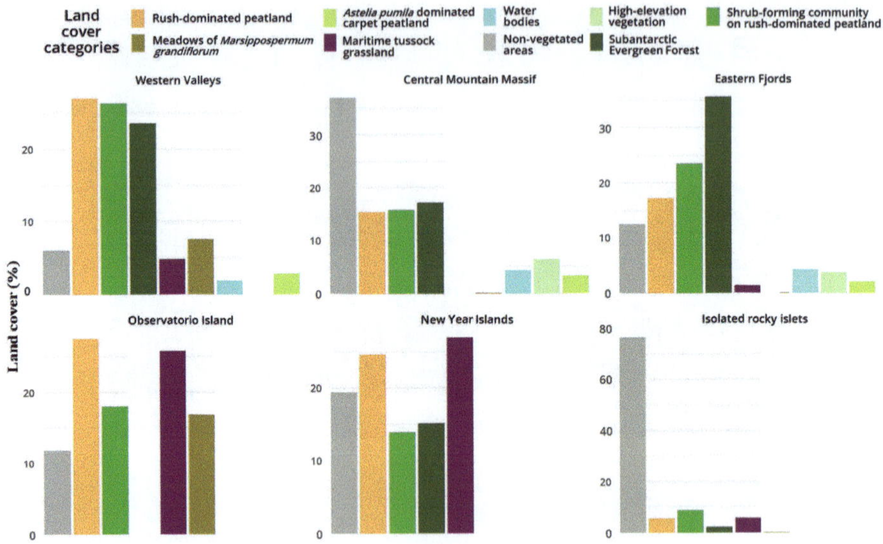

Fig. 4.10 Vegetation communities and percentage of land cover (%) in the landscape of the six sectors of the archipelago

These topographic features, together with constant wind exposure, play a key role in determining vegetation structure.

The dominant vegetation types are *rush-dominated peatlands* and the *shrub-forming community on rush-dominated peatlands*, each covering about 27% of the area (Fig. 4.10). Rush peatlands are most common on exposed slopes and hilltops, while the shrub-forming communities appear in transitional zones between peatlands and evergreen forests, often resembling the latter under harsher conditions. *Subantarctic evergreen forests* are mainly confined to the more sheltered central-northern valleys, particularly in the eastern slopes of hills. In moist lowland forests, species like *Senecio smithii*, *Marsippospermum grandiflorum*, *Anthoxanthum redolens*, and *Gunnera magellanica* may occur (Fig. 4.3b). To the west, forests intermingle with peatlands, while to the east, they transition into shrublands. This possibly reflects forest retreat under combined pressure from wind exposure and herbivory by introduced ungulates (Salom and Schiavini, Chap. 14, this volume).

Although less widespread overall, several other vegetation types reach their greatest development in this region. Noteworthy, *Marsippospermum grandiflorum meadows* are better developed here than elsewhere on the main island, covering over 7% of the landscape, particularly in the low-lying northwest (Fig. 4.7a). *Maritime tussock grasslands* are also concentrated in this region, mainly along the western and northern coasts, where they host an important breeding colony of Southern rockhopper penguins, occupying nearly 5% of the area (Fig. 4.10). This region also supports the most extensive development of *Astelia pumila-dominated peatlands*, forming contiguous patches up to 0.35 km^2 in valley bottoms and smaller patches along hill crests, often interspersed with *rush-dominated peatlands* and *shrub-forming communities on rush-dominated peatlands*.

Despite the broad valleys, *water bodies* cover less than 2% of the area, below the island average. Due to the absence of rocky peaks, *high-elevation vegetation* is scarce and restricted to small, exposed patches.

4.3.2 Central Mountain Massif

Located in the central-western sector of the main island, this region spans approximately 15 km in both east-west and north-south directions. It is bounded by the Western Valleys to the west and the Eastern Fjords to the east. Its defining feature is a rugged mountain range that includes the Bove Mountains, with peaks exceeding 800 m (Fig. 4.9b). Key coastal landmarks include Cap. Cánepa Bay to the south, and Flinders Bay, along with the Hoppner and Parry fjords, to the north (Fig. 4.2). The region's glacial history and rugged topography have also produced the island's highest density of freshwater bodies, over 100 in total, covering approximately 4% of the surface (Fig. 4.10).

Due to its mountainous terrain and harsh climatic conditions, *non-vegetated areas* dominate, covering nearly 40% of the surface (~47 km^2). In contrast, *subantarctic evergreen forests* occupy just over 17%, primarily along the more sheltered northern coast and around Cape San Antonio (Fig. 4.2). Toward the south and central interior zones, forest cover becomes sparse and fragmented. Along the exposed southern coast, where the conditions are severe these are often replaced by the *shrub-forming community on rush-dominated peatlands*, covering about 16% of the region (Fig. 4.10).

This region has the highest representation of *high-elevation vegetation*, covering 6.4% of the surface. These communities are often interspersed with *Astelia pumila-dominated peatlands* at upper altitudes. These peatlands also form mosaics with *rush-dominated peatlands* across both low valleys and elevated hilltops, depending on slope and wind exposure. Other vegetation types are limited: *maritime tussock grasslands* are rare (0.6%), occurring mainly at the northern tip of Cape San Antonio, while *Marsippospermum grandiflorum meadows* are restricted to a single patch in the western part of the cape (Fig. 4.2).

4.3.3 Eastern Fjords

This region is characterized by a deeply indented, fjord-dominated coastline (Fig. 4.9c), spanning about 37 km east-west and averaging 8 km north-south, though this varies considerably due to the irregular shore. While rugged, it has fewer *non-vegetated areas* (12%) than the Central Mountain Massif. The dominant vegetation type is the *subantarctic evergreen forest*, which gives the landscape the appearance of forested mountains. Forests reach their greatest development here, covering nearly 35% (90 km^2) of the region (Fig. 4.10), accounting for about 65% of the archipelago's total forest cover. Stands occur along both coasts, more extensively in the north, often reaching elevations of up to 450 m.

The *shrub-forming community on rush-dominated peatlands* is the second most widespread vegetation type. It typically forms a transitional zone toward higher, more exposed slopes with *rush-dominated peatlands*. Unlike in western regions, this community is more common in interior uplands. Although *water bodies* cover only 10.5 km^2, slightly less in proportion than in the Central Massif, they account for 60% of the archipelago's total freshwater surface. These are concentrated near the central mountains, along with *high-elevation vegetation*, which covers 3.7% of the region (Fig. 4.10).

Astelia pumila-dominated peatlands (2.1%) are less common, occurring mainly in the western part of the region, particularly on northern slopes. While occasionally forming contiguous patches up to 0.20 km^2, they more often appear as isolated patches within *rush-dominated peatlands*. *Maritime tussock grasslands* are sparse (1.3%), restricted to southern coastal peninsulas and, on the north coast, to the far eastern tip (Cape San Juan and parts of Cape Fourneaux), where small seabird colonies are present (Fig. 4.2). *Marsippospermum grandiflorum meadows* are virtually absent (0.1%), with only a few isolated occurrences at elevated eastern sites.

4.3.4 Observatorio Island

Observatorio Island covers 4.7 km^2 and is characterized by its low elevation which not exceed 60 m (Fig. 4.5a). This results in a complete lack of topographic shelter and likely constant wind exposure. Additionally, the proximity of Isla de los Estados may influence precipitation patterns, potentially acting as a barrier and creating dryer conditions on Observatorio Island. This ecological conditions likely limits the vegetation types that can establish, making it the only region in the archipelago entirely devoid of forest. However, studies on climatic, edaphic, and historic factors influencing vegetation distribution remain pending.

Half of the island's surface is evenly split between *rush-dominated peatlands* (27%) (Fig. 4.5c), forming a continuous matrix across the island, and *maritime tussock grasslands* (26%) (Fig. 4.5d), which fringe nearly the entire coastline except the southeastern sector (Fig. 4.2). This more sheltered sector is dominated by *shrub-forming communities on rush-dominated peatlands*, covering 18% of the area. It is the only part of the island where plants with an arboreal form occur, including scattered individuals of *D. winteri* (Fig. 4.5b).

Despite the absence of rocky summits or high elevations, the island includes a notable proportion of *non-vegetated areas*, all confined to coastal zones exposed to strong winds and wave action that inhibit plant establishment (Fig. 4.5d). Observatorio Island also supports the highest proportion of *Marsippospermum grandiflorum meadows* in the archipelago, covering 18% of its surface. These meadows occur interspersed within the *rush-dominated peatlands*, forming a mosaic-like pattern (Fig. 4.5c).

4.3.5 Año Nuevo Islands

Unlike Observatorio Island, this group supports patches of *subantarctic evergreen forest* on its eastern side, likely due to slightly higher elevations (up to 90 m a.s.l.) that provide partial shelter from prevailing winds (Fig. 4.5a). Forests are most developed on Elizalde Island, the largest in the group, where they cover 15% of the surface (Fig. 4.10). Elizalde also hosts patches of the *shrub-forming community on rush-dominated peatlands*, forming a transition between the forested east and central peatlands. *Non-vegetated areas* are concentrated along the island's western and northern shores (Fig. 4.2).

This group exhibits the highest proportional cover of *maritime tussock grassland* in the archipelago (26.9%). It is the dominant vegetation type, forming a broad coastal fringe along western shores. *Rush-dominated peatlands* closely follow in extent (24.6%), occupying the more exposed and elevated central areas. There is no representation of *water bodies, high-elevation vegetation*, or *Marsippospermum grandiflorum meadows* within this island group.

4.3.6 Isolated Rocky Islets

The more than 200 small islets that surround the main island, exhibit distinct features, which warrants separate analysis. Collectively, they cover just over 2 km^2, with an average individual size of approximately 0.01 km^2. These islets are primarily clustered along the southern coast of the main island (Fig. 4.2), where they are highly exposed to strong winds and heavy swells. As a result, despite their low elevation, *non-vegetated areas* dominate, covering 76.5% of the surface, the highest proportion in the archipelago (Fig. 4.10).

The most abundant vegetation type is the *shrub-forming community on rush-dominated peatlands*, although most of this cover is concentrated on Alexander Islet, located 500 m east of the main island, where the community is particularly well developed. *Maritime tussock grasslands* and *rush-dominated peatlands* follow, each accounting for 5.6% of the area, reflecting their strong adaptation to harsh conditions.

Somewhat unexpectedly, *Subantarctic Evergreen Forests* are present on a few islets off the northern coast, most notably the Los Tres Fernández islets in Puerto Nuevo Bay (500–1000 m from the main island), where they collectively cover less than 0.05 km^2. *Astelia pumila-dominated peatlands* are also present in sparse, small patches, comprising 1.4% of the area. *Water bodies* and *high-elevation vegetation* are absent.

4.4 Conclusion

Ecosystems are dynamic, shaped by both natural processes and external pressures such as climate change and biological invasions. On Isla de los Estados, where remoteness has helped preserve relatively intact environments, monitoring vegetation change is key to anticipating future impacts. This study, based on vegetation surveys conducted since 1999, complemented by historical data, field expertise, and institutional collaboration, provides a solid ecological baseline. The vegetation plots established during these surveys offer opportunities for future resampling to track long-term changes, evaluate the effects of introduced herbivores, and monitor vegetation responses under different climate scenarios. The vegetation map presented here represents a substantial improvement, combining ground data, satellite imagery, and expert interpretation (Shape available in Salom et al. 2025b). Together, these tools offer a practical basis for future ecological research and biodiversity monitoring, strengthening the foundation for informed management and effective conservation of the island's unique landscapes.

Competing Interests On behalf of all authors, the corresponding author declares no conflict of interest.

This work was made possible thanks to the financial support of Carlos Vairo (Museo Marítimo of Ushuaia), Fernando Zuloaga (Instituto de Botánica Darwinion), Klemens Pütz (Antarctic Research Trust), the Agencia Nacional de Promoción Científica y Tecnológica (PICT 2017 N° 897) and the Wildlife Conservation Society Argentina. We thank Argentine Army, the General Direction of Protected Areas and Biodiversity of Tierra del Fuego Province Government and the National Park Administration for their institutional support. Special thanks to the crew of the ship ARA Granville and ARA Malvinas (Argentine Army), the Orion and Explorer cruises (National Geographic and Lindblad Expeditions), and to Henk Boersma, Federico Guerrero, Darrel Day, Charlie Flesch, Guillaume, and Micky Porco Fisher from the Sarah Vorwek, Ocean Tramp, Spirit of Sidney, Fernande, Flores, Puerto Beagle vessels, respectively, for safely transporting us to and from the island. We also thank Andrea Raya Rey, Adrián Schiavini, the crew of UNTDF Medios, Máximo Lobo, Fabián Boyeras, Anabella Bonomi, Erio Curto, María Luisa Carranza, Ulises Balza, Nicolás Lois, Francisco Zunino, Andrés Capdevielle, Debora Anibaldi, Guillermo Retamar, Pablo Machado, Rodrigo Iturraspe, Patricio Blanco, Diego Ramos, Constanza Lobo, María Gowland, Iñaki Gómez, Lucas Lanusse, Fernando Encinas and Emanuel Mendieta for their assistance with logistics and/or fieldwork. We are also grateful to Vanessa Lencinas for helping with plant identification and to Alexandra Elbakyan for facilitating access to scientific literature.

References

Arana MD, Natale E, Ferreti N, Romano G, Oggero A, Posadas P, Morrone JJ (2021) Esquema biogeográfico de la República Argentina. Opera Lilloana 56:1–238

Balza U, Liljesthröm M, Pimper L, Franco-Navarro I, Cañas-Barrovecchio S, Zunino F, Domato I, Acardi S, Iturraspe R, Cano J, Raya-Rey A (2022) Status of breeding birds at Observatorio and Goffré Islands, Argentina. Polar Biol 45:667–673. https://doi.org/10.1007/s00300-022-03019-2

Baranzelli MC, Cosacov A, Espíndola A, Iglesias MR, Chan LM, Johnson LA, Sérsic AN (2018) Echoes of the whispering land: interacting roles of vicariance and selection in shaping the

evolutionary divergence of two calceolaria (Calceolariaceae) species from Patagonia and Malvinas/Falkland Islands. Evol Ecol 32:287–314. https://doi.org/10.1007/s10682-018-9938-3
Barkman JJ (1978) Synusial approaches to classification. Classification of plant communities. Springer, Dordrecht, pp 111–165
Barrera MD, Frangi JL, Richter LL, Perdomo MH, Pinedo LB (2000) Structural and functional changes in *Nothofagus pumilio* forests along an altitudinal gradient in Tierra del Fuego, Argentina. J Veg Sci 11(2):179–188
Biganzoli F (2003) Native and alien flora of Isla de los Estados (Tierra del Fuego, Argentina) and its relationships with subantarctic islands. Extended abstracts of the IBMANT/ANDEEP international symposium and workshop in 2003 [Ushuaia, Argentina] reports on Polar and Marine Research, p 507
Box EO (2002) Vegetation analogs and differences in the northern and southern hemispheres: a global comparison. Plant Ecol 163:139–154
Cabrera A (1976) Regiones fitogeográficas argentinas. In: Cabrera A (ed) Enciclopedia Argentina de Agronomía y Jardinería, 1st edn, vol 2. ACME, Buenos Aires, pp 1–85
Chytrý M, Schaminée JH, Schwabe A (2011) Vegetation survey: a new focus for applied vegetation science. Appl Veg Sci 14(4):435–439
Clements FE (1905) Research methods in ecology. University Publishing Company, Lincoln
Collado L (2015) Caracterización de la Vegetación de la Isla de los Estados-Resultados preliminares de la Campaña SDSyA 2014. Technical report submitted to the Secretary of Sustainable Development and Environment, Government of Tierra del Fuego
Congedo L (2021) Semi-automatic classification plugin: a python tool for the download and processing of remote sensing images in QGIS. J Open Source Softw 6(64):3172. https://doi.org/10.21105/joss.03172
Congedo L (2023) Semiautomaticgit/remotior_sensus. Zenodo. https://doi.org/10.5281/zenodo.10038132
Correa MN (1998) Flora patagónica, parte I. Colección Científica del INTA. Instituto Nacional de Tecnología Agropecuaria, Buenos Aires
Dengler J, Chytry M, Ewald J (2008) Phytosociology. In: Jørgensen SE, Faith BD (eds) Encyclopedia of ecology, vol 4. Elsevier, Oxford, pp 2767–2779
Dudley GR, Crow GE (1983) A contribution to the flora and vegetation of Isla de los Estados (Staten Island), Tierra del Fuego, Argentina. Antarct Res Ser, vol 37. American Geophysical Union, Washington, DC
European Space Agency (2018) Sentinel-2 products specification document. European Space Agency. https://sentinels.copernicus.eu/web/sentinel/user-guides/sentinel-2-msi/document-library. Accessed 3 Feb 2025
Frenot Y, Gloaguen JC, Massé L, Lebouvier M (2001) Human activities, ecosystem disturbance and plant invasions in subantarctic Crozet, Kerguelen and Amsterdam Islands. Biol Conserv 101(1):33–50
GRASS Development Team (2022) Geographic Resources Analysis Support System (GRASS) software, version 8.2. Open Source Geospatial Foundation. Electronic document: https://grass.osgeo.org, https://doi.org/10.5281/zenodo.5176030
Greene SW (1964) The vascular flora of South Georgia. Br Antarct Surv Sci Rep 45:1–58
Greene SW (1969) New records for South Georgian vascular plants. Br Antarct Surv Bull 22:49–59
Gremmen NJ, Smith VR (1999) New records of alien vascular plants from Marion and Prince Edward Islands, sub-Antarctic. Polar Biol 21:401–409
Groff DV, Hamley KM, Lessard TJ, Greenawalt KE, Yasuhara M, Brickle P, Gill JL (2020) Seabird establishment during regional cooling drove a terrestrial ecosystem shift 5000 years ago. Sci Adv 6(43):eabb2788. https://doi.org/10.1126/sciadv.abb2788
Guerrero PC, Contador T, Díaz A, Escobar C, Orlando J, Marín C, Medina P (2025) Southern Islands Vascular Flora (SIVFLORa) dataset: a global plant database from Southern Ocean islands. Sci Data 12:397. https://doi.org/10.1038/s41597-025-04702-9
Huertas Herrera A, Lencinas MV, Martínez Pastur G (2018) Environmental drivers of plant community assembly in Isla de los Estados at Southern Atlantic Ocean. Community Ecol 19(1):35–44. https://doi.org/10.1556/168.2018.19.1.4

Jelaska SD (2009) Vegetation mapping applications. Dev Soil Sci 33:481–496
Kent M (2011) Vegetation description and data analysis: a practical approach. Wiley, Hoboken
Küchler AW, Zonneveld IS (2012) Vegetation mapping, vol 10. Springer Science & Business Media, Berlin, Alemania
Kuhnemann O (1976) Observaciones ecológicas sobre la vegetación marina y terrestre de la Isla de los Estados (Tierra del Fuego, Argentina). Ecosur 3(6):121–248
Marchelli P, Pastorino MJ, Gallo LA (2021) Temperate subantarctic forests: a huge natural laboratory. In: Pastorino MJ, Marchelli P (eds) Low intensity breeding of native forest trees in Argentina. Springer, Cham. https://doi.org/10.1007/978-3-030-56462-9_2
Mestre L, Toro-Manríquez M, Soler R, Huertas-Herrera A, Martínez-Pastur G, Lencinas MV (2017) The influence of canopy-layer composition on understory plant diversity in southern temperate forests. For Ecosyst 4:1–13
Moore DM (1967) The vascular flora of The Falkland Islands. Br Antarct Surv Sci Rep 60:1–202
Moore DM (1983) Flora of Tierra del Fuego. Anthony Nelson & Missouri Botanical Garden, St. Louis
Nicolson M (1987) Alexander von Humboldt, Humboldtian science and the origins of the study of vegetation. Hist Sci 25:167–194
Ponce JF, Fernandez M (2014) Climatic and environmental history of Isla de los Estados, Argentina, Springerbriefs in earth system sciences. Springer, Dordrecht
Python Software Foundation (2023) Python language reference, version 2.12.8. Available at http://www.python.org. Accessed 3 Feb 2025
QGIS Development Team (2020) QGIS geographic information system, version 3.34.11. QGIS Association. https://www.qgis.org. Accessed 3 Feb 2025
Salom A, Borrelli L, Testoni D, Ramos DP, Raya Rey A, Biganzoli F (2025a) Forage selection overlap between co-occurring introduced ungulates: insights for conservation management of a Sub-Antarctic Island. Mamm Biol 105:403–414
Salom A, Cano J, Testoni D, Mansilla R, Mestre L, Collado L, Biganzoli F (2025b) Land cover map of Isla de los Estados, Año Nuevo islands (Observatorio, Goffré, Zeballos and Elizalde Islands) and adjacent islets. Figshare. Figure. https://doi.org/10.6084/m9.figshare.29391008.v1
Sancho G, Lencinas MV, Martínez Ramírez F (2023) Beyond the comfort zone: Lagenophora Cass. (Astereae) in the Malvinas/Falkland archipelago. Int Compositae Alliance 3(1):66–85
Smith RL, Clymo RS (1984) An extraordinary peat-forming community on The Falkland Islands. Nature 309(5969):617–620
Spegazzini C (1896) Plantae per Fuegiam a Carolo Spegazzini anno 1882 collectae. Juan A. Alsina, Buenos Aires
Strange IJ, Parry CJ, Parry MC, Wood RW (1988) Tussac grass in the Falklands. Falkland Islands Foundation, Brighton
Thiers B (2025) Index Herbariorum: a global directory of public herbaria and associated staff. New York Botanical Garden's Virtual Herbarium
van der Maarel E (2005) Vegetation ecology–an overview. In: Vegetation ecology, vol 3. Backwell Publishing, USA, Malden, pp 1–51
Veblen T, Donoso C, Kitzberger T, Rebertus J (1996) Ecology of southern Chilean and Argentinean Nothofagus forests. In: Veblen T, Hill RS, Read J (eds) The ecology and biogeography of Nothofagus forests. Yale University Press, New Haven, pp 293–353
Warming E (1895) Plantesamfund—Grundtræk af den økologiske plantegeografi. P.G. Philipsens, København
Whinam J (2009) Aliens in the sub-Antarctic-biosecurity and climate change. Pap Proc R Soc Tasmania 143(1):45–52
Whittaker RH (1962) Classification of natural communities. Bot Rev 28:1–239
Young SB (1973) Subantarctic rain forest of Magellanic Chile: distribution, composition, and age growth rate studies of common forest trees. Antarct Res Ser 20:307–322

Chapter 5
Diving into Kelp and Animal Forests in Isla de los Estados

Julieta Kaminsky, Gonzalo Bravo, Cecilia Paula Alonso, Mariano Rodríguez, and María Bagur

Abstract This chapter explores the underwater seascapes of Isla de los Estados, where both kelp forests and animal forests flourish. These underwater ecosystems are shaped by organisms that act as ecosystem engineers, enhancing habitat complexity and supporting high biodiversity. Kelp forests are primarily formed by brown macroalgae of the order Laminariales, while animal forests by dense aggregations of habitat-forming invertebrates. Even though animal forests also play a crucial role in ecosystem structure, they remain less studied, with significant knowledge gaps regarding their distribution and functioning. Based on a scientific diving expedition conducted in 2022, we described the underwater forests of Isla de los Estados using photo-quadrats and the Roving Diver Method. We identified over 230 taxa of mobile and sessile invertebrates, macroalgae, and fishes, revealing highly diverse communities and documenting new records of algae and invertebrates. A photographic catalog of the observed taxa was compiled, resulting in the upload of more than 1800 underwater images to the open-access platform iNaturalist. This study underscores the exceptional conservation status and ecological richness of Isla de los Estados' underwater forests. Potential threats are discussed, such as the presence of the invasive anemone *Metridium senile*, along with proposed strategies for monitoring and conservation.

Keywords Underwater seascapes · Ecosystem engineers · Benthic biodiversity · Scientific diving

J. Kaminsky (✉) · M. Bagur
CADIC-CONICET, Ushuaia, Tierra del Fuego, Argentina

G. Bravo
IBIOMAR-CONICET, Puerto Madryn, Chubut, Argentina

C. P. Alonso · M. Rodríguez
ICPA-UNTDF, Ushuaia, Tierra del Fuego, Argentina

A. Raya Rey, U. Balza (eds.), *Biodiversity and Conservation of Isla de los Estados (Chuani-sin)*, Natural and Social Sciences of Patagonia,
https://doi.org/10.1007/978-3-032-13630-5_5

5.1 Underwater Forests

Isla de los Estados (hereafter IDLE) is home to lush terrestrial forests, but beneath its surrounding waters lie equally, if not more, extraordinary underwater forests. These rich marine ecosystems, teeming with life and structural variety, remain largely unexplored. The island's geological origin and subsequent coastal geomorphological transformation (Ponce et al. 2009) have shaped a landscape of fjords and deep inlets with steep vertical walls, creating environments that range from highly exposed to relatively sheltered from wave action (Ponce and Fernández, Chap. 2, this volume). These coastal and marine settings feature rocky substrates that support the development of underwater forests. In the rocky subtidal zones of IDLE, two main types of underwater forests thrive, anchored to the substrate: kelp forests and marine animal forests (Fig. 5.1). These ecosystems are structurally complex and highly productive, formed by one or more ecosystem-engineering species that modify local physical conditions and promote the development and persistence of high biodiversity (e.g., Teagle et al. 2017; Rossi et al. 2017). Their structural complexity enhances environmental heterogeneity, enabling the formation of distinct layers and supporting characteristic species within each forest type.

Kelp forests are formed by one or more brown algae species mainly belonging to the order Laminariales. These large algae provide structure to the ecosystem and support the development of other macroalgal strata that grow within the forest (e.g., Kühnemann 1970).

Marine animal forests can be described as three-dimensional structures formed by benthic animals such as sponges, octocorals, bivalves, bryozoans, polychaetes or tunicates. These structures create new ecological niches and colonization surfaces for other organisms, thereby enhancing ecosystem functions and services (Orejas et al. 2022).

The distribution of underwater forests around IDLE is shaped by a combination of biotic and abiotic factors, with the availability of rocky substrate, seabed slope and wave exposure playing particularly important roles (Kühnemann 1976; Dayton 1985). Although both types of submerged forests can coexist in some areas, they typically follow a spatial pattern: as slope increases, animal-dominated communities become more prevalent, while kelp tend to diminish or disappear especially on vertical or overhanging walls. Given the island's fjord-like geography, vertical walls are common and frequently host rich animal forests. In contrast, flatter areas typically support dense kelp forests. This diversity of habitats makes IDLE a fascinating and highly biodiverse region, with each forest type harboring distinct communities. Its geographic proximity to the highly productive and biodiverse Marine Protected Area Namuncurá-Burdwood Bank (Schejter et al. 2016), further reinforces its importance as a hotspot for benthic biodiversity.

In this chapter, we dive into both the kelp and animal forests that thrive in the submerged ecosystems of IDLE. The descriptions presented are primarily based on a scientific expedition carried out in October 2022 (see Sect. 5.5), during which we surveyed these ecosystems through scuba diving, documenting their structure and biodiversity.

Fig. 5.1 Underwater seascapes at Isla de los Estados: kelp forests (**a** *Macrocystis* and **b**, **c** *Lessonia*) and animal forests (**d** sponges, octocorals and barnacles, **e** polychaetes, **f** sponges and ascidians in a vertical wall) supporting a wide variety of life forms

5.2 Kelp Forests

Coastal rocky ecosystems in IDLE are widely colonized by kelp forests. In contrast to other regions where top-down control by sea urchins and other herbivores strongly drives ecosystem dynamics, in Southern South America the distribution and functioning of kelp forests are primarily shaped by intraspecific interactions and abiotic factors such as wave exposure, availability of rocky substrate, light, and nutrients (see Dayton 1985; Graham et al. 2007 for reviews). Kelp forests in IDLE are structured by three species of brown Laminariales and one Fucales algae, with spatial variations in their biomass. The most abundant species, both in distribution and biomass, is the giant kelp *Macrocystis pyrifera*, forming forests between 2 and 20 m in depth (Fig. 5.2a). In the upper subtidal zone, *Lessonia flavicans* is commonly found (1–4 m depth, Fig. 5.2b), while *L. searlesiana* is present in the lower subtidal zone (10–25 m, Fig. 5.2c). In areas with high wave exposure, the upper subtidal zone is colonized by *Durvillaea antarctica*, a brown algae from the order Fucales (Fig. 5.2d).

Fig. 5.2 Canopy-forming species in Isla de los Estados: arboreal and shrub strata. (**a**) Underwater view of *Macrocystis pyrifera*, (**b**) *Lessonia flavicans*, and (**c**) *Lessonia searlesiana*. (**d**) Aerial view of *Durvillaea antarctica*, a Fucal algae found in the upper subtidal in wave exposed areas

Canopy-forming species grow attached to rocky substrates and extend through the water column to reach the ocean surface, allowing their detection through remote sensing (e.g., Bell et al. 2020; Mora-Soto et al. 2020). In recent decades, the global study of kelp forests using remote sensing has expanded significantly, primarily due to advancements in spatial, spectral, and temporal resolution of imagery (Reshitnyk et al. 2023). High-resolution satellite and aerial images can be used to map forest extent and density, detect spatial and temporal variations in biomass loss or growth, identify regional factors driving these changes, and even develop vulnerability maps under future climate scenarios (e.g., Mora-Soto et al. 2022). In Tierra del Fuego, kelp forests are present year-round, and their photosynthetic biomass exhibits high reflectance in the near-infrared spectrum, allowing precise differentiation from the surrounding seawater. Here, we present the updated distribution of kelp forests in IDLE (Fig. 5.3a), using a collection of Sentinel-2 images from 2022 to 2025 and the Random Forest classification method in Google Earth Engine (following Houskeeper et al. 2022). Most of the kelp forests are found along the northern coast, forming extensive and dense patches, while along the steep southern coast they tend to form narrow belts located inside relatively more sheltered sites protected from the prevailing south-westerly winds. The estimated forested area is ~25 km^2, an area comparable to some of the diverse terrestrial vegetation formations in the island (Salom et al., Chap. 4, this volume). Although *M. pyrifera* is the most abundant kelp in IDLE, the satellite image classification method does not allow differentiation between other species such as *L. flavicans* or *D. antarctica*. For this reason, it is important to consider that this estimate of kelp forest area is for mixed forests (Fig. 5.3b).

Sub-Antarctic kelp forests in the Fuegian archipelago are characterized by high macroalgal biodiversity (e.g., Santelices and Ojeda 1984; Ramírez 2010; Kaminsky et al. 2024; Rodríguez et al. 2024). While underwater forests in IDLE remain relatively unexplored, certain studies have described the structure and richness associated with kelp forests on the island (see Kühnemann 1976). Beneath the canopy of *M. pyrifera*, at least three additional macroalgal strata are typically found: shrub, herbaceous, and encrusting layers. During the 2022 expedition in IDLE, the richness of understory algae was described for the first time by an Argentinian team using scuba diving, taking *in situ* photographs and collecting specimens across different sampling sites to build a reference herbarium (see sampling sites in Fig. 5.3a). Identification was carried out based on available literature and regional collections, primarily through comparisons of morphological and anatomical features with records from previous studies (e.g., Boraso de Zaixso and Zaixso 2013). In this study, 37 macroalgae species were preliminarily identified (Bravo et al. *in prep*). In some locations, mixed forests occurred, where *Lessonia* spp. contributed to the shrub layer. The herbaceous layer is composed of a high diversity of red (Rhodophyta), brown (Ochrophyta), and green (Chlorophyta) macroalgae. Among the red species, the most abundant were *Cladodonta lyallii*, *Paraglossum* spp., *Plocamium secundatum*, *Myriogramme livida*, *Callophyllis variegata* and *C. atrosanguinea*, *Ptilonia magellanica*, *Pseudophycodrys phyllophora*, *Phycodrys quercifolia*, *Ballia callitricha*, *Heterosiphonia berkeleyi*, *Picconiella pectinata* and *Rhodymenia* spp. (Fig. 5.4). The most commonly found brown algae included

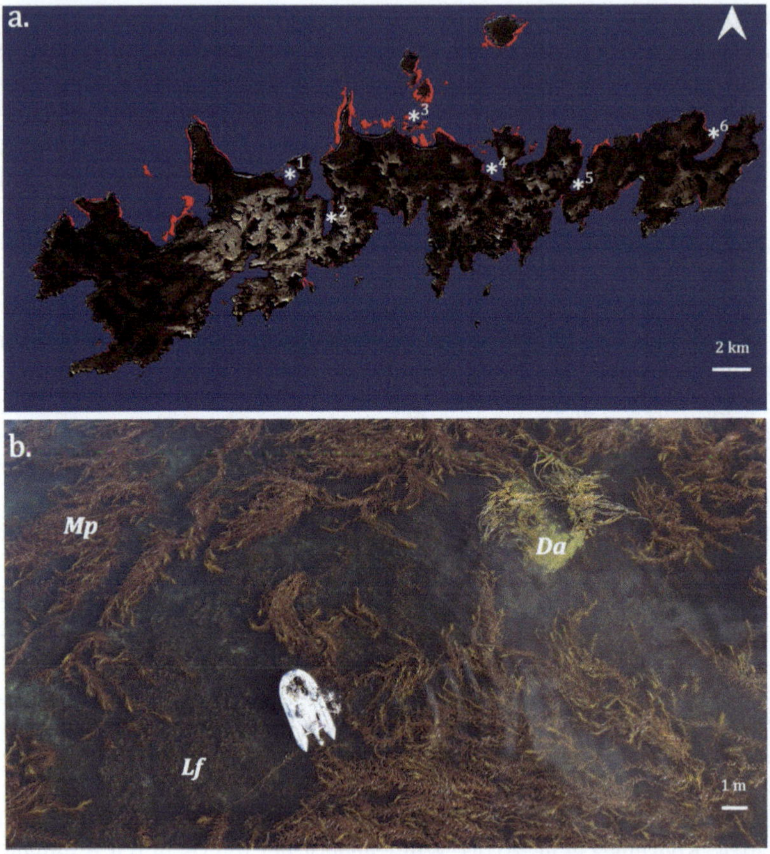

Fig. 5.3 (**a**) Kelp forests distribution in Isla de los Estados indicated by red in the map, and diving sites sampled during the expedition in 2022: (1) Puerto Hoppner, (2) Puerto Parry, (3) Rocas Moyano, (4) Puerto Basil Hall, (5) Puerto Cook and (6) San Juan de Salvamento. Map by Julieta Kaminsky and Tom Bell (WHOI). (**b**) Canopy-forming species from a drone image: *Macrocystis pyrifera*, *Lessonia flavicans* and *Durvillaea antarctica*

Dictyota falklandica, *Halopteris obovata*, *Desmarestia ligulata*, *D. confervoides* and *D. chordalis*. Among the green macroalgae, filamentous forms such as *Bryopsis* sp. or *Derbesia marina* were present, while *Codium subantarcticum* formed a thick cover of around 5–10 cm (Fig. 5.4e). The muscinal layer, found directly on the substrate, was characterized by a great diversity of encrusting coralline red algae (Fig. 5.4h). This study reports and makes available *in situ* field pictures of many of these species in their environment (see Bravo et al. 2025). Some records are of particular interest, such as the case of the brown algae *Dictyota falklandica*, which is the first record for the area (Küpper et al. 2019, Fig. 5.4g).

Fig. 5.4 Macroalgal strata associated with kelp forests in Isla de los Estados: examples of species found in the herbaceous and muscinal layers. (**a**) *Ptilonia magellanica*, (**b**) *Pseudophycodrys phyllophora*, (**c**) *Callophyllis variegata*, (**d**) *Phycodrys quercifolia*, (**e**) *Codium subantarcticum*, (**f**) *Ballia callitricha*, (**g**) *Dictyota falklandica*, (**h**) encrusting coralline red algae

5.3 Animal Forests

Marine animal forests have existed since the Paleozoic era, when stalked crinoids dominated the seafloor (Rossi et al. 2017). The concept itself, however, is not new. As early as 1869, Alfred Russel Wallace referred to "animal forests" when describing coral- and sponge-dominated reefs in *The Malay Archipelago* (Wallace 1869). More recently, a consensus-based definition was proposed through a dedicated scientific workshop (Orejas et al. 2022), describing marine animal forests as three-dimensional structures formed by benthic animals acting as autogenic ecosystem engineers, which create new ecological niches and colonization surfaces for other organisms, ultimately enhancing the provision of ecosystem functions and services. This definition builds on earlier characterizations that emphasized their biogenic nature, structural complexity, and the prevalence of long-lived suspension feeders such as sponges, corals, bivalves, hydrozoans, and bryozoans (Rossi 2013; Rossi et al. 2017; Soares et al. 2020). Like kelp forests, marine animal forests can also grow in different forms, forming vertical and functional strata. For example, Cárdenas and Montiel (2015) classified the sessile invertebrates from the Strait of Magellan in five growth forms due to their three-dimensional structures, which show analogies with macroalgal architectural forms (e.g., trees, bushes, mounds, plates and sheets; see Table 1 in Cárdenas and Montiel (2017)). Today, these complex communities are gaining increasing recognition in current discussions for their ecological significance and potential role as candidates for nature-based solutions (Orejas et al. 2022), with their restoration being explored through approaches such as biodiversity-carbon credit schemes (Shefy et al. 2025).

Marine animal forests are among the most widespread yet understudied benthic communities in the world's oceans, particularly in temperate latitudes (Rossi et al. 2017). Their broad ecological distribution is due to the wide range of environmental conditions they can occupy, from shallow coasts to mesophotic zones, although much of their extent and composition remain poorly documented. In temperate ecosystems, these assemblages typically thrive at depths between 30 and 150 m, where reduced light availability limits algal competition and favours the establishment of suspension-feeding animals (Cerrano et al. 2019; Bell et al. 2024). Nevertheless, under specific environmental scenarios such as reduced light penetration, high sedimentation, strong herbivory pressure, the presence of steep vertical walls, or areas with intense water flow, animal-dominated communities can also occur at shallower depths. Notable shallow-water examples have been described in semi-enclosed or fjord-like environments such as Lough Hyne in Ireland (Bell and Barnes 2000), Fiordland in New Zealand (Harris et al. 2021), the northern Adriatic Sea in the Mediterranean (Piazzi et al. 2023) and the Bathurst Channel in Tasmania (Barrett and Edgar 2010). In Chile, animal forests have been reported both along exposed open coasts (Campoy et al. 2023) and within fjord systems (De la Torriente et al. 2024), while additional examples from shallow open coastal areas include the Ionian Sea (Strano et al. 2020) and the Atlantic coast of Argentina (Meretta and Genzano 2025). In this context, the fjords and steep-walled inlets of IDLE (Fig. 5.5a) offer favourable conditions for the development of shallow animal forests, resembling those described in other temperate fjord systems.

Consistent with these observations, during our dives across various areas of IDLE we recorded animal forests at similarly shallow depths (down to 25 m), with the highest cover occurring in zones characterized by strong water flow, particularly within narrow channels, vertical walls, and overhang surfaces. These assemblages often exhibited marked three-dimensional complexity, sometimes structured by multiple species (Fig. 5.5). In several cases, their architectural intricacy exceeded that of animal forests previously described in shallow Atlantic coastal waters

Fig. 5.5 (**a**) Fjords and steep-walled inlets of IDLE and the sailing vessel from the scientific diving expedition carried out in October 2022. (**b-g**) Examples of different marine animal forests in fjord's narrow channels, vertical walls and overhang surfaces at Isla de los Estados

(Meretta and Genzano 2025), and occasionally approached the structural complexity reported in Antarctic systems (Gutt et al. 2017).

Among the different types of animal forests documented during our dives in IDLE, one of the most frequent was the polychaete field formed by *Chaetopterus variopedatus* (Fig. 5.5d). These fields covered extensive areas and created a complex three-dimensional habitat that supported a diverse array of invertebrate species (Fig. 5.6). Similar assemblages have been studied by De la Torriente et al. (2024) in the Magellanic region, where they reported species richness with a median of around 20 species per site. That study also highlighted the importance of three-dimensional habitats created by sessile ecosystem engineers, such as *C. variopedatus*, in enhancing local biodiversity.

Animal forests seem to be particularly frequent in IDLE, potentially more so than in other neighboring regions at comparable depths. For example, it was found that active suspension feeders were more abundant in IDLE than in the neighbouring Mitre Península (69 vs. 45% of the total benthic numerical abundance of invertebrates, Friedlander et al. (2020)). We hypothesize that one of the main factors driving this pattern is the presence of steep rocky walls throughout the fjords in IDLE, which provide an ideal substrate for sessile invertebrates. In these environments, they avoid competition with macroalgae for space, since algae, being primary producers, require flatter surfaces to maximize light capture (Cárdenas and Montiel 2017). The steep slope descending rapidly into deeper waters characteristic of the southern coast of the island, combined with strong southern winds and wave exposition that likely hinder the proper development of kelp, may explain the reduced kelp coverage observed in that area (see Fig. 5.3a). Nevertheless, it is presumed that these steep submerged walls are colonized by rich assemblages of sessile invertebrates.

Sessile invertebrates were not only more abundant but also exhibited greater diversity on the vertical and overhanging walls of IDLE than on flat surfaces, a pattern that has also been documented in other parts of Argentine Patagonia (Bravo et al. 2020) and across various regions globally (see Miller and Etter 2011). These vertical and overhanging surfaces hosted dense and highly diverse communities; while overhangs were consistently and entirely encrusted by sessile invertebrates, vertical walls also supported rich assemblages, though their coverage was more variable (Fig. 5.5). In this study, 34 sessile invertebrates species were preliminarily identified inhabiting these vertical and overhanging walls (Bravo et al. *in prep*). The most conspicuous species included: sea anemones, *Stomphia selaginella*, *Paractis laevis*, *Halcurias* sp., *Antholoba achates*, and *Metridium senile*, along with soft corals such as *Alcyonium haddoni* and a wide variety of sponges including *Cliona* sp., *Phorbas ferrugineus*, *Mycale magellanica*, and *Amphimedon maresi*. Bryozoans like *Alcyonidium australe* and *Crisia* sp., often forming bush-like colonies that added structural complexity and increased habitat heterogeneity (Fig. 5.5). These were often interspersed among colonial ascidians such as *Aplidium fuegiense*, *Aplidium* sp., *Lissoclinum* sp., *Polyzoa opuntia*, and *Sycozoa sigillinoides*, the latter occasionally covering large areas (Fig. 5.5), while solitary ascidians, including *Cnemidocarpa ohlini*, *Pyura legumen*, *Paramolgula gregaria*, and *Cnemidocarpa*

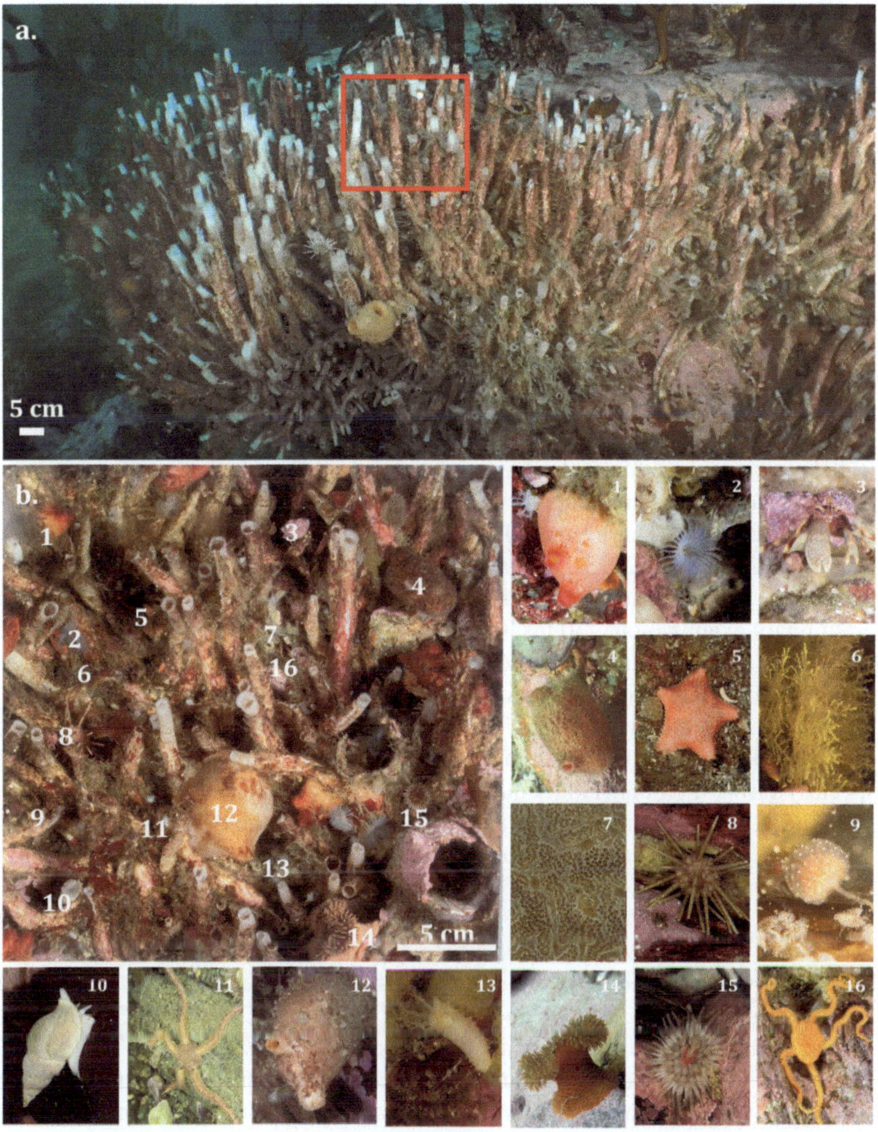

Fig. 5.6 Composition of a *Chaetopterus variopedatus* field observed in IDLE. (**a**) Wide view of the polychaete forests covering the seafloor. (**b**) Photo-quadrat (25 × 25 cm) placed over the field. Sixteen invertebrate taxa identified within the quadrat: (1) *Cnemidocarpa ohlini*, (2) *Perkinsiana antarctica*, (3) *Pagurus comptus*, (4) *Pyura legumen*, (5) *Asterina fimbriata*, (6) Bryozoa, (7) class Ascidiacea, (8) *Austrocidaris canaliculata*, (9) *Sycozoa gaimardi*, (10) *Pareuthria fuscata*, (11) *Ophiactis asperula*, (12) *Paramolgula gregaria*, (13) class Holothuroidea, (14) *Antholoba achates*, (15) *Isoparactis fionae*, (16) *Ophiomyxa vivipara*

verrucosa, were also well represented. Among the polychaetes, *Chaetopterus variopedatus* was notable for forming large tube-like structures, whereas *Perkinsiana antarctica* and *P. magalhaensis* were also observed. Large barnacles such as *Austromegabalanus psittacus* further enhanced the structural complexity of these assemblages. A variety of mobile invertebrates, including echinoderms, crustaceans, and molluscs, were frequently observed across this intricate three-dimensional matrix, interacting with the sessile fauna and contributing to the overall dynamism of the community (see next section).

5.4 Biodiversity and Mobile Species in Kelp and Animal Forests

Both kelp and animal forests host a variety of mobile animals that use these ecosystems for shelter, foraging, mating, and as nursery grounds (e.g., Rossi et al. 2017; Teagle et al. 2017). During the 2022 expedition, and following the Roving Diver Method (Bravo et al. 2023), a total of 1813 observations of benthic biodiversity from IDLE were uploaded to the iNaturalist platform (see Bravo et al. 2025). In a preliminary examination, 230 taxa of mobile and sessile invertebrates, macroalgae and fishes were identified (61 Mollusca, 34 Echinodermata, 27 Chordata, 23 Arthropoda, 23 Cnidaria, 17 Rhodophyta, 11 Annelida, 10 Bryozoa, 8 Ochrophyta, 7 Porifera, 5 Chlorophyta, 2 Brachiopoda, 1 Nemertea). Friedlander et al. (2020) enumerated a total of 158 different benthic taxa from 13 phyla (including sessile species) during their underwater quantitative surveys conducted by scuba diving in 2018 in IDLE and Península Mitre and concluded taxonomic richness was the highest at IDLE, when compared to other marine protected areas in the region (Friedlander et al. 2023).

To assess the importance of the medium- and small-scale variability in species abundance in IDLE, we conducted photo-quadrat surveys at each sampling site (Figs. 5.3a and 5.7, see Sect. 5.5 for detail), and later identified and counted 70 mobile taxa (29 Mollusca, 26 Echinodermata, 9 Arthropoda, 3 Annelida, 2 Chordata, 1 Nemertea) from the images obtained (quadrat size 25 × 25 cm, n = 783, 6 sites, Bravo et al. *in prep*). The most abundant mobile invertebrates on the island were two species of sea cucumbers (*Cladodactyla crocea* and *Chiridota pisanii*), limpets (*Nacella* spp.), a hermit crab (*Pagurus comptus*) and isopods from the Sphaeromatidae family (Fig. 5.8). These species were commonly found across most sites. Other frequently observed mobile species, though not necessarily abundant, were the pink small urchin *Pseudechinus magellanicus* and different molluscs (*Margarella violacea*, *Pareuthria fuscata*, *Fissurella* spp. and chitons, Fig. 5.8). The sea stars have been described as a diverse group in IDLE (Friedlander et al. 2020; Fraysse and Pérez, Chap. 8, this volume). We could identify ~10 different species from our diving records (without looking at any material under the stereomicroscope).

Fig. 5.7 Examples of photo-quadrats (25 × 25 cm) (**a**) with mobile invertebrates: (**b**) red sea cucumbers *Chiridota pisanii*, and (**c**) the green urchin *Arbacia dufresnii*

While sea anemones are traditionally considered sessile species, they can slowly move in different ways under certain circumstances (e.g. Parker 1916; Lawn and Ross 1982). Due to this, and the fact that they are not suspension feeders, we did not consider them as animal forest builders. Sea anemones were frequent at IDLE. We have observed at least nine different species during our dives (see Bravo et al. 2025), including *Bunodactis octoradiata* and *Antholoba achates*, which were reported by Friedlander et al. (2020). In addition, *Stomphia selaginella* and *Paractis laevis* represent new records for this location, with no prior documentation of their presence in IDLE. Worryingly, we recorded for the first time at IDLE an exotic species, *Metridium senile* (Fig. 5.9). This species has been detected at Puerto Hoppner (Fig. 5.3a), where it forms patches on the rocks with a mean density of 195.75 ± 587.97 individuals/m^2 (n = 303 quadrats; presence of *M. senile* in 67 quadrats) at 4 m depth (see Sect. 5.6 for further discussion).

Fig. 5.8 Examples of mobile species with frequent occurrences in kelp and animal forests in Isla de los Estados: (**a**) isopods from the Sphaeromatidae family, (**b**) *Chiridota pisanii*, (**c**) *Nacella* spp., (**d**) *Arbacia dufresnii*, (**e**) *Pagurus comptus*, (**f**) *Cladodactyla crocea*, (**g**) *Psolus patagonicus*, (**h**) *Pseudechinus magellanicus*, (**i**) *Margarella violacea*, (**j**) Chitonida, (**k**) *Fissurella* sp., (**l**) *Ophiactis asperula*, (**m**) Asteroidea, (**n**) *Pareuthria fuscata*, (**o**) *Crepipatella dilatata*, (**p**) *Asterina fimbriata*, (**q**) *Austrocidaris canaliculata*, (**r**) Newtoniellidae, (**s**) *Halicarcinus planatus*, (**t**) *Anasterias antarctica*. Images taken following the Roving Diver Method (see Bravo et al. 2023, 2025 for details). The numbers below indicate the total occurrence considering photo-quadrats from all sampling sites

5.5 Scientific Diving in Underwater Forests

IDLE is a remote region where navigational challenges and logistical constraints for conducting research have limited our understanding of its underwater seascapes. This is especially true for submerged ecosystems, as only a few scientific expeditions have performed *in situ* diving surveys. One of the last accounts of documenting marine macroalgae without diving is the work by Oscar Kühnemann (1976), based on a 1967 expedition by Argentine researchers from the Marine Biology Research Center (CIBIMA). This study documented the heterogeneity of both terrestrial and marine flora through trawls and by observing kelp debris and wrack accumulations along the shore. Other expeditions have documented marine fauna using nets and trawls (see e.g., Chiesa et al., Chap. 7, this volume; Rojo et al., Chap. 11, this volume; Fraysse and Pérez, Chap. 8, this volume).

Fig. 5.9 First record of the exotic sea anemone *Metridium senile* at Isla de los Estados: (**a**) photo-quadrat (25 × 25 cm) in Puerto Hoppner, (**b**) detailed image of *M. senile* on vertical walls

The first scientific study to employ scuba diving for the characterization of kelp forests in the region was conducted by Paul Dayton in 1973 (Dayton 1985). During this campaign, Dayton explored kelp forests along IDLE, as well as sites in Tierra del Fuego and the Magallanes region. He described kelp morphology and proposed ecological factors regulating the structure and functioning of these ecosystems. This pioneering work provided a comparative framework between sub-Antarctic kelp forests of southern South America and those in the Northern Hemisphere, contributing to our understanding of both shared and distinct ecological drivers. After this work, to our knowledge, more than 40 years passed without any official scientific campaign involving diving. Only in 2018, Friedlander et al. (2020) revisited several of the original sites studied by Dayton and compared historical observations with new data and remote sensing analysis, showing no significant changes in forest density or extent, suggesting a high degree of ecological persistence over time.

In October 2022, a team of five scientific divers from research institutions from Argentina embarked on an expedition aboard the *Kostat* sailing vessel to IDLE (Fig. 5.5a). All divers had extensive experience in scientific diving in the region and were familiar with the algae and invertebrate species of Argentine Patagonia. The team has contributed more than 6000 underwater photographic records to the iNaturalist platform, which, to our knowledge, represents the largest public collection of seafloor images from Argentina. The primary objectives of the campaign in IDLE were to conduct scuba dives across the island's varied underwater habitats and to produce audiovisual materials aimed at increasing public awareness -particularly among the population of Tierra del Fuego- regarding the unique marine biodiversity of the region. Over the course of the expedition, dives were conducted at six distinct sites around the island (see Fig. 5.3a), some of which had been previously surveyed by Dayton or Friedlander et al., while new sites were explored. Diving ranged from shallow to deeper areas (~2–30 m depth) to record species in both kelp and animal forests. At each site, we conducted standardized photo-quadrat sampling (25 × 25 cm) along roving transects between 100 and 150 m in length (Figs. 5.7a and 5.9a). Transects were conducted without significant changes in depth and always within *Macrocystis pyrifera* forests. Photo-quadrats were taken without visually selecting the area, by allowing the frame to descend randomly onto the seafloor. For more details on the methodology of roving photo-quadrat transects, see Bravo et al. (2021). In each sampling area, between 60 and 150 photo-quadrats were taken. In parallel, we employed the Roving Diver Method (see Bravo et al. 2023) to compile an inventory of the underwater species present in the area (see Bravo et al. 2025). By combining these two techniques, we obtained estimates of percent cover for sessile species, densities of mobile invertebrates, and a more complete record of species richness, as many rare species (or fish) were not captured in the photo-quadrats but were detected during the active search performed with the Roving Diver Method. The findings of this study (Bravo et al. *in prep*) will be published in due course. During this expedition, Baited Remote Underwater Video systems (BRUVs) were deployed at several sites around the island, providing valuable data on fish assemblages (Merlo et al. 2025). In addition, underwater video recordings were made with the intention of producing a documentary film focused on the island's marine environments.

These efforts contribute to a more comprehensive understanding of the island's underwater ecosystems, complementing earlier scientific works and enhancing outreach by showing the ecological richness of these remote and often inaccessible areas to a broader audience. The island's subtidal habitats are remarkably diverse and characterized by high species richness and abundance. Particularly notable are the frequent vertical rock walls—typical of fjord-like landscapes—which are rarely encountered in the Argentinian section of the Beagle Channel and offer exceptional opportunities for diving-based research.

Despite the challenges posed by cold water temperatures and rapidly changing weather conditions, the underwater landscapes of IDLE offer an exceptional experience for divers and researchers alike. Among the potential hazards encountered during fieldwork, we highlight the presence of leopard seals (*Hydrurga leptonyx*), which were observed resting on coastal rocks but are known to display territorial and occasionally aggressive behaviour underwater. Additionally, strong tidal currents, which vary according to the tidal cycle and can be intense in certain locations, must be taken into consideration when planning dives. The island's southern coast, which is more exposed to prevailing southern winds and features a rugged shoreline, could not be surveyed during this campaign and remains a target for future research. Given the steep bathymetric profiles—where depth increases sharply within a few meters from shore—this sector is likely to feature vertical walls of considerable scientific interest.

5.6 Ecological Importance, Threats and Conservation

Ecosystem engineers are organisms that change the abiotic environment by physically reshaping its structure. As a result, they often influence other organisms, species interactions, and ecosystem processes (Gutiérrez and Jones 2008). Both kelp and animal forests are important ecosystem engineers, altering the physical structure of coastal environments (macroalgae and animals) and, in the case of animal forests, even deep-sea habitats. As ecosystem engineers, both can modify their environment through their structures in many ways, often generating functional impacts over large areas and exerting control over different ecosystem functions such as sedimentary processes, coastal protection and habitat availability to other organisms (Gutiérrez et al. 2011). The ecosystem engineer concept emphasizes the importance of structural change. This way, kelp forests, especially those formed by canopy-forming species like *Macrocystis pyrifera* or *Lessonia* spp., and large animal forests such as mussel beds or polychaete fields, serve as major ecosystem engineers in coastal marine environments mainly by:

1. *Creating habitat structure*: Their three-dimensional physical structures provide shelter, breeding grounds, and surfaces for a diverse range of marine species, enhancing biodiversity and community stability.
2. *Modifying water flow*: Kelp forests can significantly dampen ocean currents—more so in dense interiors—affecting sedimentation patterns, larval transport, and feeding efficiency of suspension feeders. Animal forests can also influence the water flow on smaller spatial scale.
3. *Altering light availability*: Surface and subsurface canopies reduce light penetration, influencing the distribution and composition of understory algae and invertebrates, and promoting niche differentiation.
4. *Influencing nutrient dynamics*: Canopy shading and current dampening can reduce nutrient availability and light for interior kelps, affecting their growth. Edge individuals often have better access to light and nutrients. Complex biogeochemical cycling also occurs within animal forests.
5. *Affecting sedimentation*: Kelp and animal forests impact local sediment dynamics—some species trap sediments, while others sweep it away, though their influence on large-scale erosion remains uncertain. Suspension feeders such as mussels typically trap sediments through their feeding, creating habitat for infaunal species.

6. *Supporting biodiversity*: The structural complexity of kelp and animal forests boosts recruitment and abundance of fish, invertebrates, and epiphytes, promoting ecosystem diversity.

Many examples of the ways kelp and animal forests can act as ecosystem engineers have been compiled in review studies such as Schiel and Foster (2015), Rossi et al. (2017), Gutiérrez et al. (2011).

Moreover, multiple ecosystem services are provided by kelp forests irrespective of their structural changes (e.g., Eger et al. 2023). These include: (1) their role as primary producers; as highly productive ecosystems they form the base of rich marine food webs, mainly through detrital pathways, and (2) their potential contribution to carbon fixation and sequestration (blue carbon), since kelp absorbs CO_2, helping to mitigate climate change (e.g., Krumhansl and Scheibling 2012; Pessarrodona et al. 2023).

In many regions around the world, underwater forests are increasingly being affected by multiple anthropogenic pressures that may negatively impact their biodiversity and functioning at both local and regional scales (Rossi 2013; Eger et al. 2024). The main stressors include rising global temperatures and the increased occurrence of extreme events such as marine heatwaves and storms, as well as pollution linked to urban coastal development and maritime traffic (including transportation, tourism, and fishing). Additional key pressures include the unregulated harvesting of canopy-forming species and their associated biota, and the spread of invasive species. Recent studies have shown shifts in canopy-forming species distribution and biomass loss in some areas (see Krumhansl et al. 2016). While some canopy-forming species, such as *M. pyrifera*, exhibit high phenotypic plasticity that allows them to adjust their morphology and physiology in response to environmental changes (Graham et al. 2007), these adaptive strategies come with trade-offs that could undermine forest resilience, potentially leading to degradation and loss.

Even though there is a global trend of kelp forest decline, recent studies conducted in sub-Antarctic Tierra del Fuego have shown no signs of such decline. On the contrary, they suggest that this region may serve as a climatic refuge for underwater forests (Friedlander et al. 2020, 2023; Mora-Soto et al. 2021, 2022). In terms of extreme events, both marine heatwaves and cold spells have been recorded in Tierra del Fuego, yet recent remote sensing monitoring indicated that these events have not significantly altered the distribution or area covered by kelp forests (Mora-Soto et al. 2022). However, rising temperatures may accelerate glacial melt, increasing runoff and sediment input. These sediments may reduce light availability and degrade substrate quality, thereby impacting photosynthesis and the recruitment of kelp and associated species (Palacios et al. 2021; Kaminsky et al. 2024; Coral-Santacruz et al. 2024). Further studies could assess long-term effects of these factors to identify early warning signs of changes in kelp forest extent and density. To our knowledge, no studies to date have assessed the impact of heatwaves on animal forests in Tierra del Fuego.

Despite being a remote and difficult-to-access region, IDLE has seen an increase in cruise ship tourism in recent years, with vessels using the islands as shelter or as intermediate stops (Instituto Fueguino de Turismo 2024). According to tourism sector statistics, between 2016 and 2020, 96% of cruise tourists who visited IDLE were from abroad. This highlights—and at the same time explains—the limited local and national awareness of the island's rich ecosystems. Nearly 98% of surveyed visitors valued the island's pristine nature and protected status as a key reason for their trip. Given the site's remoteness, monitoring and regulating visitor landings is challenging; if tourism continues to grow, it could result in increased impacts on the area. The traffic of small boats during landing operations may cause physical damage to coastal habitats, such as the cutting or detachment of macroalgae. Additionally, the rise in maritime traffic heightens the risk of introducing non-native species. During this expedition, we recorded the presence of the sea anemone *Metridium senile* specifically in the dives conducted at Puerto Hoppner, a site frequently used by sailing vessels due to its narrow entrance, allowing access only to small boats, and its sheltered inner bay, which offers excellent anchorage during strong wind events. *M. senile* is considered an exotic species in the region and has been expanding along the coasts of Argentina and Chile (Gimenez et al. 2023). It was first reported in the Beagle Channel in 2019 (Häussermann et al. 2022) and appears to colonize vertical rocky walls rapidly, likely aided by its ability for asexual reproduction through laceration, with maximum laceration rates occurring every 12 hours (Anthony and Svane 1995). Studies along the Chilean coast have shown that *M. senile* has recently increased in abundance in areas previously occupied by sea urchins (*Loxechinus albus*), potentially causing significant economic impacts on this important fishery (Molinet et al. 2023). In the study by Molinet et al. (2023), it was shown that the abundance of *M. senile* increased significantly, from 67 individuals/m^2 in 2016 to an average of 1764 individuals/m^2 in 2021. In our scientific campaign, we recorded a mean density of 195 individuals/m^2 in Puerto Hoppner, which could indicate the onset of an expansion in the region. Given this emerging trend, we recommend investing time and resources into monitoring the *M. senile* population in IDLE to assess its potential growth and the possible ecological impacts it may have on the area.

Another potential pressure that could affect underwater forests is the expansion of aquaculture operations, such as salmon farming, an activity widely established in ecosystems throughout northern Patagonia in Chile. A historical summary of salmon farming in Chile, its main threats and critical knowledge gaps that need to be addressed to fully understand the environmental effects of this industry are reviewed in Quiñones et al. (2019). Specifically in kelp forests, salmonid fish farms' effluents may cause changes in hard-bottom macroalgal assemblages and in the composition of the epiphytic communities living on kelp thalli (Haugland et al. 2021; Oh et al. 2015). Furthermore, the emission of hydrogen peroxide in shallow areas, a chemotherapeutant frequently used to treat infections caused by salmon sea lice, might produce reduced photosynthetic efficiency that may lead to growth problems and mortality in kelp (Haugland et al. 2019). Although in Tierra del Fuego (Argentina) the law explicitly prohibits marine salmon farming since 2021, there are ongoing pressures to establish such operations, and in the Chilean part of the archipelago they are not banned. While there is broad local consensus against expanding these farms, remote locations in Chile are still seeing new installations. Given that Fuegian archipelago is shared between Argentina and Chile, it is essential to promote complementary conservation strategies between both countries to maintain the high conservation status of these underwater seascapes.

In Argentina, different conservation strategies have been implemented over recent decades to protect the coastal ecosystems of Tierra del Fuego. For instance, the creation of the Nature Reserve in IDLE in 1991, its designation as an Ecological, Historical, and Touristic Reserve in the Provincial Constitution and the recent declaration of the Peninsula Mitre Nature Reserve in 2022 all aim to preserve both terrestrial and coastal-marine environments in the region. Additionally, in 2024, the harvest of natural kelp forests formed by *Macrocystis pyrifera* was officially banned. These policies emphasize not only the importance of promoting scientific research to generate up-to-date information but also the need to raise public awareness about these ecosystems and their contributions to society—such as through environmental education. In this context, it is particularly important to plan research strategies that support the monitoring of both kelp forests and associated fauna in remote, logistically challenging areas like IDLE, for example complementing scientific diving with remote sensing studies.

Tierra del Fuego is home to some of the most pristine kelp and marine forests on the planet, representing both a significant opportunity and a considerable challenge. In some regions, kelp and animal forest restoration efforts are underway, though they often come with high costs (see Shefy et al. 2025; Eger et al. 2022). Evidence shows that restoration tends to fail in the absence of sufficient ecological knowledge, effective strategies to address the drivers of degradation, long-term planning, and adequate resources. Thus, conserving these forests is far more cost-effective than investing in their future restoration.

In addition to its 17,000 hectares of native terrestrial forest, Isla de los Estados also harbors lesser-known underwater kelp and animal forests, each hosting unique varieties of flora and fauna. These marine forests add yet another layer to the island's singular beauty and ecological richness, underscoring our responsibility, and that of future generations, to ensure its preservation.

Acknowledgments We are grateful to Tom W. Bell (Woods Hole Oceanographic Institution) for supervision and advice on remote sensing analyses. We would like to thank Miguel Porco Fisher, the captain from the Kostat vessel (Puerto Beagle company), for logistical support during the diving expedition to Isla de los Estados. Thanks to all the specialists who contributed to species identifications on the iNaturalist project: Camila Vasquez (Actinaria), Alicia Boraso (macroalgae), Cintia Fraysse (Asteroidea), Mariano Martínez (Holothuroidea), Diego Urteaga (Polyplacophora), Ignacio Chiesa (Amphipoda), Facundo Llompart, and Eloísa Giménez (fish). We thank the staff of LADE (Líneas Aéreas del Estado) for facilitating the transport of large volumes of diving and photographic equipment at no additional cost, in support of Argentine science.

Competing Interests This research was supported by the Argentinean National Science Council (CONICET) and the National University of Tierra del Fuego (UNTDF). This project was funded by a subsidy from Tierra del Fuego Provincial Secretariat for Science and Technology to pursue the scientific diving expedition to Isla de los Estados (Res. N°42/22) and a technical assistance agreement (STAN) between CADIC-CONICET and Arakur SA. We would like to acknowledge

CONICET Documental for their advice on the acquisition of audiovisual material, SIT-Study Abroad, Jumará Films, and The Life Aquatic Producciones for their funding support. The authors have no conflicts of interest to declare that are relevant to the content of this chapter.

Ethical Approval The fieldwork was conducted with permission from Tierra del Fuego Province (Res. N°345/21, N°117/22, N°310/22).

References

Anthony K, Svane I (1995) Effects of substratum instability on locomotion and pedal laceration in *Metridium senile* (Anthozoa:Actiniaria). Mar Ecol Prog Ser 124:171–180. https://doi.org/10.3354/meps124171

Barrett NS, Edgar GJ (2010) Distribution of benthic communities in the fjord-like Bathurst Channel ecosystem, South-Western Tasmania, a globally anomalous estuarine protected area. Aquat Conserv Mar Freshw Ecosyst 20:397–406. https://doi.org/10.1002/aqc.1085

Bell JJ, Barnes DKA (2000) The influences of bathymetry and flow regime upon the morphology of sublittoral sponge communities. J Mar Biol Assoc U K 80:707–718. https://doi.org/10.1017/S0025315400002538

Bell TW, Allen JG, Cavanaugh KC, Siegel DA (2020) Three decades of variability in California's giant kelp forests from the Landsat satellites. Remote Sens Environ 238:110811. https://doi.org/10.1016/j.rse.2018.06.039

Bell JJ, Micaroni V, Harris B, Strano F, Broadribb M, Rogers A (2024) Global status, impacts, and management of rocky temperate mesophotic ecosystems. Conserv Biol 38:e13945. https://doi.org/10.1111/cobi.13945

Boraso de Zaixso AL, Zaixso JM (2013) Elementos para el estudio de las macroalgas de Argentina, Comodoro Rivadavia. 204 pp. ISBN 978-987-1937-14-1

Bravo G, Livore JP, Bigatti G (2020) The importance of surface orientation in biodiversity monitoring protocols: the case of patagonian rocky reefs. Front Mar Sci 7:578595. https://doi.org/10.3389/fmars.2020.578595

Bravo G, Moity N, Londoño-Cruz E, Muller-Karger F, Bigatti G, Klein E, Choi F, Parmalee L, Helmuth B, Montes E (2021) Robots versus humans: automated annotation accurately quantifies essential ocean variables of rocky intertidal functional groups and habitat state. Front Mar Sci 8:691313. https://doi.org/10.3389/fmars.2021.691313

Bravo G, Kaminsky J, Bagur M, Alonso CP, Rodríguez M, Fraysse C, Lovrich G, Bigatti G (2023) Roving diver survey as a rapid and cost-effective methodology to register species richness in sub-Antarctic kelp forests. Diversity 15:354. https://doi.org/10.3390/d15030354

Bravo G, Kaminsky J, Bagur M, Alonso CP, Rodríguez M. iNaturalist Project: "Biodiversidad Submarina Isla de los Estados". Available online: https://www.inaturalist.org/projects/biodiversidad-submarina-isla-de-los-estados. Accessed 5 May 2025

Campoy AN, Pérez-Matus A, Wieters EA, Alarcón-Ireland R, Garmendia V, Beldade R, Navarrete SA, Fernández M (2023) The hidden diversity of temperate mesophotic ecosystems from Central Chile (southeastern Pacific Ocean) assessed through towed underwater videos. Diversity 15:360. https://doi.org/10.3390/d15030360

Cárdenas CA, Montiel A (2015) The influence of depth and substrate inclination on sessile assemblages in subantarctic rocky reefs (Magellan region). Polar Biol 38:1631–1644. https://doi.org/10.1007/s00300-015-1729-5

Cárdenas CA, Montiel A (2017) Coexistence in cold waters: animal forests in seaweed-dominated habitats in southern high latitudes. In: Rossi S, Bramanti L, Gori A, Orejas C (eds) Marine animal forests. Springer International Publishing. https://doi.org/10.1007/978-3-319-17001-5_50-1

Cerrano C, Bastari A, Calcinai B, Di Camillo C, Pica D, Puce S, Valisano L, Torsani F (2019) Temperate mesophotic ecosystems: gaps and perspectives of an emerging conservation challenge for the Mediterranean Sea. Eur Zool J 86:370–388. https://doi.org/10.1080/24750263.2019.1677790

Coral-Santacruz D, Méndez F, Marambio J, Haye PA, Bahamonde F, Mansilla A (2024) Effects of glacial melting on physiological performance of Macrocystis pyrifera in the Fjord of the mountains, Magellanic Sub-Antarctic ecoregion, Chile. J Appl Phycol 36:3637–3648. https://doi.org/10.1007/s10811-024-03362-3

Dayton PK (1985) The structure and regulation of some South American kelp communities. Ecol Monogr 55:447–468. https://doi.org/10.2307/2937131

De la Torriente A, Espinoza-León IM, Valenzuela-Lobos LA, Antolinez A, Serrano A (2024) The southernmost Errina Antarctica hydrocoral savannah in Patagonian waters. Sci Rep 14:9579. https://doi.org/10.1038/s41598-024-60207-2

Eger AM, Marzinelli EM, Christie H, Fagerli CW, Fujita D, Gonzalez AP, Hong SW, Kim JH, Lee LC, McHugh TA, Nishihara GN, Tatsumi M, Steinberg PD, Vergés A (2022) Global kelp forest restoration: past lessons, present status, and future directions. Biol Rev 97:1449–1475. https://doi.org/10.1111/brv.12850

Eger AM, Marzinelli EM, Beas-Luna R, Blain CO, Blamey LK, Byrnes JEK, Carnell PE, Choi CG, Hessing-Lewis M, Kim KY, Kumagai NH, Lorda J, Moore P, Nakamura Y, Pérez-Matus A, Pontier O, Smale D, Steinberg PD, Vergés A (2023) The value of ecosystem services in global marine kelp forests. Nat Commun 14:1894. https://doi.org/10.1038/s41467-023-37385-0

Eger AM, Eddy N, McHugh TA, Arafeh-Dalmau N, Wernberg T, Krumhansl K, Verbeek J, Branigan S, Kuwae T, Caselle JE, Giraldo Ospina A, Vergés A (2024) State of the world's kelp forests. One Earth 7:1927–1931. https://doi.org/10.1016/j.oneear.2024.10.008

Friedlander AM, Ballesteros E, Bell TW, Caselle JE, Campagna C, Goodell W, Hüne M, Muñoz A, Salinas-de-León P, Sala E, Dayton PK (2020) Kelp forests at the end of the earth: 45 years later. PLoS One 15:e0229259. https://doi.org/10.1371/journal.pone.0229259

Friedlander AM, Ballesteros E, Caselle JE, Hüne M, Adler AM, Sala (2023) Patterns and drivers of benthic macroinvertebrate assemblages in the kelp forests of Southern Patagonia. PLoS One 18:e0279200. https://doi.org/10.1371/journal.pone.0279200

Gimenez LH, Battini N, González-Muñoz R, Glon H (2023) Invader in disguise for decades: the plumose sea anemone *Metridium senile* in the Southwestern Atlantic Ocean. Biol Invasions 25:2159–2173. https://doi.org/10.1007/s10530-023-03031-5

Graham MH, Vásquez JA, Buschmann AH (2007) Global ecology of the giant kelp *Macrocystis*: from ecotypes to ecosystems. Oceanogr Mar Biol Annu Rev 45:39–88. https://doi.org/10.1201/9781420050943

Gutiérrez JL, Jones CG (2008) Ecosystem engineers. In: Encyclopedia of life sciences. Wiley, Chichester. https://doi.org/10.1002/9780470015902.a0021226

Gutiérrez JL, Jones CG, Byers JE, Arkema KK, Berkenbusch K, Commito JA, Duarte CM, Hacker SD, Lambrinos JG, Hendriks IE, Hogarth PJ, Palomo MG, Wild C (2011) Physical ecosystem engineers and the functioning of estuaries and coasts. In: Wolanski E, McLusky D (eds) Treatise on estuarine and coastal science. Academic Press, London, pp 53–81. https://doi.org/10.1016/B978-0-12-374711-2.00705-1

Gutt J, Cummings V, Dayton P, Isla E, Jentsch A, Schiaparelli S (2017) Antarctic marine animal forests: three-dimensional communities in Southern Ocean ecosystems. In: Rossi S, Bramanti L, Gori A, Orejas C (eds) Marine animal forests. Springer International Publishing. https://doi.org/10.1007/978-3-319-21012-4_8

Harris B, Davy SK, Bell JJ (2021) Benthic community composition of temperate mesophotic ecosystems (TMEs) in New Zealand: sponge domination and contribution to habitat complexity. Mar Ecol Prog Ser 671:21–43. https://doi.org/10.3354/meps13758

Haugland BT, Rastrick SPS, Agnalt AL, Husa V, Kutti T, Samuelsen OB (2019) Mortality and reduced photosynthetic performance in sugar kelp Saccharina latissima caused by the salmon-lice therapeutant hydrogen peroxide. Aquacult Environ Interact 11:1–17. https://doi.org/10.3354/aei00292

Haugland BT, Armitage CS, Kutti T, Husa V, Skogen MD, Bekkby T, Carvajalino-Fernández MA, Bannister RJ, White CA, Norderhaug KM, Fredriksen S (2021) Large-scale salmon farming in Norway impacts the epiphytic community of Laminaria hyperborea. Aquacult Environ Interact 13:81–100. https://doi.org/10.3354/aei00392

Häussermann V, Molinet C, Díaz Gómez M, Försterra G, Henríquez J, Espinoza Cea K, Matamala Ascencio T, Hüne M, Cárdenas CA, Glon H, Barahona Toledo N, Subiabre Mena D (2022) Recent massive invasions of the circumboreal sea anemone *Metridium senile* in North and South Patagonia. Biol Invasions 24:3665–3674. https://doi.org/10.1007/s10530-022-02878-4

Houskeeper HF, Rosenthal IS, Cavanaugh KC, Pawlak C, Trouille L, Byrnes JEK, Bell TW, Cavanaugh KC (2022) Automated satellite remote sensing of giant kelp at The Falkland Islands (Islas Malvinas). PLoS One 17:e0257933. https://doi.org/10.1371/journal.pone.0257933

Instituto Fueguino de Turismo (2024) Información Estadística: Informe Temporada de Cruceros. Available online: https://infuetur.gob.ar/. Accessed 5 May 2025

Kaminsky J, Bagur M, Boraso A, Rodríguez M, Buschmann AH, Quartino ML, Schloss IR (2024) Giant kelp recruitment declines and changes in the understory algae in a sub-Antarctic urban ecosystem. Hydrobiologia 851:4783–4802. https://doi.org/10.1007/s10750-024-05623-1

Krumhansl KA, Scheibling RE (2012) Production and fate of kelp detritus. Mar Ecol Prog Ser 467:281–302. https://doi.org/10.3354/meps09940

Krumhansl KA, Okamoto DK, Rassweiler A, Novak M, Bolton JJ, Cavanaugh KC, Connell SD et al (2016) Global patterns of kelp forest change over the past half century. Proc Natl Acad Sci USA 113:13785–13790. https://doi.org/10.1073/pnas.1606102113

Kühnemann O (1970) Algunas consideraciones sobre los bosques de *Macrocystis pyrifera*. Physis 29:273–296

Kühnemann O (1976) Observaciones ecológicas sobre la vegetación marina y terrestre de la Isla de los Estados (Tierra del Fuego, Argentina). Ecosur 6:121–248

Küpper FC, Peters AF, Kytinou E, Asensi AO, Vieira C, Macaya EC, De Clerck O (2019) *Dictyota falklandica* sp. nov. (Dictyotales, Phaeophyceae) from The Falkland Islands and southernmost South America. Phycologia 58:640–647. https://doi.org/10.1080/00318884.2019.1648990

Lawn ID, Ross DM (1982) The behavioural physiology of the swimming sea anemone *Boloceroides mcmurrichi*. Proc R Soc Lond B 216:315–334. https://doi.org/10.1098/rspb.1982.0077

Meretta PE, Genzano G (2025) A temperate gorgonian coral garden in the shallow sublittoral off Buenos Aires, Argentina. Reg Stud Mar Sci 81:103971. https://doi.org/10.1016/j.rsma.2024.103971

Merlo PJ, De Wysiecki A, Venerus LA, Trobbiani GA, Irigoyen AJ (2025) Latitudinal and habitat-driven diversity of coastal fishes in the southernmost Southwest Atlantic Ocean. J Biogeogr 52:e15156. https://doi.org/10.1111/jbi.15156

Miller R, Etter R (2011) Rock walls: small-scale diversity hotspots in the subtidal Gulf of Maine. Mar Ecol Prog Ser 425:153–165. https://doi.org/10.3354/meps09025

Molinet C, Häussermann V, Astorga M, Barahona N, Espinoza K, Diaz M, Díaz P, Henríquez J, Matamala T, Soto D (2023) Population expansion of the invasive sea anemone *Metridium senile* in the spatial mesoscale of a sea urchin bed in North-Western Patagonia. Biol Invasions 25:1101–1118. https://doi.org/10.1007/s10530-022-02965-6

Mora-Soto A, Palacios M, Macaya E, Gómez I, Huovinen P, Pérez-Matus A, Young M, Golding N, Toro M, Yaqub M, Macias-Fauria M (2020) A high-resolution global map of giant kelp (*Macrocystis pyrifera*) forests and intertidal green algae (Ulvophyceae) with Sentinel-2 imagery. Remote Sens 12:694. https://doi.org/10.3390/rs12040694

Mora-Soto A, Capsey A, Friedlander AM, Palacios M, Brewin PE, Golding N, Dayton P et al (2021) One of the least disturbed marine coastal ecosystems on earth: spatial and temporal persistence of Darwin's sub-Antarctic giant kelp forests. J Biogeogr 48:2562–2577. https://doi.org/10.1111/jbi.14221

Mora-Soto A, Aguirre C, Iriarte JL, Palacios M, Macaya EC, Macias-Fauria M (2022) A song of wind and ice: increased frequency of marine cold-spells in Southwestern Patagonia and their possible effects on giant kelp forests. J Geophys Res 127:e2021JC017801. https://doi.org/10.1029/2021JC017801

Oh ES, Edgar GJ, Kirkpatrick JB, Stuart-Smith RD, Barrett NS (2015) Broad-scale impacts of salmon farms on temperate macroalgal assemblages on rocky reefs. Mar Pollut Bull 98(1-2):201–209. https://doi.org/10.1016/j.marpolbul.2015.06.049

Orejas C, Carreiro-Silva M, Mohn C, Reimer JD, Samaai T, Allcock AL, Rossi S (2022) Marine animal forests of the world: definition and characteristics. Res Ideas Outcomes 8:e96274. https://doi.org/10.3897/rio.8.e96274

Palacios M, Osman D, Ramírez J, Huovinen P, Gómez I (2021) Photobiology of the giant kelp *Macrocystis pyrifera* in the land-terminating glacier fjord Yendegaia (Tierra del Fuego): a look into the future? Sci Total Environ 751:141810. https://doi.org/10.1016/j.scitotenv.2020.141810

Parker GH (1916) Locomotion of sea-anemones. Proc Natl Acad Sci USA 2:449–450. https://doi.org/10.1073/pnas.2.8.449

Pessarrodona A, Franco-Santos RM, Wright LS, Vanderklift MA, Howard J, Pidgeon E, Wernberg T, Filbee-Dexter K (2023) Carbon sequestration and climate change mitigation using macroalgae: a state of knowledge review. Biol Rev 98:1945–1971. https://doi.org/10.1111/brv.12990

Piazzi L, Turicchia E, Rindi F, Falace A, Gennaro P, Abbiati M, Bandelj V, Calcinai B, Ciriaco S, Costantini F, Gianni F, Kaleb S, Puce S, Ponti M (2023) NAMBER: a biotic index for assessing the ecological quality of mesophotic biogenic reefs in the northern Adriatic Sea. Aquatic Conserv Mar Freshw Ecosyst 33:298–311. https://doi.org/10.1002/aqc.3922

Ponce JF, Rabassa JO, Martinez OE (2009) Fiordos en Isla de los Estados: descripción morfométrica y génesis de los únicos fiordos en la Patagonia Argentina. Rev Assoc Geol 65:638–647. https://revista.geologica.org.ar/raga/article/view/865

Quiñones RA, Fuentes M, Montes RM, Soto D, León-Muñoz J (2019) Environmental issues in Chilean salmon farming: a review. Rev Aquac 11:375–402. https://doi.org/10.1111/raq.12337

Ramírez ME (2010) Benthic marine flora from southern South America and Antarctica. An Inst Patagon 38:57–71. https://doi.org/10.4067/S0718-686X2010000100003

Reshitnyk L, Saccomanno V, Bell T, Cavanaugh K (2023) Mapping canopy-forming kelps in the Northeast Pacific: a guidebook for decision-makers and practitioners. The Hakai Institute, Heriot Bay. https://doi.org/10.21966/7ze4-x883

Rodríguez JP, Gibelli V, Bahamonde F, Marambio J, Calderón M, Bustamante D, Rozzi R, Mansilla A (2024) Primer catálogo de macroalgas marinas bentónicas del canal Beagle occidental: Relevancia del nuevo Herbario de Criptógamas Subantárticas (HCS) UMAG-CHIC, Chile. An Inst Patagon 52:1–41

Rossi S (2013) The destruction of the 'animal forests' in the oceans: towards an over-simplification of the benthic ecosystems. Ocean Coast Manag 84:77–85. https://doi.org/10.1016/j.ocecoaman.2013.07.004

Rossi S, Bramanti L, Gori A, Orejas C (2017) An overview of the animal forests of the world. In: Rossi S, Bramanti L, Gori A, Orejas C (eds) Marine animal forests. Springer International Publishing. https://doi.org/10.1007/978-3-319-17001-5_1-1

Santelices B, Ojeda FP (1984) Effects of canopy removal on the understory algal community structure of coastal forests of *Macrocystis pyrifera* from southern South America. Mar Ecol Prog Ser 14:165–173. https://doi.org/10.3354/meps014165

Schejter L, Rimondino C, Chiesa I, Díaz de Astarloa JM, Doti B, Elías R, Escolar M, Genzano G, López-Gappa J, Tatián M, Zelaya DG, Cristobo J, Perez CD, Cordeiro RT, Bremec CS (2016) Namuncurá marine protected area: an oceanic hot spot of benthic biodiversity at Burdwood Bank, Argentina. Polar Biol 39:2373–2386. https://doi.org/10.1007/s00300-016-1913-2

Schiel DR, Foster MS (2015) The biology and ecology of giant kelp forests. University of California Press, Oakland. https://doi.org/10.1525/9780520961098

Shefy D, Rossi S, Rinkevich B (2025) Not just carbon: biodiversity credits for restoration of the marine animal forests. Ecosyst People 21:2488350. https://doi.org/10.1080/26395916.2025.2488350

Soares MDO, Araújo JTD, Ferreira SMC, Santos BA, Boavida JRH, Costantini F, Rossi S (2020) Why do mesophotic coral ecosystems have to be protected? Sci Total Environ 726:138456. https://doi.org/10.1016/j.scitotenv.2020.138456

Strano F, Micaroni V, Costa G, Bertocci I, Bertolino M (2020) Shallow-water sponge grounds along the Apulian coast (central Mediterranean Sea). Mar Biodivers 50:7. https://doi.org/10.1007/s12526-019-01026-x

Teagle H, Hawkins SJ, Moore PJ, Smale DA (2017) The role of kelp species as biogenic habitat formers in coastal marine ecosystems. J Exp Mar Biol Ecol 492:81–98. https://doi.org/10.1016/j.jembe.2017.01.017

Wallace A (1869) The Malay archipelago. Macmillan and Company

Chapter 6
Mosses of Isla de los Estados

Adolfina Savoretti and María Paz Martinoli

Abstract Bryophytes, which include mosses, liverworts, and hornworts, are present in most terrestrial ecosystems and are especially dominant in cold climates. Isla de los Estados has been an important site for bryophyte collection since the late eighteenth century, with contributions from notable botanists such as Menzies, Skottsberg, Castellanos, and Matteri. Although a considerable portion of the collected specimens has been identified and published, many remain undetermined and dispersed across various herbaria, making it difficult to produce a comprehensive species list. This study aims to provide a preliminary list of moss species from the island by reviewing historical collections, databases, and herbarium records. A total of 765 records were compiled, identifying 66 species across 37 families and 36 genera, along with 30 additional genera identified to that level. This work represents an initial step toward documenting the bryophyte diversity of Isla de los Estados.

Keywords Bryophyte diversity · Moss taxonomy · Herbarium collections · Historical expeditions · Floristic inventory

6.1 Introduction

Bryophytes include mosses (Division Bryophyta), liverworts (Division Marchantiophyta), and hornworts (Division Anthocerotophyta). Collectively, this group comprises approximately 20,000 species (Shaw et al. 2011). They inhabit all ecosystems on the planet, except for marine ecosystems (Goffinet and Buck 2013;

A. Savoretti (✉)
CIT-Tierra del Fuego-CONICET, Río Grande, Tierra del Fuego, Argentina

ICPA UNTDF, Ushuaia, Tierra del Fuego, Argentina
e-mail: masavoretti@untdf.edu.ar

M. P. Martinoli
CADIC-CONICET, Ushuaia, Tierra del Fuego, Argentina

A. Raya Rey, U. Balza (eds.), *Biodiversity and Conservation of Isla de los Estados (Chuani-sin)*, Natural and Social Sciences of Patagonia,
https://doi.org/10.1007/978-3-032-13630-5_6

Schofield 1985). The three groups share a similar life cycle in which they require water for fertilization, the gametophyte is perennial and dominant in terms of both size and longevity, the sporophyte is monosporangiate, unbranched, and its complete development occurs in connection with the gametophyte—characteristics that make them very different from the rest of terrestrial plants (Schofield 1985).

Among bryophytes, mosses constitute the most numerous Division in terms of species diversity, with approximately 12,500–13,000 species (Gradstein et al. 2001; Goffinet et al. 2009). In Argentina, 990 moss species are recorded (Matteri 2003). Tierra del Fuego Province is the Argentine jurisdiction where the most studies on the diversity of this group have been conducted and, consequently, where the highest number of species is recorded, with a total of 256 moss species (Matteri 2003). Numerous floristic studies of the Fuegian area have been published (Matteri 1988, 1989; Matteri and Schiavone 1991; Matteri and Schiavone 2002, Cottet and Messuti 2023), as well as important taxonomic works compiled in different volumes of the Cryptogamic Flora (Matteri 1975, 1985; Schiavone 1993; Ochyra and Matteri 2001).

Among the existing moss records for to the region are those from Isla de los Estados. However, to date, there is no work that compiles all available information specifically for this island. This gap makes it difficult to assess the extent of current knowledge and to gain comprehensive understanding of the island's moss diversity. The primary aim of this study is to create a preliminary list of moss species found on Isla de los Estados by reviewing historical collections, through literature search, online database searches, and visits to Argentine herbaria. A secondary objective is to compile information on historical collections, includying both identified and unidentified material. This work represents an initial step toward organizing and identifying historical material from the island, contributing to the development of a comprehensive list of its bryophyte species.

6.2 Materials and Methods

The information required for compiling the species list was obtained based on the review of herbarium specimens, the record book of the Herbarium of the Argentine Museum of Natural Sciences "Bernardino Rivadavia" (BA), and the Herbarium of the Spegazzini Institute (LPS). Additionally, a search was conducted in GBIF (https://doi.org/10.15468/dl.thaq5x). Literature from the Fuegian region was consulted (e.g., Kühnemann and Goncalves Carrales, 1975; Matteri 1975, 1985, 1988, 1993, 2003; Matteri and Hyvönen 1992; Matteri and Schiavone 2002; Ochyra and Matteri 2001). A review of the status of specific names was made using the Missouri Botanical Garden database (https://tropicos.org) and the updated classification of moss genera proposed by Bernard Goffinet and William R. Buck was followed (https://bryology.eeb.uconn.edu/classification). QGIS 3.22 software was used for map preparation.

6.3 Results

6.3.1 Expeditions and Collectors

The recorded moss specimens were collected during several expeditions to Isla de los Estados (Fig. 6.1).

The first specimens for this area were collected by Archibald Menzies in 1787 during the expedition called "Prince of Wales" commanded by Lieutenant Colnett (1786–1789) (Imshaug 1972). Thereafter, in an Italian-Argentine expedition led by Giacomo Bove in 1882, Carlos Spegazzini (Viegas Barros and Malvestitti 2019) collected moss specimens among other plant and fungi groups (Imshaug 1972). Carl Skottsberg visited the island in January 1902 at the Swedish South-Polar Expedition (Svenska Sydpolarexpeditionen 1901–1903) as the expedition's botanist (Official narrative Swedish south polar expedition; Lewander 2002). Alberto Castellanos was on Isla de los Estados from December 1933 to March 1934, sent by the Argentine Museum of Natural Sciences "Bernardino Rivadavia" with the objective of studying the island's flora (Castellanos 1935). This expedition is recorded in the BA herbarium entry books as "Excursión del Museo."

Additional records of specimens collected by P. Pastorino in August 1966 have been found in BA. No information about the expedition or the collector's first name was possible to find. These specimens are the only ones collected during winter, and due to the considerable number of records, they are worth mentioning.

Gabriela G. Hässel de Menéndez was on the island in 1967, in a trip organized by the Bernardino Rivadavia Museum of Natural Sciences in conjunction with the

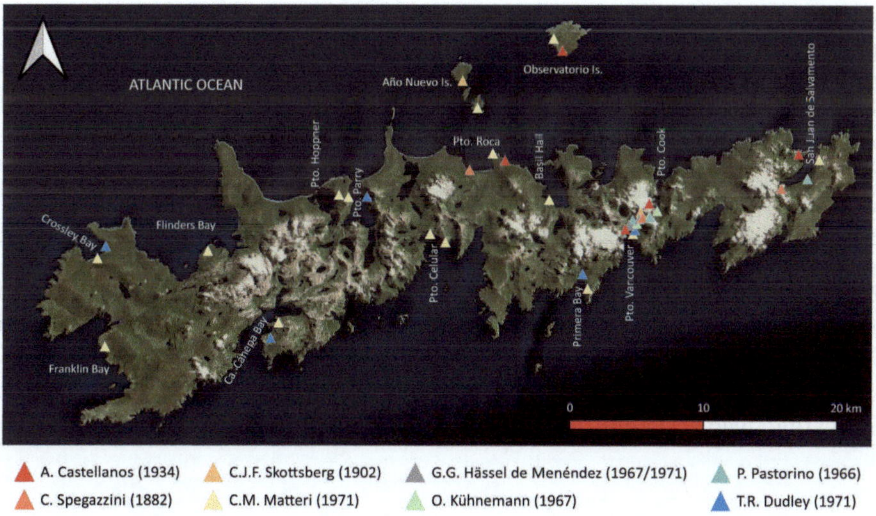

Fig. 6.1 Map of Isla de los Estados indicating the collection sites and names of bryophyte collectors from the nineteenth and twentieth centuries

National Council of Scientific Research and Technology (CONICET), with support from the Argentine Army (Hässel de Menéndez and Solari 1975). That same year, Oscar Kühnemann collected bryophytes, lichens, and algae on the island (Hässel de Menéndez 2003; Kühnemann and Gonçalves Carralves 1975).

Celina M. Matteri, Gabriela G. Hässel de Menéndez and Theodore R. Dudley shared the "R.V. Hero" Botanical Expedition in 1971 during October and November (Imshaug 1972). Among all these expeditions, the work of Celina M. Matteri stands out. As a bryologist and moss specialist, she collected many specimens and identified many of them. Notably, she is the only collector to have sampled mosses in all localities on Isla de los Estados where moss collection has been recorded, including Basil Hall, Flinders Bay, Franklin Bay, Puerto Hoppner, and Puerto Celular (Fig. 6.1). Some of her specimens are cited in various fascicles of the Cryptogamic Flora of Tierra del Fuego, which also includes the description of a new moss species, *Distichophyllum submersum* Matteri (1975). To our knowledge, this was the most recent bryological expedition to Isla de los Estados.

6.3.2 Herbaria

Most of the material collected on Isla de los Estados is housed in BA. However, material deposited in other herbaria has also been recorded.

List of herbaria where moss specimens from Isla de los Estados are deposited (herbarium codes according to *Index Herbariorum*, Thiers, 2025):

BA: Museo Argentino de Ciencias Naturales "Bernardino Rivadavia"
FLAS: University of Florida Herbarium—Bryophytes
H-Br: V. F. Brotherus Herbarium
L: Naturalis Biodiversity Center
LPS: Museo de La Plata, Instituto Spegazzini
MNHN-PC: Muséum national d'Histoire naturelle
MO: Missouri Botanical Garden
NY: The New York Botanical Garden
S-Bryophytes: Bryophytes Swedish Museum of Natural History
TENN: University of Tennessee—Knoxville

Many specimens have been examined and identified by recognized botanists and specialists from different moss groups. Some of them are: Jules Cardot, Heikki Roivainen Dale Vitt, Ryszard Ochyra, Halina Bednarek-Ochyra Bruce H. Allen, Allan J. Fife, Dana Griffin III, Celina M. Matteri, Steven Churchill, Esther Fuertes, Guillermo Suárez, Harumi Ochi, Juan Larraín and Lisa Pokorny.

6.3.3 Moss Diversity of Isla de los Estados

Based on the information obtained, 765 records were registered. Among these, 37 families, 36 genera, 66 species and 3 varieties are recognized. Additionally, 30 other genera were recorded in specimens determined to this taxonomic level. Other samples have been determined only to the Family level (ca. 100 records). The number of families per locality is shown in Fig. 6.2.

The list of mosses presents on Isla de los Estados is organized alphabetically by Family. For each taxon, the locality or localities where it has been collected are specified. Localities are mentioned through initials as detailed below: Observatory Island (OI), Alferez Grofe Island (AGI), Año Nuevo Island (ANI), Basil Hall Bay (BHB), Crossley Bay (CB), Puerto Vancouver (PV), Hoppner Bay (HB), Puerto Cook (PC), Puerto Parry (PP), Puerto Celular (PCe), Puerto Roca (PR), Franklin Bay (FrB), Flinders Bay (FB), Capitán Cánepa Bay (CCB), San Juan de Salvamento (SJDS), Primera Bay (PB).

List of Mosses from Isla de los Estados

Amblystegiaceae

- *Drepanocladus* (Müll. Hal.) G. Roth: PR, PB, FB

Sanionia uncinata (Hedw.) Loeske: PV

Vittia pachyloma (Mont.) Ochyra: SJDS

Andreaeaceae

- *Andreaea* Hedw.: PCe, CCB, SJDS, PB, FB, BHB, CB, PV, PC

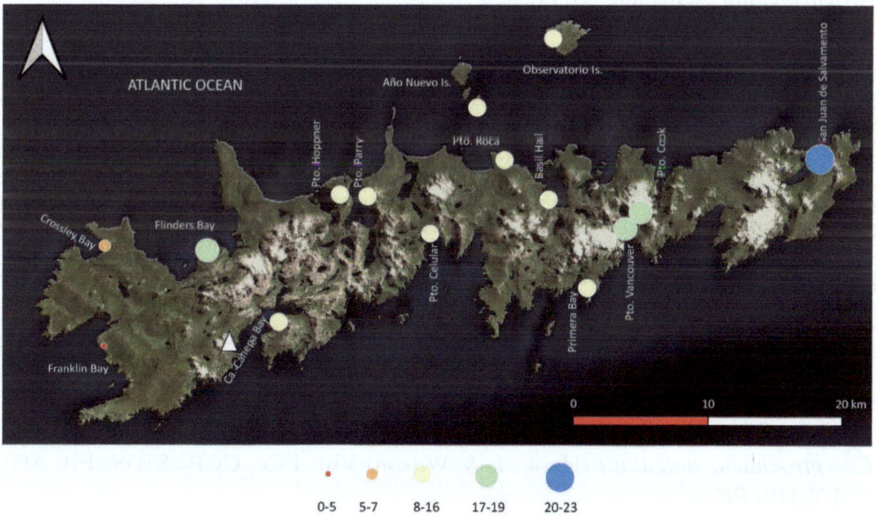

Fig. 6.2 Map of Isla de los Estados showing bryophyte family diversity across bays (intervals are based on natural breaks). The white triangle marks the site where Hässel de Menéndez collected the Type specimen of *Distichophyllum submersum* Matteri

Andreaea acutifolia Hook.f. & Wilson: PC
Andreaea gainii Cardot: PP
Andreaea mutabilis Hook. f. & Wilson: PC
Andreaea regularis Müll. Hal.: PC
Andreaea wilsonii Hook. f.: PP

Bartramiaceae
• *Bartramia patens* Brid.: PCe, PR,FrB, PB, OI, PC
Bartramia patens Brid. var. *patens*: FB
Breutelia aureola (Besch. ex Müll. Hal.) Besch.: PP, PCe, CCB, SJDS, FB, PV, HB, PC
Breutelia dumosa Mitt.: PP, SJDS, PB
Breutelia integrifolia (Taylor) A. Jaeger: PB, FB, HB, PC
Breutelia plicata Mitt.: PP, FB, BY
Conostomum magellanicum Sull.: CCB
Conostomum pentastichum (Brid.) Lindb.: CCB, SJDS, FB, BHB, PV, PC
Philonotis vagans (Hook.f. & Wilson) Mitt.: SJDS, PB

Brachytheciaceae
• *Brachythecium* Schimp.: PCe, FrB, PB, AGI, PC
Brachythecium paradoxum (Hook. f. & Wilson) A. Jaeger: FB, BHB, PV, HB

Bryaceae
• *Bryum* Hedw.: SJDS, PB, FB, AGI, PC
Bryum argenteum Hedw.: AGI
Bryum clavatum (Schimp.) Müll. Hal.: SJDS
Brachymenium Schwägr.: SJDS

Calliergonaceae
• *Warnstorfia fluitans* (Hedw.) Loeske: FrB, SJDS

Catagoniaceae
• *Catagonium* Müll. Hal.: PP,CCB, FB, PV, HB
Catagonium nitens (Brid.) Cardot: PR
Catagonium nitidum (Hook. f. & Wilson) Broth.: OI, AGI, PC

Daltoniaceae
• *Achrophyllum* Vitt & Crosby: PP, PR, SJDS, PB, OI, AGI, PV, PCe
Achrophyllum haesselianum (Matteri) Matteri): PP, SJDS, PB, FB, Pce
Achrophyllum magellanicum (Besch.) Matteri: PP, SJDS, FB, OI, PV, PC
Achrophyllum magellanicum var. *oligodontum* (Matteri) Matteri: FB
Calyptrochaeta Desv.: PR, CCB, AGI, BHB
Calyptrochaeta apiculata (Hook. f. & Wilson) Vitt: PCe, CCB, SJDS, FB, AGI, PV, HB, PC
Calyptrochaeta apiculata var. *spathulata* (Matteri) Matteri: PB
Distichophyllum Dozy & Molk.: PP, CCB, SJDS, PB, BHB, PV, HB, PC, Pce
Distichophyllum dicksonii (Hook. & Grev.) Mitt.: Pto Parry, PR, SJDS, BHB, PV, HB, PC

Distichophyllum eremitae (A. Jaeger) Paris: PP, SJDS, FB, AGI, PV, HB
Distichophyllum flaccidum (Hook. f. & Wilson) Mitt.: PP, HB
Distichophyllum pulchellum (Hampe) Mitt.: FrB
**Distichophyllum submersum* Matteri: PP, *CCB, SJDS, FB, PV
Holotypus: Cascada del Arroyo Capitán Fernández, G.G Hässel de Menéndez (BA18312)

Dicranaceae
• *Chorisodontium* (Mitt.) Broth: PR, SJDS, BHB
Chorisodontium spegazzinii (Müll. Hal.) Roiv.: PC
Dicranoloma (Renauld) Renauld: PP, PR, CCB, SJDS, PB, FB, OI, AGI, BHB, CB, PV, HB, PC, PCe
Dicranoloma australe (Besch. ex Müll. Hal.) Paris: SJDS
Dicranoloma robustum (Hook. f. & Wilson) Paris: SJDS
Platyneuron (Cardot) Broth.: AGI, BHB, CB, HB, Pce, SJDS, FB, OI
Platyneuron praealtum (Mitt.) Ochyra & Bedn.-Ochyra: AGI, CB
Dicranum Hedw.: PC

Dicranellaceae
• *Dicranella* (Müll. Hal.) Schimp.: SJDS

Distichiaceae
• *Distichium* Bruch & Schimp.: CCB, SJDS, BHB

Ditrichaceae
• *Ceratodon* Brid.: CB, PV, PC
Ditrichum Timm ex Hampe: PP, CCB, FB, BHB, PC

Fontinalaceae
• *Fontinalis* Hedw.: PC

Funariaceae
• *Funaria* Hedw.: FB, CB, PV

Grimmiaceae
• *Grimmia* Hedw.: PP, CCB, SJDS, PB, FB, BHB, PV, HB, Pce
Racomitrium Brid.: PP, PR, CCB, SJDS, PB, BHB, PV, HB, PC
Racomitrium geronticum Müll. Hal.: PV, PC
Racomitrium heterostichoides Cardot: PC
Racomitrium lanuginosum (Hedw.) Brid.: PP, PR, CCB, SJDS, FB, BHB, PV, HB, Pce
Racomitrium pachydictyon Cardot: PC
Racomitrium willii (Müll. Hal.) Kindb.: PC
Schistidium Bruch & Schimp.: PB

Hymenolomataceae
• *Hymenoloma* Dusén: CCB

Hypnaceae

• *Hypnum* Hedw.: OI

Lembophyllaceae
• *Acrocladium* Mitt.: SJDS
Acrocladium auriculatum (Mont.) Mitt.: PP, SJDS, AGI, PC
Camptochaete gracilis (Hook. f. & Wilson) Paris: SJDS, CCB
Isothecium Brid.: PCe
Rigodium pseudothuidium Dusén: PR

Leptostomataceae
• *Leptostomum* R. Br.: SJDS, PB, BHB, Pce
Leptostomum menziesii R.Br.: PP, PR, FrB, PB, FB, PV, PC

Lepyrodontaceae
• *Lepyrodon* Hampe: PP, PB, OI
Lepyrodon lagurus (Hook.) Mitt.: PR

Leucobryaceae
• *Campylopus* Brid.: SJDS, OI, AGI, PC

Mniaceae
• *Mielichhoferia* Nees & Hornsch.: SJDS
Mnium Hedw.: PV, PC
Pohlia Hedw.: HB
Pohlia wahlenbergii (F. Weber & D. Mohr) A.L. Andrews: SJDS

Neckeraceae
• *Bryobuckia arbusculans* (Müll. Hal.) Enroth: FB

Oedipodiaceae
• *Oedipodium* Schwägr.: PV

Orthodontiaceae
• *Leptotheca* Schwägr.: PP, SJDS, HB
Leptotheca gaudichaudii Schwägr.: BHB
Orthodontium Wilson: SJDS, OI

Orthotrichaceae
• *Macromitrium* Brid.: HB
Muelleriella Dusén: PCe, CCB, SJDS, PB, FB, AGI, BHB, PV, HB
Muelleriella crassifolia (Hook. f. & Wilson) Dusén: PR, OI, PC
Ulota D. Mohr: PP, PV, PC
Zygodon Hook. & Taylor: PV, PC

Plagiotheciaceae
• *Plagiothecium conostegium* Herzog: PCe, SJDS, OI, AGI

Polytrichaceae
• *Atrichopsis* Cardot: PP
Atrichopsis trichodon (Hook. f. & Wilson) N.E. Bell & Hyvönen: PC

Polytrichum Hedw.: SJDS, CB

Pottiaceae

• *Barbula* Hedw.: PC, AGI, PC
Hymenostylium Brid.: PB
Pottia (Rchb.) Fürnr.: PB
Syntrichia brachyclada (Cardot) R.H.Zander: FB
Syntrichia subpapillosa (Cardot & Broth.) Matteri: FB, PCe
Syntrichia princeps (De Not.) Mitt.: FB
Tortula Hedw.: PR, CCB, SJDS, FB, AGI
Tortula anderssonii Ångstr.: HB
Tortula chrysopila (Müll. Hal.) Paris: SJDS, PB, OI

Ptychomniaceae

• *Cladomniopsis crenato-obtusa* M. Fleisch.: SJDS, FB
Ptychomnion cygnisetum (Müll. Hal.) Kindb.: Pto Cook, PP, PCe, PR, CCB, SJDS, PB, FB, BHB, CB, PV, HB, PC
Ptychomnion densifolium (Brid.) A. Jaeger: SJDS, PB, FB, OI, AGI, CB, PV, HB, PC

Pylaisiadelphaceae

• *Isopterygium* Mitt.: PP

Rhacocarpaceae

• *Rhacocarpus* Lindb.: PP, CCB, SJDS, BHB, PV, HB

Rhizogoniaceae

• *Goniobryum subbasilare* (Hook.) Lindb.: PR
Pyrrhobryum Mitt.: PP, SJDS
Rhizogonium Brid.: PC, PCe, SJDS, FB
Rhizogonium mnioides (Hook.) Wilson.: PP, SJDS

Seligeriaceae

• *Blindia* Bruch & Schimp.: PP, PR, FB, HB, PC
Blindia magellanica Schimp. ex Müll. Hal.: PB
Blindia robusta Hampe: CCB

Sematophyllaceae

• *Rhaphidorrhynchium* Besch. ex M. Fleisch: PP

Splachnaceae

• *Tayloria* Hook.: HB
Tayloria dubyi Broth.: SJDS, AGI, CB, HB
Tayloria magellanica (Brid.) Mitt.: PP, SJDS, FB, PV, HB

Sphagnaceae

• *Sphagnum* L.: PCe, PR, SJDS, PB, FB, CB, CCB
Sphagnum falcatulum Besch.: SJDS, FB, PC
Sphagnum fimbriatum Wilson: SJDS, PC
Sphagnum magellanicum Brid.: SJDS

6.4 Final Remarks

This preliminary list of mosses from Isla de los Estados reveals that the island harbors approximately 25% of the total moss diversity currently documented for the Argentine portion of Tierra del Fuego. This figure highlights the island's significant contribution to the regional bryoflora, despite its limited accessibility and the relatively low sampling effort over time.

A substantial portion of the existing knowledge about mosses on Isla de los Estados is the result of the work of Celina M. Matteri. Her dual role as both an active field collector and a specialist in moss taxonomy has been instrumental. Through her expeditions and subsequent taxonomic work, Matteri has provided the most comprehensive and consistent dataset for the island to date. Her collections span all known localities with moss records, and her detailed study of the specimens has laid the groundwork for future bryological research in the region.

The difficulty of accessing Isla de los Estados—combined with the relatively small number of bryologists in Argentina compared to researchers working on other taxonomic groups—underscores the importance of these expeditions. Each record from the island represents not only a data point but also the outcome of significant logistical effort and taxonomic expertise. In this context, the work carried out by bryologists on historical and recent collections becomes especially valuable.

Given the high number of specimens identified only to the Family level, there remains much to be done. The first author of this study, in collaboration with the Herbarium of the Museo Argentino de Ciencias Naturales, will undertake a detailed revision of this undetermined material. This effort will not only contribute to refining the moss checklist for Isla de los Estados but also to advancing our understanding of the island's unique and underexplored bryophyte diversity.

Competing Interest To Diego Gutiérrez (General Curator BA), Agustina Yañez (Associated Curator BA) Celeste Alvarenga (Technical Staff BA), Fabricio Valdez (LPS), as well as to the Argentine Herbarium Network. We are also grateful to Anabela Plos, node manager of GBIF Argentina. Special thanks to Juan Larraín and Christian Zanotti for their continuous support and valuable recommendations. We also thank Martín Castro, Paula Sotelano, Soledad Jiménez, Oriana Hernández Herrero, and Damián A. Fernández for their valuable contributions during the preparation of the manuscript.

References

Castellanos A (1935) Observaciones de algunas aves de Tierra del Fuego e Isla de los Estados. Hornero 006(01):022–035

Cottet AC, Messuti MI (2023) Nuevos registros de distribución para la flora de musgos (Bryophyta) de la Patagonia, Argentina. Bol Soc Argent Bot 58:215–227

Goffinet B, Buck WR (2013) The evolution of body form in bryophytes. Ann Plant Rev 45:51–90

Goffinet B, Buck WR. Classification of the bryophyta. On-line version available at http://bryology.eeb.uconn.edu/classification/. Checked on May 2025

Goffinet BW, Buck R, Shaw AJ (2009) Morphology, anatomy, and classifycation of the Bryophyta. In: Goffinet B, Shaw AJ (eds) Bryophyte biology. Cambridge University Press, Nueva York, pp 55–138

Gradstein SR, Churchill SP, Salazar-Allen N (2001) Guide to the bryophytes of tropical America, Memoirs of the New York Botanical Garden, vol 86. New York Botanical Garden Press, Bronx, pp 1–577

Hässel de Menéndez GG (2003) Obituarios, Oscar Kühnemann (22-XII-1911/24-XII-1999). Bol Soc Arg Bot 38(3–4):357–359

Hässel de Menéndez GG, Solari SS (1975) Orden Calobryales. Orden Jungermanniales. In: Guarrera SA, Gamundi de Amos I, Rabinovich de Halperin D (dir) Flora Criptogámica de Tierra del Fuego, vol 15, no 2. Fundación para la educación la ciencia y la cultura, Buenos Aires

Imshaug HA (1972) R/V Hero Cruise 71-5 to Isla de los Estados. Antarct J US 7(2):42–44

Kühnemann O, Goncalves Carrales MF (1975) Monografía preliminar para la flora criptogámica fueguina: Los musgos de la familia Ptychomniaceae (Bryophyta). Darwin 19(2-4):583–617

Lewander L (2002) The representations of the Swedish Antarctic expedition, 1901–03. Polar Rec 38(205):97–114

Matteri CM (1975) Orden Hookeriales. In: Guarrera SA, Gamundi de Amos I, Rabinovich de Halperin D Flora Criptogámica de Tierra del Fuego. Fundación para la educación la ciencia y la cultura, Buenos Aires, vol 14 (9)

Matteri CM (1985). Bartramiaceae. In: Guarrera SA, Gamundi de Amos I, Rabinovich de Halperin D Flora Criptogámica de Tierra del Fuego. Fundación para la educación la ciencia y la cultura, Buenos Aires, vol 14 (7)

Matteri CM (1988) Distribution maps of mosses from Tierra del Fuego, Argentina 1. Lindbergia 14:176–178

Matteri CM (1989) Additions to the Moss Flora of Tierra del Fuego (Argentina). Bryologist 92(2):174–177

Matteri CM (1993) Distribution maps of mosses from Tierra del Fuego, Argentina 5. Lindbergia 18:138–140

Matteri CM (2003) Los musgos (Bryophyta) de Argentina. Bryoph Divers Evol 24(1):33–100

Matteri CM, Hyvönen J (1992) Distribution maps of mosses from Tierra del Fuego, Argentina 2. Lindbergia 17:17–19

Matteri CM, Schiavone MM (1991) La Vegetación Muscinal de Bahía Buen Suceso, Tierra del Fuego, Argentina. Bryologist 94(4):368–376

Matteri CM, Schiavone MM (2002) Catálogo de los musgos (Bryophyta) de la región Fueguina en Argentina y Chile. Rev Mus Argent Cienc Nat 4(2):111–138

Ochyra R, Matteri CM (2001) Amblystegiaceae. In: Guarrera SA, Gamundi de Amos IJ, Matteri CM Flora Criptogámica de Tierra del Fuego. Fundación para la educación la ciencia y la cultura, Buenos Aires, vol 14 (10)

Schiavone MM (1993) Orden polytrichinales. In: Guarrera SA, Gamundi de Amos IJ, Matteri CM Flora Criptogámica de Tierra del Fuego. Fundación para la educación la ciencia y la cultura, Buenos Aires, vol 14 (12)

Schofield WB (1985) Introduction to bryology, vol 192. Macmillan Publ. Co., New York

Shaw AJ, Szövényi P, Shaw B (2011) Bryophyte diversity and evolution: window into the early evolution of land plants. Am J Bot 98(3):352–369

Thiers B (regularly updated, accessed 2025) Index Herbariorum: a global directory of public herbaria and associated staff. New York Botanical Garden's Virtual Herbarium. http://sweetgum.nybg.org/ih

Viegas Barros JP, Malvestitti M (2019) Un manuscrito de Carlos Spegazzini con datos inéditos sobre la lengua haush. Indiana 36(2):101–128

Chapter 7
Marine Isopods and Amphipods (Crustacea: Peracarida) Recorded from Isla de los Estados and Surrounding Waters

Ignacio Luis Chiesa, Emanuel Pereira, Sofía Calderón López, Julián Santiago, and Brenda Lía Doti

Abstract Isopods and amphipods are ubiquitous crustaceans in benthic marine environments. Both orders exhibit high biodiversity in the Southern Ocean and the Magellan Region; however, no previous study has synthesized the fauna recorded from Isla de los Estados (Argentina). To address this gap, we present a comprehensive list of marine isopods and amphipods reported from this island and its surrounding waters. Data were obtained from the literature and the Global Biodiversity Information Facility (GBIF) database. In total, 29 species of isopods and 22 species of amphipods were recorded across 18 sampling stations carried out during 10 different surveys. Most of the reported species exhibit a Magellanic distribution pattern. The biodiversity reported here is relatively low compared to adjacent areas of the Magellan Region, suggesting that Isla de los Estados and its marine surroundings remain undersampled. Further surveys, particularly in the deeper waters south of the island, are likely to reveal additional species from both crustacean orders.

I. L. Chiesa (✉)
Centro Austral de Investigaciones Científicas (CADIC-CONICET), Ushuaia, Tierra del Fuego, Argentina
e-mail: chiesa.ignacio@conicet.gov.ar; ignacio.nacho.chiesa@gmail.com

E. Pereira · S. C. López · B. L. Doti (✉)
Universidad de Buenos Aires, Facultad de Ciencias Exactas y Naturales, Departamento de Biodiversidad y Biología Experimental (DBBE), Buenos Aires, Argentina

CONICET – Universidad de Buenos Aires, Instituto de Biodiversidad y Biología Experimental y Aplicada (IBBEA), Buenos Aires, Argentina
e-mail: bdoti@bg.fcen.uba.ar

J. Santiago
Universidad de Buenos Aires, Facultad de Ciencias Exactas y Naturales, Departamento de Biodiversidad y Biología Experimental (DBBE), Buenos Aires, Argentina

A. Raya Rey, U. Balza (eds.), *Biodiversity and Conservation of Isla de los Estados (Chuani-sin)*, Natural and Social Sciences of Patagonia,
https://doi.org/10.1007/978-3-032-13630-5_7

Keywords Peracarida diversity · Marine bentos · Magellan region · Amphipod and isopod species richness

7.1 Introduction

Isopods and amphipods are among the most important groups of marine benthic invertebrates in terms of both diversity and abundance (Hessler et al. 1979; Brandt et al. 2007; Lörz et al. 2021). Species of these taxa inhabit coastal, shelf, and deep-sea communities across the world's oceans (Poore and Bruce 2012; Momtazi and Saeedi 2024). They occupy a wide range of habitats, including hard or soft substrates, algae, and even other invertebrates, such as corals, sponges, and decapods, which provide shelter, food, and mating opportunities (Barnard and Karaman 1991; Schejter et al. 2012; Rumbold et al. 2021). Isopods and amphipods exhibit diverse feeding strategies, including detritivory, suspension feeding, predation, and scavenging (Guerra-García et al. 2014; Poore and Bruce 2012). These groups also play key roles in benthic and demersal food webs, not only serving as prey for higher trophic levels (other crustaceans, fish, marine mammals and seabirds) but also contributing to nutrient cycling (Robertson and Mann 1980; Keable 1995; Dauby et al. 2003; Ritter and Bourne 2024). Like all peracarid crustaceans, isopods and amphipods lack a free-living larval stage, making them excellent models for biogeographical studies (Arfianti and Costello 2020).

Despite their ubiquity and ecological importance, knowledge of isopods and amphipods fauna from the Argentine continental shelf remains limited and fragmented. For amphipods, two key catalogs compile the known records from the Argentine continental shelf and the Antarctic and Subantarctic waters (see López Gappa et al. 2006; De Broyer et al. 2007). Whereas, studies of isopods have primarily focused on specific families or suborders, for example Bastida and Torti (1973) reviewed the Serolidae from the Argentine Sea, Doti et al. (2014) addressed the shallow-water Asellota, and Pereira (2022) examined the Valvifera from both shallow and deep waters. Additionally, Doti et al. (2020a) provided a comprehensive review of deep-water amphipods, isopods, and cumaceans off the Argentine coast. However, no previous study has specifically synthesized the isopod and amphipod species recorded from Isla de los Estados (Argentina) and its surrounding waters, although some records have been included in broader regional catalogs.

This chapter aims to fill the gap providing a synthesis of current knowledge on marine benthic isopod and amphipod species from Isla de los Estados. It compiles all known records from the island and adjacent waters, including information on surveys, sampling stations, bathymetric ranges, and brief notes on species distributions.

7.2 Material and Methods

Isla de los Estados lies approximately 35 km east of Isla Grande de Tierra del Fuego. The island is characterized by steep relief, rocky coastlines, and numerous deep fjords and coves. Its surrounding waters include several smaller islands and islets (Ponce and Fernández 2026, Chap. 2, this book). Biogeographically, Isla de los Estados is part of the Magellan Region (López Gappa et al. 2006; De Broyer et al. 2007; Doti et al. 2014).

The study area encompasses the intertidal zone and marine waters surrounding Isla de los Estados. Its boundaries were defined by extending 12 km outward from the island's extreme cardinal points, covering the area between 54°32′10″S to 55°01′45″S latitude and 64°56′40″W to 63°36′35″W longitude (Fig. 7.1).

Previous records of benthic marine isopods and amphipods from this area were compiled from the literature. Additionally, the Global Biodiversity Information Facility (GBIF) database was consulted, and only records verified by trained taxonomists were included. A few isopod species are newly reported here for Isla de los Estados. All species are listed according to the current taxonomic classification (WoRMS, accessed May 2025), or their most recent taxonomic revision.

The localities/stations considered within the study area are listed in Table 7.1, and plotted in Figs. 7.1 and 7.2, which show the distribution of Isopoda and Amphipoda records, respectively. For those localities/stations lacking precise

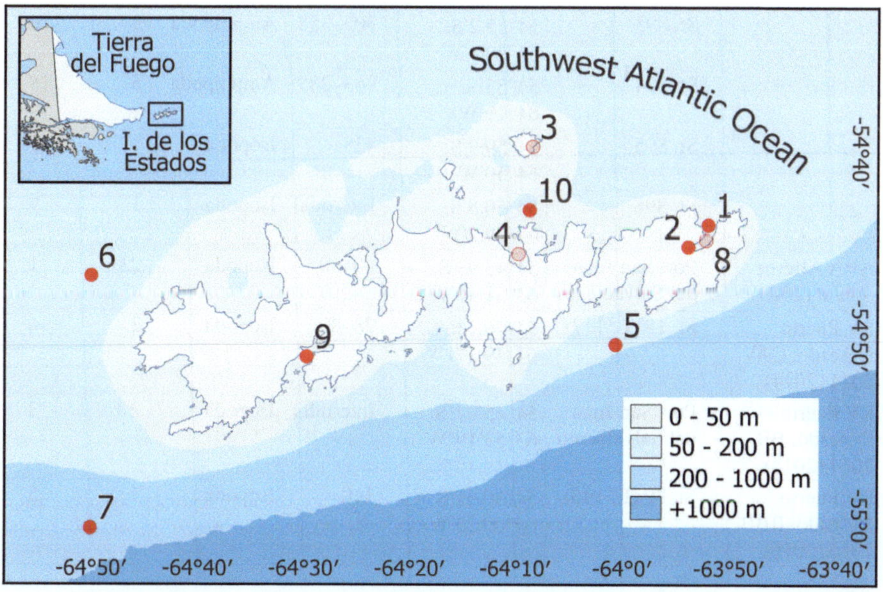

Fig. 7.1 Localities/stations with isopod records from the study area. Filled circles correspond to locations/stations with coordinates provided in the original publication, empty circles indicate localities/stations without coordinates in the original publication

Table 7.1 Surveys and locations/stations data of isopods and amphipods recorded from the study area

Survey and/or ships	Stations/ locations	Geographic coordinates	Depth (m)	Order/s	No. of species	Sta. code
Swedish "Eugenie" Expedition (1852)	York Bay	Not given	7–11	Amphipoda	1	14
Italian Antarctic Expedition (1881–1882)	Port Cook	Not given	Intertidal	Amphipoda	2	12
	Port Roca	Not given	Intertidal	Amphipoda	3	11
Belgian Antarctic Expedition (1897–1899)	Port San Juan del Salvamento	Not given	Intertidal	Amphipoda	1	13
Swedish Antarctic Expedition (1901–1903)	St. 3	54°43′S, 64°08′W	36	Isopoda/ Amphipoda	1 (I)/3 (A)	10
	York Bay	Not given	7–12	Amphipoda	5	15
	Observatorio Island	Not given	Intertidal	Isopoda	2	3
Not given (see Bastida and Torti 1973)	Port Basil Hall	Not given	Not given	Isopoda	1	4
	San Juan del Salvamento Bay	Not given	Not given	Isopoda	1	8
RV Hero, Cruise 715 (1971)	St. 882	54°50′S, 64°10′W	76	Amphipoda	1	17
	St. 893	54°55.2′S, 64°20.4′W	303–358	Amphipoda	3	16
	St. 894	54°55′S, 64°19′W	263–285	Amphipoda	5	18
	St. 895	54°59.9′S, 64°50′W	438	Isopoda	1	7
	St. 898	54°50.8′S, 64°29.2′W	Intertidal	Isopoda	1	9
ARA Alférez Sobral (2003)	Port San Juan del Salvamento	54°43.93′S, 63°51.42′W	40–60	Isopoda	23	1
RV Puerto Deseado, CAV 2014 (2014)	St. 18 (L. 117)	54°46.136′S, 64°49.171′W	78–80	Isopoda	2	6
RV Puerto Deseado, BB 2014 (2014)	Port San Juan del Salvamento	54°45′07″S, 63°53′19″W	Intertidal	Isopoda	2	2
RV Puerto Deseado, BBB 2016 (2016)	St. 39 (L. 140)	54°50.41′S, 64°00.25′W	160	Isopoda	1	5

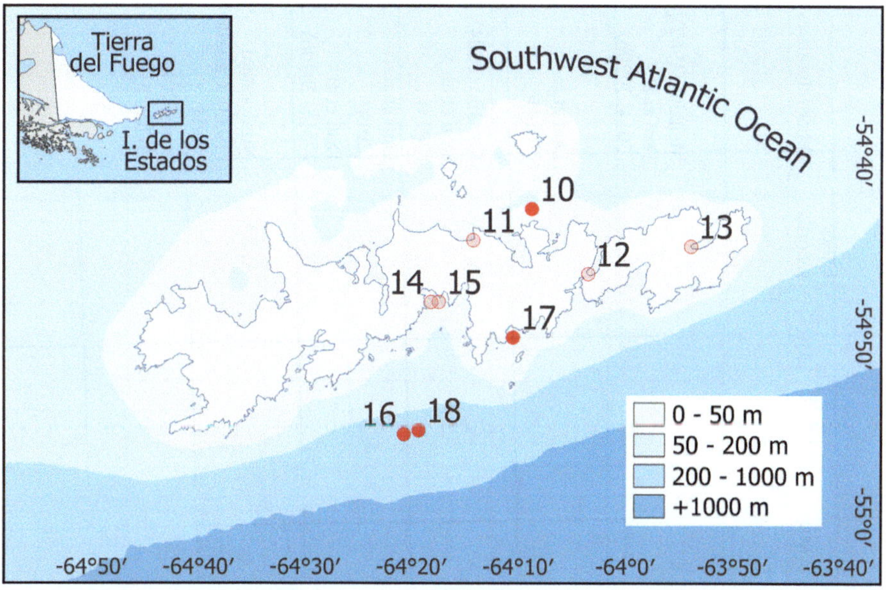

Fig. 7.2 Localities/stations with amphipod records from the study area. Filled circles correspond to locations/stations with coordinates provided in the original publication, empty circles indicate localities/stations without coordinates in the original publication

coordinates in the original publication, approximate latitude and longitude were estimated using Google Earth for mapping purposes. All localities/stations were mapped using QGIS 3.34 software (QGIS, 2024), with geographic base layers obtained from the Instituto Geográfico Nacional de la República Argentina (IGN, accessed May 2025). Table 7.2 presents the isopod and amphipod species recorded in the study area, along with their geographic distribution within the Magellan Region, and adjacent waters, as follows: A, Antarctic; SA, Subantarctic; O, other localities.

Specimens collected during surveys by the ARA *Alférez Sobral* and the RV *Puerto Deseado* and reported herein have been deposited in the Carcinological Collection of the Departamento de Biodiversidad y Biología Experimental, Facultad de Ciencias Exactas y Naturales, Universidad de Buenos Aires (DBBE, FCEN, UBA).

7.3 Results

A total of 18 localities/stations with records of isopods and amphipods were reported within the study area. Amphipods were recorded at nine stations and isopods at 10 (Fig. 7.3). Only one station, Station 3 from the Swedish Antarctic Expedition, yielded records of both taxa (see Table 7.1).

Table 7.2 Isopod and amphipod species recorded from the study area

	Sta. code	References for IDLE records	Other Magellan records							Other records	Other references
			MI	BC	SM	BB	AMR	PMR	SMR		
Isopoda											
Asellota											
Acanthaspidiidae											
Ianthopsis laevis Menzies, 1962	1	Doti et al. (2014)		x	x	x	x	x			Menzies (1962), Doti et al. (2014, 2020b)
Janiridae											
Iais pubescens (Dana, 1852)	1, 2, 3	Nordenstam (1933), Doti et al. (2014)	x	x	x		x				Doti et al. (2014)
Austrofilius furcatus Hodgson, 1910	1	Doti et al. (2014)	x	x			x		x	A, SA	Castelló (2004), Doti et al. (2014)
Ianiropsis cf. *chilensis* Menzies, 1963	1	Doti et al. (2014)					x				Doti et al. (2014)
Iathrippa menziesi Sivertsen & Holthuis, 1980	1	Doti et al. (2014)		x	x	x	x				Doti et al. (2014, 2020b)
Neojaera antarctica (Pfeffer, 1887)	1	Doti et al. (2014)	x	x	x	x	x			A, SA	Nordenstam (1933), Castelló (2004), Doti et al. (2014, 2020a, b)

Joeropsidiidae											
Joeropsis curvicornis (Nicolet, 1849)	1	Doti et al. (2014)	x	x	x	x	x			O	Doti et al. (2014, 2020a, b)
Joeropsis intermedius Nordenstam, 1933	1	Doti et al. (2014)	x	x	x	x	x	x		A	Castelló (2004), Doti et al. (2014, 2020a, b)
Munnidae											
Munna sp.	1	Doti et al. (2014)		x	x		x				Doti et al. (2014)
Paramunnidae											
Abyssianira argentinensis Menzies, 1962	1	Doti et al. (2014)		x		x	x				Doti et al. (2014, 2020a, b)
Abyssianira sp. A	1	Doti et al. (2014)		x		x	x				Doti et al. (2014, 2020b)
Allorostrata ovalis Winkler, 1994	1	Doti et al. (2014)		x	x	x	x				Doti et al. (2014, 2020b)
Antennulosignum elegans Nordenstam, 1933	1	Doti et al. (2014)	x	x			x				Doti et al. (2014)
Austronanus dentatus (Nordenstam, 1933)	1	Doti et al. (2014)	x	x		x	x				Doti et al. (2014, 2020b)
Meridiosignum menziesi (Winkler, 1994)	1	Doti et al. (2014)		x	x	x	x				Doti et al. (2014, 2020b)
Paramunna integra Nordenstam, 1933	1	Doti et al. (2014)	x	x		x	x				Doti et al. (2014, 2020b)

(continued)

Table 7.2 (continued)

	Sta. code	References for IDLE records	Other Magellan records							Other records	Other references
			MI	BC	SM	BB	AMR	PMR	SMR		
Pleurosignum magnum Vanhöffen, 1914	1	Doti et al. (2014)	x	x		x	x			A	Vanhöffen (1914), Doti et al. (2014, 2020b)
Sporonana sp.	1	Doti et al. (2014)		x	x	x	x				Doti et al. (2014, 2020b)
Santiidae											
Santia hispida (Vanhöffen, 1914)	1	Doti et al. (2014)	x	x	x		x		x	A	Vanhöffen (1914), Doti et al. (2014)
Stenetriidae											
incertae sedis drakensis (Schultz, 1982)	7	Schultz (1982)									
Sphaeromatidea											
Serolidae											
Acanthoserolis schythei (Lütken, 1859)	4, 8	Bastida and Torti (1973)	x	x	x			x	x	A, SA	Sheppard (1933), Doti et al. (2020a, b)
Thysanoserolis elliptica (Sheppard, 1933)	1	Present study	x		x		x				Doti (2009)
Sphaeromatidae											
Exosphaeroma gigas (Leach, 1818)	1, 2, 3	Nordenstam (1933)	x	x	x			x		A, SA, O	Brandt (1998)
Moruloidea darwinii (Cunningham, 1871)	5, 6	Calderón López and Doti (2025)	x	x	x	x	x		x	SA	Calderón López and Doti (2025)

Valvifera											
Antarcturidae											
Litarcturus americanus (Beddard, 1886)	6	Present study	x	x		x	x				Pereira (2022)
Arcturidae											
Neastacilla estadoensis (Schultz, 1981)	9	Schultz (1981)					x				Pereira (2022)
Idoteidae											
Edotia bilobata Nordenstam, 1933	10	Pereira and Doti (2017)	x	x			x				Pereira (2022)
Rectarcturidae											
Rectarcturus kophameli (Ohlin, 1901)	1	Pereira (2022)		x	x	x	x			O	Pereira (2022)
Xenarcturidae											
Xenarcturus spinulosus Sheppard, 1957	1	Pereira (2022)	x		x	x	x				Pereira (2022)
Amphipoda											
Amaryllididae											
Erikus sp.	15	Schellenberg (1931)	x	x	x		x	x	x		López Gappa et al. (2006), De Broyer et al. (2007)

(continued)

Table 7.2 (continued)

	Sta. code	References for IDLE records	Other Magellan records							Other records	Other references
			MI	BC	SM	BB	AMR	FMR	SMR		
Caprellidae											
Caprella ungulina Mayer, 1903	14	Mayer (1903)		x	x		x		x	C	De Broyer et al. (2006), Rumbold et al. (2021)
Paracaprella tenuis Mayer, 1903	17	GBIF (Guerra García)								O	Guerra García (2002)
Corophiidae											
Haplocheira barbimana robusta Barnard, 1932	10	Schellenberg (1931)	x	x	x	x	x	x	x		López Gappa et al. (2006), De Broyer et al. (2007)
Epimeriidae											
Epimeria (*Metepimeria*) *acanthurus* (Schellenberg, 1931)	18	Watling and Holman (1981)	x			x		x	x		López Gappa et al. (2006), De Broyer et al. (2007), d'Udekem and Verheye (2017)
Eusiridae											
Eusirus antarcticus Thomson, 1880	15	Schellenberg (1931)	x	x	x		x	x		A, SA	López Gappa et al. (2006), De Broyer et al. (2007)

Gammarellidae											
Austroregia regis (Stebbing, 1914)	15	Schellenberg (1931)	x	x	x			x			López Gappa et al. (2006), De Broyer et al. (2007)
Iphimediidae											
Iphimedia multidentata (Schellenberg, 1931)	18	Watling and Holman (1981)	x	x	x	x		x			López Gappa et al. (2006), De Broyer et al. (2007)
Ischyroceridae											
Jassa alonsoae Conlan, 1990	15	Schellenberg (1931)	x	x	x	x	x			SA, O	López Gappa et al. (2006), De Broyer et al. (2007), Conlan et al. (2021)
Liljeborgiidae											
Liljeborgia macrodon Schellenberg, 1931	18	Holman and Watling (1983)		x	x			x	x		López Gappa et al. (2006), d'Udekem (2008)
Liljeborgia octodentata Schellenberg, 1931	18	Holman and Watling (1983)	x	x	x		x	x	x		López Gappa et al. (2006), d'Udekem (2008)

(continued)

Table 7.2 (continued)

	Sta. code	References for IDLE records	Other Magellan records							Other records	Other references
			MI	BC	SM	BB	AMR	PMR	SMR		
Photidae											
Gammaropsis (*Paranaenia*) *dentifera* (Haswell, 1879)	10	Schellenberg (1931)	x	x	x					SA, O	López Gappa et al. (2006), De Broyer et al. (2007)
Phoxocephalidae											
Phoxorgia sinuata (Barnard, 1932)	10	Schellenberg (1931)		x	x		x	x	x	SA, O	López Gappa et al. (2006), Alonso et al. (2008)
Pontogeneiidae											
Gondogeneia gracilicauda (Schellenberg, 1931)	15	Schellenberg (1931)	x	x	x						López Gappa et al. (2006), De Broyer et al. (2007)
Paramoera fissicauda (Dana, 1852)	11, 12	Ruffo (1947)	x	x	x		x	x		A, SA, O	López Gappa et al. (2006), De Broyer et al. (2007)
Sebidae											
Seba saundersii Stebbing, 1875	16	Holman and Watling (1983)	x	x	x		x			A, SA	López Gappa et al. (2006), De Broyer et al. (2007)
Seba subantarctica Schellenberg, 1931	16	Holman and Watling (1983)	x	x	x	x			x	SA	López Gappa et al. (2006), De Broyer et al. (2007)

Seba typica (Chilton, 1884)	16	Holman and Watling (1983)		x							López Gappa et al. (2006), De Broyer et al. (2007)
Stegocephalidae											
Andaniotes linearis Barnard, 1932	18	Watling and Holman (1981)	x	x		x	x	x	x	A, SA	López Gappa et al. (2006), De Broyer et al. (2007)
Talitridae											
Gondwanorchestia scutigerula (Dana, 1852)	11, 12	Ruffo (1947)	x	x	x		x	x	x	SA	López Gappa et al. (2006), De Broyer et al. (2007), Lowry et al. (2020)
Protorchestia nitida (Dana, 1852)	13	Ruffo (1949)		x	x			x	x		López Gappa et al. (2006), De Broyer et al. (2007)
Uristidae											
Uristes schellenbergi (Lowry & Bullock, 1976)	11	Ruffo (1947)	x	x	x		x	x		O	López Gappa et al. (2006), De Broyer et al. (2007), Pérez-Schultheiss (2016)

Abbreviations: *MI* Malvinas Islands, *BC* Beagle Channel, *SM* Strait of Magellan, *BB* Burdwood Bank, *AMR* Atlantic sector of Magellan Region, *PMR* Pacific sector of Magellan Region, *SMR* southern sector of Magellan Region, *A* Antarctic, *SA* Subantarctic, *C* cosmopolitan, *O* any region other than "A", "SA" or Magellan Region

Fig. 7.3 Some marine isopods and amphipods from Isla de los Estados and surroundings. Isopods, (**a**) *Exosphaeroma gigas*; (**b**) *Acanthoserolis schythei*; (**c**) *Rectarcturus kophameli*; (**d**) *Xenarcturus spinulosus*. Amphipods; (**e**) *Caprella ungulina*; (**f**) *Austroregia* sp.; (**g**) *Uristes schellenbergi*

Species of these two peracarid orders were collected during 10 expeditions that explored Isla de los Estados and its surrounding waters since the XIX century. Most of these campaigns included the island as an opportunistic sampling site in the context of broader Antarctic or Subantarctic expeditions. The earliest of these was the Swedish "Eugenie" Expedition in 1852 (Table 7.1).

Most localities/stations (13) were located near the coast or within sheltered bays, and their depth data therefore correspond to shallow waters, ranging from the intertidal zone to 76 m depth. The remaining five stations, three of which were conducted by the RV *Hero*, were located in deeper waters, between 78 and 358 m depths (see Figs. 7.1 and 7.2; Table 7.1). Considering both taxonomic groups, the highest number of stations (5) was sampled during Cruise 715 of the RV *Hero* in 1971. For amphipods, the highest species richness (9 spp.) was also collected by the RV *Hero*, whereas the ARA *Alférez Sobral* yielded the highest richness for isopods (23 spp.) (Table 7.1).

A total of 22 amphipods species belonging to 16 families and 19 genera were recorded in the study area (Table 7.2). The family Sebidae has the highest species richness (3 spp.), while Caprellidae, Liljeborgiidae, Pontogeneiidae and Talitridae were represented by two species each (Table 7.2). Most amphipod species were recorded at a single locality; only two species, *Gondwanorchestia scutigerula* and *Paramoera fissicauda*, were found at two stations (Ports Roca and Cook). Most amphipods recorded around Isla de los Estados are also distributed throughout other areas of the Magellan Region: 16 species have been cited from the Malvinas Islands, 18 from the Strait of Magellan, and 20 species from the Beagle Channel (Table 7.2). Nearly half (10) of the amphipod species cited from Isla de los Estados are known only from the Magellan region, while nine species also occur in Antarctic and/or Subantarctic waters (Table 7.2). Most amphipods (14 spp.) were collected from shallow waters stations/localities, whereas eight species were sampled by the RV *Hero* at two deeper sites (263–285 and 303–358 m depth) (Table 7.2).

Regarding isopods, a total of 29 species were recorded in the study area. The suborder Asellota was the most diverse with 20 species, followed by Valvifera (5 spp.), and Sphaeromatidea (4 spp.) (Table 7.2). Two species, *Thysanoserolis elliptica* (Sheppard, 1933) and *Litarcturus americanus* (Beddard 1886), are newly recorded here from Isla de los Estados. Only two species, *Exosphaeroma gigas* and its symbiont *Iais pubescens,* were recorded at three localities. Two additional species, *Acanthoserolis schythei* and *Moruloidea darwinii*, were found at two sites, while all remaining species were recorded at a single locality (Table 7.2). With the exception of Stenetriidae *incertis sedis drakensis,* which was found only once at St. 895 (RV *Hero*), all isopod species recorded from the study area have also been reported for other localities within the Magellan Region (Table 7.2). Eight species also occur in Antarctic and/or Subantarctic waters, and three species have been recorded in regions outside the ones mentioned above (Table 7.2).

7.4 Discussion

In the Southwest Atlantic, López Gappa et al. (2006) and Doti et al. (2014) found, respectively, that amphipods and shallow-water Asellota isopod diversity increases towards higher latitudes. Notably, several localities in the southern Magellan Region, such as Puerto Deseado, Malvinas Islands, the Strait of Magellan, and the Beagle Channel have been identified as areas of high species richness (see Chiesa et al. 2005; Doti et al. 2005, 2014; López Gappa et al. 2006; Alonso 2012). Even in the Marine Protected Area Namuncurá-Burdwood Bank (MPANBB) the isopod species richness (43 spp.) is higher than that recorded from Isla de los Estados and surrounding waters (29 spp., see Doti et al. 2020b). This relatively low number of species appears to be the result of limited sampling efforts in the area, rather than reflecting actual low diversity. Although Isla de los Estados has been visited by several foreign expeditions since the XIX century, these surveys often had broader geographical objectives. Indeed, half of the expeditions listed in Table 7.1 had Antarctic waters as primarily target. Consequently, many stations near Isla de los Estados were likely sampled opportunistically during transits to other destinations or when vessels sought shelter in bays to avoid the strong winds that prevail in the southern Southwest Atlantic. We are aware that three incidental samples from Isla de los Estados were collected under such circumstances, during expeditions originally focused on the MPANBB (CAV 2014, BB 2014, and BBB2016, see Table 7.1).

In addition to limited sampling, most of the existing records originate from taxonomic studies, which often emphasize descriptions of new species and may not include full species lists for each sample. As a result, most stations/locations in the study area reported only a few species or even just one. For isopods, only one station (ARA *Alférez Sobral*) was analyzed in the context of a comprehensive biodiversity study, which led to the identification of 23 species (see Doti et al. 2014). This taxonomic bias is also evident in the low number of stations that recorded both isopods and amphipods: only one out of 18 stations recorded both taxa. Regarding amphipods, few stations received more thorough taxonomic attention and thus yielded greater species richness. For example, the York Bay station (sampled during the Swedish Antarctic Expedition) was studied by Alfred Schellenberg, while Sta. 894 (sampled by the RV *Hero*) was examined by Heather Holman and Les Watling (see Schellenberg 1931; Watling and Holman 1981; Holman and Watling 1983, respectively). These two stations showed the highest amphipod richness in the study area, with five species each.

The waters surrounding Isla de los Estados have been identified as a key habitat for numerous top predators, with frequent sightings of cetaceans (Dellabianca et al. 2023). Several seabird species, including penguins, cormorants, and petrels, establish breeding colonies along the island's rugged coastline (Schiavini et al. 2005; Balza 2021; Raya Rey et al. 2022; Dodino et al. 2024). The island also hosts three breeding colonies of pinnipeds (Crespo et al. 2015; Milano et al. 2020). In addition, a total of 33 decapod and 46 fish species have been recorded in its surrounding waters (Diez et al. 2016; Di Salvatore et al. 2023; Rojo et al. 2026, Chap. 11, this book).

Isopods and amphipods play a crucial role as trophic intermediaries, linking lower and higher levels of the food web (Dauvin 2024). Expanded and targeted benthic sampling, combined with comprehensive taxonomic studies, is expected to significantly increase species records of these peracarid groups around Isla de los Estados. Improved knowledge of the region's biodiversity will enhance our understanding of local trophic interactions. Moreover, the island's geographical position provides valuable opportunities for phylogeographic and dispersal studies, particularly through rafting mechanisms. Our results also highlight a notably low sampling effort along the southern slope of the island. Given the importance of deep-water benthic communities as a biodiversity hotspot, this underexplored area emerges as a strategic priority for future research.

Competing Interests We are grateful to Andrea Raya Rey and Ulises Balza (CADIC) for the invitation to participate in the present chapter of this book.

This work was partially funded by the Universidad de Buenos Aires (UBACyT 20020220400064BA) and the Consejo Nacional de Investigaciones Científicas y Técnicas (PIP 11220200102070CO). The authors have no conflicts of interest to declare that are relevant to the content of this chapter.

References

Alonso GM (2012) Amphipod crustaceans (Corophiidea and Gammaridea) associated with holdfasts of *Macrocystis pyrifera* from the Beagle Channel (Argentina) and additional records from the Southwestern Atlantic. J Nat Hist 46(29–30):1799–1894

Alonso GM, Rauschert M, de Broyer C (2008) A catalogue of the Antarctic and sub-Antarctic Phoxocephalidae (Crustacea: Amphipoda: Gammaridea) with taxonomic, distribution and ecological data. Zootaxa 1752(1):1–40

Arfianti T, Costello MJ (2020) Global biogeography of marine amphipod crustaceans: latitude, regionalization, and beta diversity. Mar Ecol Prog Ser 638:83–94

Balza U (2021) Ecología y conservación del carancho austral (*Phalcoboenus australis*) en Isla de los Estados. Dissertation, Universidad de Buenos Aires

Barnard JL, Karaman GS (1991) The families and genera of marine gammaridean Amphipoda (except marine gammaroids). Rec Aust Mus 13(1, 2):1–866

Bastida R, Torti MR (1973) Los isópodos Serolidae de la Argentina. Clave para su reconocimiento. Physis Sec A 32:19–46

Brandt A (1998) Sphaeromatidae (Crustacea, Isopoda) of the Beagle Channel and description of *Cymodopsis beageli* n. sp. Beaufortia 48:137–162

Brandt A, Brix S, Brökeland W et al (2007) First insights into the biodiversity and biogeography of the Southern Ocean deep sea. Nature 447:307–310

Calderón López S, Doti B (2025) Review of the genus *Moruloidea* (Crustacea, Isopoda, Sphaeromatidae) from the Southwest Atlantic, with description of two new species from the marine-protected area Namuncurá-Burdwood Bank. Polar Biol 48(1):24

Castelló J (2004) Isopods (Crustacea, Isopoda) from the Spanish "Bentart-94/95" expeditions to the South Shetland Islands (subAntarctic). Polar Biol 28:1–14

Chiesa IL, Alonso GM, Zelaya DG (2005) Species richness and faunistic affinities of the Gammaridea and Corophiidea (Amphipoda) from shallow waters of southern Tierra del Fuego, Argentina: preliminary results. Sci Mar 69(2):167–174

Conlan KE, Desiderato A, Beermann J (2021) *Jassa* (Crustacea: Amphipoda): a new morphological and molecular assessment of the genus. Zootaxa 4939(1):1–191

Crespo EA, Schiavini AC, García NA et al (2015) Status, population trend and genetic structure of South American fur seals, *Arctocephalus australis*, in southwestern Atlantic waters. Mar Mamm Sci 31(3):866–890

d'Udekem d'Acoz C (2008) Shelf and abyssal *Liljeborgia* Bate, 1861 of the Southern Ocean (Crustacea, Amphipoda, Liljeborgiidae). Bull Inst R Sci Nat Bel Biol 77:45–286

d'Udekem d'Acoz C, Verheye ML (2017) *Epimeria* of the Southern Ocean with notes on their relatives (crustacea, Amphipoda, Eusiroidea). Eur J Taxon 359:1–553

Dauby P, Nyssen F, Broyer CD (2003) Amphipods as food sources for higher trophic levels in the Southern Ocean: a synthesis. In: Huiskes AHL, Gieskes WWC, Rozema J et al (eds) Antarctic biology in a global context. Backhuys Publishers, Leiden, pp 129–134

Dauvin JC (2024) Overview of predation by birds, cephalopods, fish and marine mammals on marine benthic amphipods. J Mar Sci Eng 12(3):403

De Broyer C, Lowry JK, Jażdżewski K, Robert H (2007) Catalogue of the Gammaridean and Corophiidean Amphipoda (Crustacea) of the Southern Ocean with distribution and ecological data. In: De Broyer C (ed) Census of Antarctic marine life: synopsis of the Amphipoda of the Southern Ocean. Bull Inst R Sci Nat Belg Biol 77(1):1–35

Dellabianca NA, Torres MA, Ordoñez C et al (2023) Marine protected areas in the southern southwest Atlantic: insights from marine top predator communities. Aquatic Conserv Mar Freshw Ecosyst 33(5):472–487

Di Salvatore P, Albano MJ, Diez MJ et al (2023) Decapods of the southern tip of South America and the Marine Protected Area Namuncurá–Burdwood Bank: a nearshore–offshore comparison. Diversity 15(11):1143

Diez MJ, Cabreira AG, Madirolas A, Lovrich GA (2016) Hydroacoustical evidence of the expansion of pelagic swarms of *Munida gregaria* (Decapoda, Munididae) in the Beagle Channel and the Argentine Patagonian shelf, and its relationship with habitat features. J Sea Res 114:1–12

Dodino S, Balza U, Riccialdelli L et al (2024) Pre-molt dispersal and use of marine protected areas by Southern Rockhopper Penguins (*Eudyptes chrysocome*) at the southernmost oceanic regions of South America. Prog Oceanogr 229:103369

Doti BL (2009) Sistemática y biogeografía de los isópodos Asellota (Crustacea) del Mar Argentino. Dissertation, Universidad de Buenos Aires

Doti BL, Roccatagliata D, Zelaya DG (2005) The shallow-water Asellota (Crustacea: Isopoda) from the Beagle Channel: preliminary taxonomic and zoogeographical results. Sci Mar 69(2):159–166

Doti BL, Roccatagliata D, López Gappa J (2014) An inverse latitudinal biodiversity pattern in asellote isopods (Crustacea, Peracarida) from the Southwest Atlantic between 35° and 56°S. Mar Biodivers 44(1):115–125

Doti BL, Chiesa IL, Roccatagliata D (2020a) Biodiversity of the deep-sea isopods, Cumaceans, and amphipods (Crustacea: Peracarida) recorded off the Argentine coast. In: Hendrickx ME (ed) Deep-Sea pycnogonids and crustaceans of the Americas. Springer Nature, Cham, pp 157–191

Doti BL, Chiesa IL, Roccatagliata D (2020b) Biodiversity of isopoda and Cumacea (Peracarida, crustacea) from the marine protected area Namuncurá-Burdwood Bank, South-West Atlantic. Polar Biol 43(10):1519–1534

Guerra-García J (2002) Redescription of five rare caprellids (Crustacea: Amphipoda: Caprellidea) collected from Tanzanian coasts. Helgol Mar Res 55:221–231

Guerra-García J, De Figueroa JT, Navarro-Barranco C et al (2014) Dietary analysis of the marine Amphipoda (Crustacea: Peracarida) from the Iberian Peninsula. J Sea Res 85:508–517

Hessler RR, Wilson GD, Thistle D (1979) The deep-sea isopods: a biogeographic and phylogenetic overview. Sarsia 64(1–2):67–75

Holman H, Watling L (1983) Amphipoda from the Southern Ocean: families Colomastigidae, Dexaminidae, Leucothoidae, Liljeborgiidae, and Sebidae. Antarct Res Ser 38:215–262

Keable SJ (1995) Structure of the marine invertebrate scavenging guild of a tropical reef ecosystem: field studies at Lizard Island, Queensland, Australia. J Nat Hist 29:27–45

López Gappa J, Alonso GM, Landoni NA (2006) Biodiversity of benthic Amphipoda (crustacea: Peracarida) in the Southwest Atlantic between 35°S and 56°S. Zootaxa 1342:1–66

Lörz AN, Kaiser S, Oldeland J et al (2021) Biogeography, diversity and environmental relationships of shelf and deep-sea benthic Amphipoda around Iceland. PeerJ 9:e11898

Lowry JK, Myers AA, Perez-Schultheiss J (2020) *Gondwanorchestia tristanensis* gen. nov. sp. nov., a new southern hemisphere genus and species of talitrid amphipod (Amphipoda, Senticaudata, Talitridae). Zootaxa 4748(2):375–381

Mayer P (1903) Die Caprellidae der "Siboga" Expedition. Siboga Exped 34:1–160

Menzies RJ (1962) The zoogeography, ecology and systematics of the Chilean marine isopods. Lunds Univ Årsskr NF Avd 2 57(11):1–162

Milano VN, Grandi MF, Schiavini AC et al (2020) Sea lions (*Otaria flavescens*) from the end of the world: insights of a recovery. Polar Biol 43:695–706

Momtazi F, Saeedi H (2024) Exploring latitudinal gradients and environmental drivers of amphipod biodiversity patterns regarding depth and habitat variations. Sci Rep 14(1):30547

Nordenstam Å (1933) Marine isopoda of the families Serolidae, Idotheidae, Pseudidotheidae, Arcturidae, Parasellidae and Stenetriidae mainly from the South Atlantic. Further Zool Res Swed Antarct Exped 1901–1903(3):1–284

Pereira E (2022) Sistemática filogenética y biogeografía de isópodos Valvifera (Crustacea: Peracarida) de la Plataforma Continental y el Talud de Argentina. Dissertation,. Universidad de Buenos Aires

Pereira E, Doti B (2017) *Edotia abyssalis* n. sp. from the Southwest Atlantic Ocean, first record of the genus (isopoda, Valvifera, Idoteidae) in the deep sea. Zool Anz 268:19–31

Pérez-Schultheiss J (2016) Sinopsis de la superfamilia Lysianassoidea (Amphipoda: Gammaridea) en Chile. Bol M Nac Hist Nat Chile 65:193–246

Ponce JF, Fernández M (2026) Chapter 2: Geological and paleoenvironmetal evolution of Isla de los Estados. In: Raya Rey A, Balza U (eds) Biodiversity and conservation of Isla de los Estados (Chuanisin)—a XXI Century Terra Incognita. Springer Nature, Cham

Poore GCB, Bruce NL (2012) Global diversity of marine isopods (Except Asellota and Crustacean Symbionts). PLoS One 7(8):e43529

Raya Rey A, Balza U, Domato I, Zunino F (2022) New Magellanic penguin *Spheniscus magellanicus* colony in a subantarctic Island. Polar Biol 45(10):1553–1558

Ritter CJ, Bourne DG (2024) Marine amphipods as integral members of global ocean ecosystems. J Exp Mar Biol Ecol 572:151985

Robertson AI, Mann KH (1980) The role of isopods and amphipods in the initial fragmentation of eelgrass detritus in Nova Scotia, Canada. Mar Biol 59(1):63–69

Rojo JH, Bruno DO, Llompart F, Chalde T, Delpiani M (2026) Chapter 11: Fish diversity in Isla de los Estados and surrounding areas. In: Raya Rey A, Balza U (eds) Biodiversity and conservation of Isla de los Estados (Chuanisin)—a XXI Century Terra Incognita. Springer Nature, Cham

Ruffo S (1947) Studi sui crostacei anfipodi. XV. Su alcune specie di Anfipodi dell´ Argentina e della Terra del Fuoco. Boll Lab Entomol Agrar Portici 7:326–332

Ruffo S (1949) Amphipodes (II). Résultats du Voyage de la Belgica en 1897–99. Rapports Scientifiques Zoologie, 1–58

Rumbold CE, Chiesa IL, Farías NE (2021) New epibiotic association in the deep-sea: the amphipod *Caprella ungulina* and the Patagonian lobsterette *Thymops birsteini* in the South-Western Atlantic. J Mar Biol Assoc U K 101(8):1171–1179

Schejter L, Chiesa IL, Doti BL, Bremec C (2012) *Mycale (Aegogropila) magellanica* (Porifera: Demospongiae) in the southwestern Atlantic Ocean: endobiotic fauna and new distributional information. Sci Mar 76(4):753–761

Schellenberg A (1931) Gammariden und Caprelliden des Magellangebietes, Südgeorgiens und der Westantarktis, Further zoological results of the Swedish Antarctic expedition 1901–1903, vol 2, pp 1–290

Schiavini A, Yorio P, Gandini P et al (2005) Los pingüinos de las costas argentinas: estado poblacional y conservación. El Hornero 20(1):5–23

Schultz GA (1981) Arcturidae from the Antarctic and Subantarctic seas (Isopoda, Valvifera) part 1. Biol Antarct Seas X. Antarct Res Ser 32:63–94

Schultz GA (1982) Species of Protallocoxoidea and Stenetrioidea (Isopoda, Asellota) from the Antarctic and southern seas. Biol Antarct Seas X. Antarct Res Ser 32:17–62

Sheppard EM (1933) Isopod crustacea, part I. The family Serolidae. Discover Rep 7:253–362

Vanhöffen E (1914) Die Isopoden der Deutschen Südpolar-Expedition 1901–1903. Zoologie 7(4):447–598

Watling L, Holman H (1981) Additional acanthonotozomatid, paramphithoid and stegocephalid Amphipoda from the Southern Ocean. Proc Biol Soc Wash 94:181–227

Chapter 8
Biodiversity and Development Patterns of Asteroids from Isla de los Estados and Surrounding Areas

Cintia P. Fraysse and Analia F. Pérez

Abstract Echinoderms are typically the most abundant and diverse members of the benthic communities. Within this Phylum, asteroids are especially diverse and abundant in Antarctic and Subantarctic waters, as well as in the South Atlantic Ocean, accounting for approximately 12% of total species richness in these regions. Asteroids play key ecological roles and occupy various trophic levels within food webs, including that of top predators. The developmental patterns are highly diverse, ranging from planktotrophic pelagic larvae to lecithotrophic larvae, retained larvae through brooding, and even direct development without a larval stage. Studying these patterns provides valuable insights into species dispersal and connectivity across geographic regions, as developmental modes are associated with differences in larval duration in the water column, which directly influences dispersal potential and population connectivity. In recent years, the number of marine and oceanographic studies funded by Argentine public institutions has increased; however, many areas of the South Atlantic remain poorly studied. The geographical isolation of Isla de los Estados (54° 16′ S, 64° 17′ W), separated from the main island of Tierra del Fuego by the Le Maire Strait, along with its adverse sea conditions, has historically limited scientific exploration in the region. As a result, our understanding of this remote region and its marine fauna remains limited. This chapter presents the diversity of asteroid species found in the waters surrounding Isla de los Estados, based on data collected during oceanographic surveys and a comprehensive literature review. It also compares the northern and southern zones of the island in terms of species composition and developmental patterns. Continued research in this region is essential to deepen our understanding of high-latitude ecosystems and

C. P. Fraysse (✉)
Laboratorio de Ecología, Fisiología y Evolución de Organismos Acuáticos, Centro Austral de Investigaciones Científicas (CADIC), Consejo Nacional de Investigaciones Científicas y Técnicas (CONICET), Ushuaia, Tierra del Fuego, Argentina

A. F. Pérez
Laboratorio de Invertebrados Marinos, Centro de Ciencias Naturales, Ambientales y Antropológicas (CCNAA), Universidad Maimónides, Consejo Nacional de Investigaciones Científicas y Técnicas (CONICET), Ciudad Autónoma de Buenos Aires, Argentina

A. Raya Rey, U. Balza (eds.), *Biodiversity and Conservation of Isla de los Estados (Chuani-sin)*, Natural and Social Sciences of Patagonia,
https://doi.org/10.1007/978-3-032-13630-5_8

their connectivity, ultimately contributing to the development of effective conservation strategies and the protection of these species and the marine ecosystems they inhabit.

Keywords Equinoderms · South Atlantic Ocean · Geographical isolation · Species richness

8.1 Introduction

8.1.1 Asteroids in High-Latitude Environments

Echinoderms are typically among the most abundant and diverse organisms found in benthic communities, having successfully colonized all ocean depths, including Antarctic waters (De Broyer et al. 2025). Within this group, asteroids exhibit particularly high diversity in the macrobenthos, representing approximately 12% of overall species richness in Antarctic waters and the South Atlantic Ocean (Mah and Blake 2012; Fraysse et al. 2018, 2024). Interestingly, the benthic marine fauna of Antarctica is better studied and understood than that of sub-Antarctic waters. It is estimated that around 300 asteroid species inhabit Antarctic waters (Moreau et al. 2019), including several that play key roles in benthic ecosystems (McClintock et al. 1988, 2008; Moreau et al. 2019). In recent years, Argentine public institutions have increased funding for marine and oceanographic research projects (Schejter et al. 2016, 2020a, b; Fraysse et al. 2018; Bremec et al. 2019; Gordillo et al. 2019; Doti et al. 2020). Despite these efforts, many areas of the South Atlantic remain insufficiently studied.

8.1.2 Some Insights into the Oceanography of Isla de los Estados

Isla de los Estados (54° 16′ S, 64° 17′ W) is located in the southernmost region of Argentina and belongs to the province of Tierra del Fuego, Antártida, e Islas del Atlántico Sur. The island rests on the Argentine continental shelf and is separated from Isla Grande de Tierra del Fuego by the Le Maire Strait (Ponce et al. 2011). Covering an area of 496 km^2, the island extends up to 62 km in length (east to west) with an average width of 6 km (Ponce et al. 2011). It represents the southeasternmost extension of the Andes Mountain range (Ponce and Fernández 2014).

Information about the oceanographic conditions of the water masses surrounding Isla de los Estados remains limited. In contrast, the Antarctic Circumpolar Current (ACC), the largest ocean current and the primary driver of inter-ocean water exchange, has been extensively studied. Flowing eastward, the ACC transports cold, relatively fresh Subantarctic waters (34 PSU) (Peterson and Whitworth III 1989; Piola and Gordon 1989), which are nutrient-rich (Acha et al. 2004; Romero et al. 2006) from the Drake Passage along the upper western slope of the Argentine basin (Guihou et al. 2020). A branch of the ACC, the Cape Horn Current, bathes the southern part of Isla de los Estados. This current then splits into two main branches: one flowing northward through the Le Maire Strait, while the main branch continues eastward, bathing the eastern coast of Isla de los Estados (Matano et al. 2019; Guihou et al. 2020; Palma et al. 2021; National Geographic 2023; Di Salvatore et al. 2023). These branches contribute to the formation of the Malvinas Current (Palma et al. 2021). As a result, Isla de los Estados is influenced by the subsection of the ACC, which delivers cold Subantarctic waters (Kühnemann 1976). Despite this, detailed knowledge of the region's oceanographic dynamics and transport processes remains scarce. What is known indicates a strong influence from currents along the shelf break and the Subantarctic Front (Orsi et al. 1995; Guihou et al. 2020).

8.1.3 Development Patterns

Studying sea star development patterns is essential for understanding their ecology and evolution (McEdward and Morgan 2001; Moreau et al. 2017, 2019). These patterns provide insights into dispersal and connectivity across geographic regions, as different developmental patterns are associated with varying larval durations, which in turn influence the dispersal potential and population connectivity (Moreau et al. 2019; Fraysse et al. 2024).

Asteroids exhibit a wide range of developmental modes, from indirect to direct development. Their larval stages can be pelagic or benthic, with planktotrophic (feeding) or lecithotrophic (non-feeding) nutrition (Mileikovsky 1971; Chia 1974; McEdward and Miner 2001). Most asteroid species produce pelagic larvae, which may be either planktotrophic or lecithotrophic, allowing for broad dispersal (Hyman 1955; Bosch and Pearse 1990; Chia and Walker 1991; Byrne and Barker 1991). In contrast, a smaller number of species exhibit non-pelagic (benthic) lecithotrophic larvae that are brooded by the parents (retained larvae). In some cases, the larval stage is entirely absent, with the embryo developing directly into a juvenile. These diverse reproductive strategies have been extensively studied, especially regarding parental care and incubation (Chia 1968; Himmelman et al. 1982; Strathmann and Strathmann 1982; McClary and Mladenov 1990; Pearse and Bosch 1994; Byrne 2005; Gillespie and McClintock 2007; Gil et al. 2011; Rivadeneira et al. 2017; Pérez et al. 2017; Fraysse et al. 2020a, b). For example, *Leptasterias (Hexasterias) polaris* (Himmelman et al. 1982; Hamel and Mercier 1995) and *Asterina phylactica* (Strathmann et al. 1984) protect their embryos beneath their bodies. In contrast, *Leptasterias hexactis* (Chia 1968) and *Anasterias antarctica* (Salvat 1985; Gil et al. 2011; Pérez et al. 2015; Fraysse et al. 2021) construct a nidamental chamber by arching their arms and retaining the lecithotrophic larva on their mouth. In *Ctenodiscus australis* and *Leptychaster spp.* (Lieberkind 1926; Rivadeneira et al.

2017), brooding occurs among the paxillae on the aboral surface. In contrast, species in the family Pterasteridae brood their young inside an aboral nidamental chamber (Hyman 1955; McClary and Mladenov 1990; Fraysse et al. 2020a, b). Internal brooding has also been documented in *Leptasterias groenlandica*, which retains embryos inside the stomach (Lieberkind 1920) and in several species of the genus *Patiriella*, which brood their offspring in the intragonadal space (Komatsu et al. 1990; Byrne 1996, 2005; Dartnall et al. 2003).

Due to the remote location and geographic isolation of Isla de los Estados, along with the harsh oceanographic conditions, few scientific studies have been conducted in the area, limiting our understanding of this unique region and its fauna. This chapter describes the asteroids community, from the waters surrounding Isla de los Estados, providing an updated checklist of species. It also compares the northern and southern zones of the island in terms of species composition and developmental patterns, presenting evidence that these zones represent ecologically distinct environments. Continued research is crucial for improving our understanding of high latitude ecosystems and their biogeographic connectivity, ultimately enhancing the conservation and protection of these species and the marine ecosystems they inhabit.

8.2 Materials and Methods

8.2.1 Study Area and Sampling

The study area encompassed Isla de los Estados and its surrounding waters, with a particular focus on the northern and southern zones of the island. An updated species occurrence checklist was compiled using data from multiple sources: specimens collected during two Oceanographic Research Cruises, "BBB ABR 2016" and "BOPD 2017", conducted aboard the oceanographic vessel ARA "Puerto Deseado" (Fraysse et al. 2018, 2024), records from a recent technical report in the area (Di Gallo 2025), and occurrence data retrieved from the GBIF dataset for the region (GBIF 2025a)

Sampling during the research cruises was carried out using a bottom otter trawl to collect epibenthic organisms and a Rauschert dredge to target infaunal species. In total, nine sampling stations were conducted, covering a variety of seabed types. All sampling was performed in a qualitative manner. Both cruises took place in austral autumn, during which environmental parameters (salinity, temperature, and depth) were recorded with a CTD probe. Salinity ranged from 32.70 to 33.17 PSU, temperature ranged from 8.53 to 10.19 °C, and depth ranged from 153 to 418 m. During the research cruises, specimens were photographed and subsequently preserved in 10% formaldehyde for taxonomic identification to the lowest possible level.

8.2.2 *Taxonomic and Reproductive Determination*

Taxonomic identifications were achieved following Sladen (1889), Koehler (1912), Verril (1914), Fisher (1940), Bernasconi (1962, 1970), Tablado (1982), Clark and Downey (1992), Mah et al. (2014) and Arntz and Rauschert (2015). Taxonomy and species distributions were subsequently cross-checked using the Register of Antarctic Marine Species (Clarke and Johnston 2003; De Broyer et al. 2025) to correct possible synonymies or spelling errors and to compare previously known distributions with those observed in this study. Additionally, the developmental mode of each species, classified as internal brooding, oral brooding, free-lecithotrophic larvae, or free-planktotrophic larvae, was determined based on MacBride (1920), Lieberkind (1926), Fisher (1940), Hyman (1955), Bernasconi (1962, 1970), Bosch and Pearse (1990), Pearse et al. (1991), Pearse and Bosch (1994), and Fraysse et al. (2020a, b).

8.2.3 *Data Analysis*

The Shannon-Wiener diversity index (H′) and Pielou's evenness index (J′) were calculated to assess diversity and the relative distribution of species abundances at each site. Evenness was computed as the ratio between Shannon's index and the natural logarithm of species richness ($J' = H'/\ln(S)$), where S is the total number of species recorded per site. These indices provide insights into both the richness and the uniformity of species distribution. Calculations were performed using the vegan 2.6–8 package (Oksanen et al. 2024) implemented in R 4.4.2 (R Core Team 2025).

To conduct the cluster analysis, species records from Antarctica and the Beagle Channel were incorporated into the dataset (Pérez et al. 2024; Fraysse et al. 2024; GBIF 2025b). Due to unequal sampling effort and the use of various collection methods, presence/absence (binary) data were used to build the species similarity matrix. The Sørensen index was applied to construct this matrix (Clarke et al. 2006). Similarities in species composition across regions were analyzed through hierarchical clustering using the group-average linkage method. The resulting dendrogram was generated with Primer V6.0 software, with a similarity threshold of 50% used to define regional groupings.

8.3 Results

8.3.1 *Updated Checklist*

A total of 142 Asteroidea individuals were recorded in Isla de los Estados (Fig. 8.1). These individuals belong to 12 families, 22 genera, and 29 Asteroidea species (Table 8.1). Five orders were registered, with Forcipulatida being the most representative in Isla de los Estados, followed by Valvatida. *Cryptasterias brachiata* is a new record species for the area.

Fig. 8.1 Aboral and oral views of the Asteroidea species recorded in Isla de los Estados. Scale bar: 2 cm

Table 8.1 Updated checklist of Asteroidea present in the study area

Order	Family	Species	Year of record	Area	Depth (m)	IE zone		Development	Reference
						North	South		
Forcipulatida	Asteriidae	*Anasterias Antarctica*	1969, 1971, 2016, 2022	Basil Hall Port Capitán Canepa Bay Celular Port Colnett Cape Cook Port Crossley Bay Flinders Bay Hoppner Port Observatorio Island Offshore northern zone Parry Port Roca Port San Antonio Cape San Juan de Salvamento Vancouver Port	Intertidal-46	26	8	Oral brooding	GBIF, BBBABR16
		Anasterias asterinoides	1971	Hoppner Port	Intertidal	1	0	Oral brooding	GBIF
		Anasterias pedicellaris	1902, 2016	Offshore northern zone	36–183	1	1	Oral brooding	GBIF, BBBABR16
		Anasterias spirabilis	1971	Cook Port	23	1	0	Oral brooding	GBIF
		Cryptasterias brachiata	2024	Offshore eastern zone				Oral brooding	Di Gallo (2025)
		Diplasterias brandti	2016, 2017, 2022, 2024	Hoppner Port Observatorio Island Offshore eastern zone Offshore southern zone	35–418	2	3	Oral brooding	GBIF, BBBABR16, BOPD2017, Di Gallo (2025)
		Diplasterias brucei	–	Observatorio Island	35	1	0	Oral brooding	GBIF
		Asteriidae *indet*	–	Lobo Port Offshore northern zone	33–194	1	1		GBIF
	Heliasteridae	*Labidiaster radiosus*	2022–2024	Capitán Canepa Bay Hoppner Port	20	1	1		GBIF
	Pedicellasteridae	*Anteliaster australis*	1971	Offshore northern zone	60	1	0	Oral brooding	GBIF

(continued)

Table 8.1 (continued)

Order	Family	Species	Year of record	Area	Depth (m)	IE zone		Development	Reference
						North	South		
	Stichasteridae	*Cosmasterias lurida*	1969, 1971, 2016, 2022, 2023, 2024	Capitán Canepa Bay Celular Port Cook Port Flinders Bay Hoppner Port Offshore eastern zone Offshore northern zone Parry Port San Juan de Salvamento	12–183	12	3	Planktotrophic pelagic larva	GBIF, BBBABR16, Di Gallo (2025)
		Granaster nutrix	–	Observatorio Island	–	1	0	Internal brooding	GBIF
		Smilasterias triremis	–	Offshore southern zone	149	0	1	Brooder	GBIF
		Stichasteridae *indet*	2017–	Observatorio island Offshore soutthern zone	–	1	1		GBIF, BOPD2017
Paxillosida	Astropectinidae	*Bathybiaster loripes*	–	Observatorio Island		1	0	Lecithotrophic pelagic larva	GBIF
		Leptychaster kerguelenensis	1902	Offshore northern zone	36	1	0	Brooder	GBIF
		Psilaster charcoti	1969	Flinders Bay	31–40	1	0	Lecithotrophic pelagic larva	GBIF
Spinulosida	Echinasteridae	*Henricia obesa*	1971, 2016, 2022	Basil Hall Port Offshore southern zone Parry Port	Intertidal-159	2	1	Lecithotrophic pelagic larva	GBIF, BBBABR16
		Echinasteridae *indet*	1971, 2016	Black Port Offshore northern zone Offshore soutthern zone	33–159	2	2		GBIF, BBBABR16

Valvatida	Asterinidae	*Asterina fimbriata*	1971, 1973, 2022, 2024	Año Nuevo Port Basil Hall Port Black Port Cook Port Crossley Bay Hoppner Port Offshore northern zone Parry Port Roca Port San Juan de Salvamento	Intertidal-136	14	1	Oral brooding	GBIF
		Cycethra frigida	1971	Cook Port Roca Port	Intertidal	2	0	Lecithotrophic pelagic larva	GBIF
		Cycethra verrucosa	1971, 2024	Observatorio island Offshore northern zone San Juan de Salvamento	Intertidal-78	3	0	Lecithotrophic pelagic larva	GBIF
		Asterinidae *indet*	1973	Crossley Bay	–	1	0		GBIF
	Odontasteridae	*Diplodontias singularis*	1971, 2017	Flinders Bay Offshore northern zone Offshore southern zone Basil Hall Port	Intertidal-153	4	1	Planktotrophic pelagic larva	GBIF, BOPD2017
		Odontaster meridionalis	–	Observatorio Island	–	1	0	Planktotrophic pelagic larva	GBIF
		Odontaster penicillatus	1973, 2024	Black Port Crossley Bay Observatorio Island San Juan de Salvamento	Intertidal-33	6	1	Planktotrophic pelagic larva	GBIF, Di Gallo (2025)
		Odontaster validus	–	Observatorio Island	12	1	0	Planktotrophic pelagic larva	GBIF
	Poraniidae	*Glabraster antarctica*	1971, 2022, 2024	Basil Hall Port Offshore eastern zone Offshore southern zone San Juan de Salvamento	1–149	4	1	Facultative lecithotrophic pelagic larva	GBIF, Di Gallo (2025)
	Solasteridae	*Paralophaster antarcticus*	1971	Basil Hall Port Roca Port	1	2	0		GBIF
		Solaster regularis	2017, 2024	Capitán Canepa Bay Offshore southern zone	20–284	2	0	Lecithotrophic pelagic larva	GBIF, BOPD2017

(continued)

Table 8.1 (continued)

Order	Family	Species	Year of record	Area	Depth (m)	IE zone		Development	Reference
						North	South		
Velatida	Korethrasteridae	*Peribolaster folliculatus*	1971, 2016, 2017	Offshore northern zone Offshore southern zone	136–284	1	2	Planktotrophic pelagic larva	GBIF, BBBABR16, BOPD2017
	Pterasteridae	*Diplopteraster verrucosus*	2016, 2017, 2024	Capitán Canepa Bay Offshore southern zone	20–418	0	4	Internal brooding	GBIF, BBBABR16, BOPD2017
		Pteraster gibber	2022	San Juan de Salvamento	–	1	0	Internal brooding	GBIF
Asteroidea indet			1973–	Blossom Bay Hoppner Port Lobo Port Observatorio Island Offshore northern zone Offshore southern zone Toriaschi Bay	10–245	5	6		GBIF

The year, narrow area of the record, registered depth, and developmental pattern are noted when available. Also, the number of records per zone within Isla de los Estados (north or south) is noted

8.3.2 Diversity

The northern zone of Isla de los Estados (North IE) exhibited twice the species richness observed in the southern zone (South IE). In terms of abundance distribution, the Shannon-Wiener diversity index (H′) was higher in North IE, while Pielou's evenness index (J′) was greater in South IE (Table 8.2). These results suggest that the community in North IE is likely dominated by a few species, leading to a more uneven distribution of abundance. In contrast, despite its lower species richness, South IE displays a more even community structure.

8.3.3 Reproductive Patterns

In the northern zone, brooding species—those with retained larvae and direct development—account for 11 species, compared to 12 species with free-living larval development (Fig. 8.2). Nevertheless, oral brooding species remain the predominant developmental strategy in this zone. In contrast, in the southern zone, the number of species exhibiting oral brooding is equal to that of those with planktotrophic larvae. In both zones, the number of species that brood internally is low. *Smilasterias triremis* and *Leptychaster kerguelenensis* are classified as brooders in the literature, although no detailed description of the brooding type was found (Hyman 1955; O'Loughlin and O'Hara 1990). A congeneric species, *S. multipara*, has been described as a gastric brooder, retaining its young inside its cardiac stomach (O'Loughlin and O'Hara 1990).

8.3.4 Connectivity with Surrounding Areas

The hierarchical cluster analysis revealed clear patterns of similarity between zones. The SIMPROF test identified statistically significant clusters ($p < 0.05$), indicating that the observed groupings are not due to random variation. Notably, the northern zone of Isla de los Estados (North IE) clustered with the Beagle Channel, while the southern zone of the island (South IE) and Antarctica remained singletons, indicating distinct species compositions in those areas (Fig. 8.3).

Table 8.2 Calculated Shannon-Wiener diversity index (H′), Richness, and Pielou's evenness index (J′) for both zones of Isla de los Estados (North and South)

Zone	Diversity (H′)	Richness	Evenness (J′)
South IE	2.255	13	0.879
North IE	2.559	26	0.785

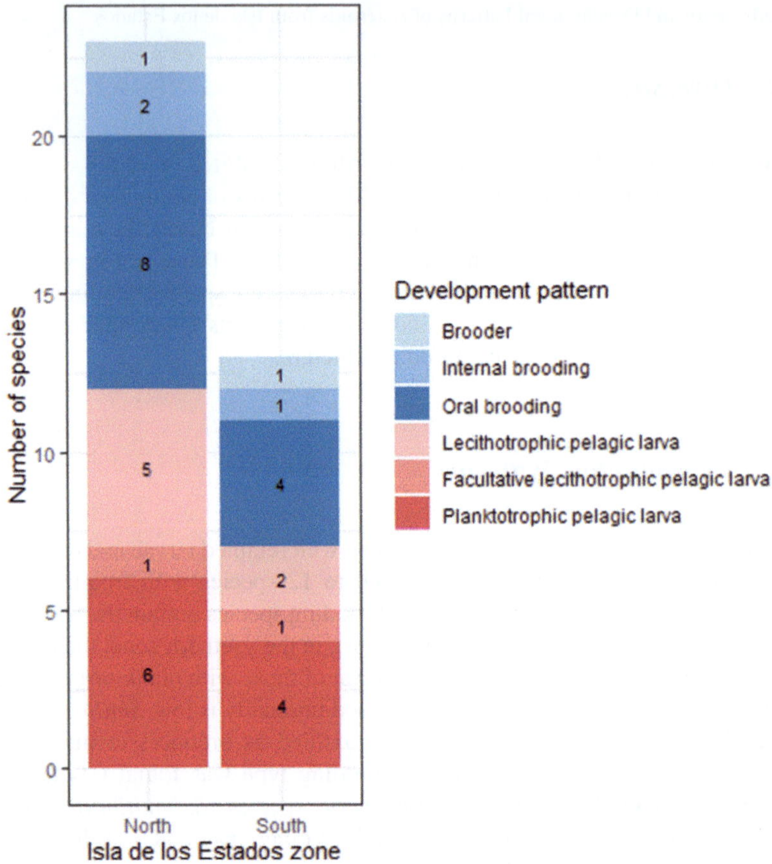

Fig. 8.2 The number of species for each Isla de los Estados zone (north and south) and their developmental pattern. The number in the column represents the richness of each developmental pattern

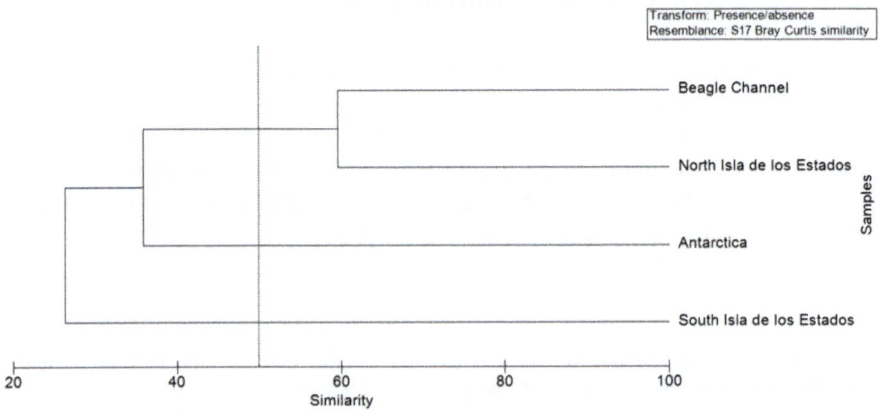

Fig. 8.3 Hierarchical cluster analysis performed using group-average linkage based on the Bray-Curtis similarity of species composition across sampling areas. Significant clusters identified by the SIMPROF test ($p < 0.05$) are indicated on the dendrogram. A threshold of 50% similarity is shown. For Antarctica and the Beagle Channel GBIF data set records (GBIF 2025b) and previous research studies (Fraysse et al. 2024) were consulted

8.4 Discussion

A total of 142 Asteroidea individuals were recorded in Isla de los Estados, comprising 29 species, including *Cryptasterias brachiata, which represents* a new record for the area (Table 8.1).

The northern zone of Isla de los Estados (North IE) exhibited twice the species richness and a higher abundance index than the southern zone (South IE), while South IE showed a higher evenness index. This suggests that in North IE, a few abundant species dominate the community, resulting in an uneven distribution of abundance. In contrast, South IE, despite the lower species richness and abundance observed in the North IE, may be associated with a more stable environment, likely influenced by milder oceanic currents. Conversely, the stronger and more variable hydrodynamic conditions in the southern zone (Guihou et al. 2020; National Geographic 2023) may act as environmental filters, limiting community assembly to species capable of withstanding high physical stress. Environmental filters refer to specific factors that constrain the range of functional traits within a local community (Díaz et al. 2007). This process occurs primarily through ecological sorting, where the representation of genotypes in a community occurs without modifying the genotypes themselves. Over evolutionary timescales, however, such filtering may lead to stabilizing or directional selection (Díaz et al. 2007). In this context, environmental filtering may reduce species richness but increase evenness, as no single species consistently gains an advantage under persistently harsh conditions. These stress-tolerant assemblages may reflect a functionally redundant and potentially resilient community structure. The concept of 'functional redundancy' is critical for maintaining ecosystem function in the face of environmental disturbances (Mori et al. 2013). Nonetheless, it is important to acknowledge a potential sampling bias in these results, as the number of oceanographic surveys conducted in the South IE was lower due to the logistical challenges associated with operating in the area.

When reproductive strategies are considered, the northern zone hosts more brooding species, particularly those with oral brooding (Fig. 8.2). This form of protected development is commonly associated with stable environments (Fraysse et al. 2024). In contrast, internal brooding is rare in both zones, consistent with its typical occurrence in deeper environments, where external cues such as light and seasonal food availability are minimal (Lieberkind 1926; Chia 1968; Fraysse et al. 2024). The island may act as a barrier, buffering North IE from the harsher conditions in the south, thereby facilitating larval retention. However, in the more dynamic South IE dominated by strong currents (Palma et al. 2021), no clear selection pattern emerges between larval retention and planktonic development; the number of species employing each strategy is approximately equal. Most brooding species in both zones belong to the family Asteriidae, which has been widely reported across Atlantic regions (Clark and Downey 1992; Fraysse et al. 2018), suggesting a possible stepping-stone dispersal mechanism (Moreau et al. 2017). Non-pelagic development in these species may have resulted from selective pressures linked to isolation following vicariant events, rather than adaptation to high latitude environments, leading to their proliferation within relatively few clades (Pearse et al. 2009). Differences in dispersal potential, low in brooders and high in species with pelagic larvae, may explain the diversity of biogeographic patterns observed (Raguá-Gil et al. 2004; Moreau et al. 2017). Species with planktonic larvae are also present in Antarctic waters, underscoring the influence of the Cape Horn current (Moreau et al. 2017, 2019; Fraysse et al. 2024).

Cluster analysis further supports this pattern, suggesting that North IE and South IE form statistically distinct groups. North IE clusters together with the Beagle Channel, and both differ significantly from South IE and Antarctica. The number of species shared by South IE and Antarctica highlights the strong influence of the Cape Horn current on larval dispersal to South IE. In contrast, North IE is likely influenced by the Malvinas Current, a sub-branch of the ACC (Piola and Gordon 1989) which flows northward through the Le Maire Strait, and connects North IE to the Beagle Channel.

In summary, North IE shares species composition with the Beagle Channel and is characterized by higher species richness and abundance, lower evenness, and a more significant proportion of brooding species, especially those exhibiting oral brooding. In contrast, South IE shows a stronger influence of the Cape Horn current, greater community evenness, and a balanced distribution of developmental strategies without a clear dominance of either larval retention or free-swimming forms. In both regions, internal brooding remains a rare occurrence.

Continued research is essential to deepen our understanding of high-latitude ecosystems and their connectivity, which will support the development of more effective conservation and management strategies for these species and their marine habitats.

Acknowledgements We are grateful to Sonia Rimbau, technician of Laboratorio de Ecología, Fisiología y Evolución de Organismos Acuáticos (CADIC-CONICET), and Dr. Pablo Di Salvatore for their technical assistance during the samplings. We are also grateful to the crew of the RV "ARA Puerto Deseado" for their support during the sampling procedures.

Funding Statement This work was funded by Fundación Científica Felipe Fiorellino (Universidad Maimónides), and grants from ANPIDTyI (PICT INV1-0100) and PADI foundation.

Competing Interests The authors declare that they have no conflict of interest.

References

Acha EM, Mianzan HW, Guerrero RA, Favero M, Bava J (2004) Marine fronts at the continental shelves of austral South America. J Mar Syst 44(1–2):83–105. https://doi.org/10.1016/j.jmarsys.2003.09.005

Arntz WE, Rauschert M (2015) Antarctic macrobenthos: a field guide of the invertebrates living at the antarctic seafloor. Arntz & Rauschert Selbstverlag

Bernasconi I (1962) Asteroideos Argentinos III. Familia Odontasteridae. Rev Mus Argent Cienc Nat 9:1–25

Bernasconi I (1970) Equinodermos antárticos. II. Asteroideos. 3. Asteroideos de la extremidad norte de la Península Antártica. Rev Mus Argent Cienc Nat 9:211–281

Bosch I, Pearse JS (1990) Developmental types of shallow-water asteroids of McMurdo Sound, Antarctica. Mar Biol 104(1):41–46. https://doi.org/10.1007/BF01313155

Bremec C, Elías R, Calla S, Genzano G, Puente-Tapia A, Schejter L (2019) Polychaetes from Burdwood Bank: "Namuncurá I" marine protected area and slope, SW Atlantic Ocean. Rev Biol Trop 67(S5):S119–S135. https://doi.org/10.15517/rbt.v67iS5.38937

Byrne M (1996) Viviparity and intragonadal cannibalism in the diminutive sea stars *Patiriella vivipara* and *P. parvivipara* (family Asterinidae). Mar Biol 125:551–567. https://doi.org/10.1007/BF00353268

Byrne M (2005) Viviparity in the sea star *Cryptasterina hystera* (Asterinidae)—conserved and modified features in reproduction and development. Biol Bull 208(2):81–91. https://doi.org/10.2307/3593116

Byrne M, Barker MF (1991) Embryogenesis and larval development of the Asteroid *Patiriella regularis* viewed by light and scanning electron microscopy. Biol Bull 180(3):332–345. https://doi.org/10.2307/1542335

Chia F-S (1968) The embryology of a brooding starfish, *Leptasterias hexactis* (Stimpson). Acta Zool 49(3):321–364. https://doi.org/10.1111/j.1463-6395.1968.tb00157.x

Chia F-S (1974) Classification and adaptive significance of developmental patterns in marine invertebrates. Thalass Jugosl 10:121–130

Chia F-S, Walker CW (1991) Echinodermata: asteroidea. In: Giese AC, Pearse JS, Pearse VB (eds) Reproduction of marine invertebrates, Echinoderms and lophophorates, vol 6. The Boxwood Press, pp 301–340

Clark AM, Downey ME (1992) Starfishes of the Atlantic. Chapman & Hall, London

Clarke A, Johnston NM (2003) Antarctic marine benthic diversity. Oceanogr Mar Biol 41:47–114

Clarke R, Somerfield P, Chapman G (2006) On resemblance measures for ecological studies, including taxonomic dissimilarities and a zero-adjusted Bray–Curtis coefficient for denuded assemblages. J Exp Mar Biol Ecol 330(1):55–80. https://doi.org/10.1016/j.jembe.2005.12.017

Dartnall AJ, Byrne M, Collins J, Hart MW (2003) A new viviparous species of asterinid (Echinodermata, Asteroidea, Asterinidae) and a new genus to accommodate the species of pan-tropical exiguoid sea stars. Zootaxa 359:1–14

De Broyer C, Clarke A, Koubbi P, Pakhomov E, Scott F, Vanden Berghe E, Danis B (2025) Register of Antarctic Marine Species (RAMS). http://www.marinespecies.org/rams

Di Gallo AC (2025) Informe de Práctica Profesionales Supervisadas de la Licenciatura en Biología: Programa de Observadores a bordo de la provincia de Tierra del Fuego Antártida e Islas del Atlántico Sur (TDFAeIAS). Universidad Nacional de Tierra del Fuego y Secretaría de Pesca y Acuicultura, Ministerio de Producción y Ambiente de Tierra del Fuego, p 30

Di Salvatore P, Albano MJ, Diez MJ, Tapella F, Pérez-Barros P, Lovrich GA (2023) Decapods of the southern tip of South America and the marine protected area Namuncurá–Burdwood Bank: a nearshore–offshore comparison. Diversity 15(11):1143. https://doi.org/10.3390/d15111143

Díaz S, Lavorel S, Chapin FS III, Tecco PA, Gurvich DE, Grigulis K (2007) Functional diversity—at the crossroads between ecosystem functioning and environmental filters. In: Canadell JG, Pataki D, Pitelka L (eds) Terrestrial ecosystems in a changing world. Springer, Berlin

Doti BL, Chiesa IL, Roccatagliata D (2020) Biodiversity of isopoda and Cumacea (Peracarida, crustacea) from the marine protected area Namuncurá-Burdwood Bank, South-West Atlantic. Polar Biol 43(10):1519–1534. https://doi.org/10.1007/s00300-020-02725-z

Fisher WK (1940) Asteroidea. Discov Rep 20:69–306

Fraysse CP, Calcagno J, Pérez AF (2018) Asteroidea of the southern tip of South America, including Namuncurá marine protected area at Burdwood Bank and Tierra del Fuego Province, Argentina. Polar Biol 41:2423–2433. https://doi.org/10.1007/s00300-018-2377-3

Fraysse CP, Boy CC, Becker Y, Calcagno JA, Pérez AF (2020a) Brooding in the southern ocean: the case of the pterasterid sea star *Diplopteraster verrucosus* (Sladen, 1882). Biol Bull 239:1–12. https://doi.org/10.1086/709664

Fraysse CP, Pérez AF, Calcagno J, Boy CC (2020b) Energetics and development modes of Asteroidea (Echinodermata) from the Southwestern Atlantic Ocean including Burdwood Bank/MPA Namuncurá. Polar Biol 43:175–186. https://doi.org/10.1007/s00300-020-02621-6

Fraysse CP, Boy CC, Pérez AF (2021) Reproductive traits of the intertidal sea star *Anasterias Antarctica* (Echinodermata: Asteroidea) from the Beagle Channel, Argentina. Mar Biol 168(12):178. https://doi.org/10.1007/s00227-021-03987-9

Fraysse CP, Boy CC, Ojeda M, Rodriguez M, Rojo JH, Pérez AF (2024) Distribution and development patterns in sea stars (Echinodermata: Asteroidea) of the subantarctic southern Atlantic. Polar Biol 47:1107–1120. https://doi.org/10.1007/s00300-024-03288-z

GBIF (2025a) Occurrence download (p. 58970) [Darwin Core Archive,text/tab-separated-values,application/zip]. The Global Biodiversity Information Facility. https://doi.org/10.15468/DL.3UZ5XM

GBIF (2025b) Occurrence download (p. 225996) [Text/tab-separated-values,application/zip]. The Global Biodiversity Information Facility. https://doi.org/10.15468/DL.QJV3U4

Gil DG, Escudero G, Zaixso HE (2011) Brooding and development of *Anasterias minuta* (Asteroidea: Forcipulata) in Patagonia, Argentina. Mar Biol 158(11):2589–2602. https://doi.org/10.1007/s00227-011-1760-1

Gillespie JM, McClintock JB (2007) Brooding in echinoderms: how can modern experimental techniques add to our historical perspective? J Exp Mar Biol Ecol 342:191–201. https://doi.org/10.1016/j.jembe.2006.10.055

Gordillo S, Bayer MS, de Aranzamendi MC, Taverna A, Morán GA (2019) The ecological role of brachiopods in the Namuncurá MPA/Burdwood Bank, off southern South America. Mar Biodivers 49(4):1667–1681. https://doi.org/10.1007/s12526-019-00935-1

Guihou K, Piola AR, Palma ED, Chidichimo MP (2020) Dynamical connections between large marine ecosystems of austral South America based on numerical simulations. Ocean Sci 16(2):271–290. https://doi.org/10.5194/os-16-271-2020

Hamel JF, Mercier A (1995) Prespawning behavior, spawning, and development of the brooding starfish *Leptasterias polaris*. Biol Bull 188(1):32–45. https://doi.org/10.2307/1542065

Himmelman J, Lavergne Y, Cardinal A, Martel G, Jalbert P (1982) Brooding behaviour of the northern sea star *Leptasterias polaris*. Mar Biol 68(3):235–240. https://doi.org/10.1007/BF00409590

Hyman LH (1955) The invertebrates: echinodermata. The coelomate bilateria, vol IV. McGraw-Hill Book Co, New York

Koehler R (1912) Échinodermes (astéries, ophiures et échinides). En Deuxième Expédition Antarctique Française (1908–1910) Commandée par le Dr Jean Charcot. Masson et Cie, p 270

Komatsu M, Kano YT, Oguro C (1990) Development of a true ovoviviparous sea star, *Asterina pseudoexigua pacifica* Hayashi. Biol Bull 179:2454–2263. https://doi.org/10.2307/1542316

Kühnemann O (1976) Observaciones ecológicas sobre la vegetación marina y terrestre de la Isla de los Estados (Tierra del Fuego, Argentina). Ecosur 3:121–248

Lieberkind I (1920) On a star-fish (*Asterias groenlandica*) which hatches its young in its stomach. Vidensk Medd Dan Naturhist Foren 72:121–126

Lieberkind I (1926) *Ctenodiscus australis* Lütken. A brood-protecting asteroid. Vid Dansk Nat Hist Foren 82:184–196

MacBride EW (1920) Echinoderma (part II) and Enteropneusta: larvae of Echinoderma and Enteropneusta. Order of the Trustees of the British Museum

Mah C, Blake D (2012) Global diversity and phylogeny of the Asteroidea (Echinodermata). PLoS One 7(4):e35644. https://doi.org/10.1371/journal.pone.0035644

Mah C, Neill K, Eléaume M, Foltz D (2014) New species and global revision of *Hippasteria* (Hippasterinae: Goniasteridae; Asteroidea; Echinodermata): *Hippasteria* revision. Zool J Linnean Soc 171(2):422–456. https://doi.org/10.1111/zoj.12131

Matano RP, Palma ED, Combes V (2019) The Burdwood Bank circulation. J Geophys Res Oceans 124(10):6904–6926. https://doi.org/10.1029/2019JC015001

McClary DJ, Mladenov PV (1990) Brooding biology of the sea star *Pteraster militaris* (O.F. Müller): energetic and histological evidence for nutrient translocation to brooded juveniles. J Exp Mar Biol Ecol 142(3):183–199. https://doi.org/10.1016/0022-0981(90)90090-Y

McClintock J, Pearse J, Bosch I (1988) Population structure and energetics of the shallow-water antarctic sea star *Odontaster validus* in contrasting habitats. Mar Biol 99(2):235–246. https://doi.org/10.1007/BF00391986

McClintock J, Angus R, Ho C, Amsler C, Baker B (2008) Intraspecific agonistic arm-fencing behavior in the Antarctic keystone sea star *Odontaster validus* influences prey acquisition. Mar Ecol Prog Ser 371:297–300. https://doi.org/10.3354/meps07710

McEdward LR, Miner BG (2001) Larval and life-cycle patterns in echinoderms. Can J Zool 79(7):1125–1170. https://doi.org/10.1139/z00-218

McEdward LR, Morgan KH (2001) Interspecific relationships between egg size and the level of parental investment per offspring in echinoderms. Biol Bull 200(1):33–50. https://doi.org/10.2307/1543083

Mileikovsky SA (1971) Types of larval development in marine bottom invertebrates, their distribution and ecological significance: a re-evaluation. Mar Biol 10(3):193–213. https://doi.org/10.1007/BF00352809

Moreau C, Saucède T, Jossart Q, Agüera A, Brayard A, Danis B (2017) Reproductive strategy as a piece of the biogeographic puzzle: a case study using Antarctic sea stars (Echinodermata, Asteroidea). J Biogeogr 44(4):848–860. https://doi.org/10.1111/jbi.12965

Moreau C, Danis B, Jossart Q, Eléaume M, Sands C, Achaz G, Agüera A, Saucède T (2019) Is reproductive strategy a key factor in understanding the evolutionary history of Southern Ocean Asteroidea (Echinodermata)? Ecol Evol 9(15):8465–8478. https://doi.org/10.1002/ece3.5280

Mori AS, Furukawa T, Sasaki T (2013) Response diversity determines the resilience of ecosystems to environmental change. Biol Rev 88(2):349–364. https://doi.org/10.1111/brv.12004

National Geographic (2023) Ocean currents in motion (earthstar geographics | NOAA (AOML, CIMAS, PMEL)) [Map]. Universidad de Miami. https://www.arcgis.com/apps/instant/atlas/index.html?appid=0cd1cdee853c413a84bfe4b9a6931f0d&webmap=c69ed6f0677d4eb98d362f6d176dadf9

O'Loughlin PM, O'Hara TD (1990) A review of the genus *Smilasterias* (Echinodermata, Asteroidea), with descriptions of two new species from southeastern Australia, one a gastric brooder, and a new species from Macquarie Island. Mem Mus Vic 50(2):307–323. https://doi.org/10.24199/j.mmv.1990.50.05

Oksanen J, Simpson GL, Blanchette G, Kindt R, Legendre P, Minchin PR, O'Hara RB, Solymos P, Stevens MH, Szoecs E, Wagner H, Barbour M, Bedward M, Bolker B, Borcard D, Carvalho G, Chirico M, De Caceres M, Durand S, et al (2024) vegan: Community Ecology Package [Software]. https://CRAN.R-project.org/package=vegan

Orsi AH, Whitworth T, Nowlin WD (1995) On the meridional extent and fronts of the Antarctic circumpolar current. Deep-Sea Res I Oceanogr Res Pap 42(5):641–673. https://doi.org/10.1016/0967-0637(95)00021-W

Palma ED, Matano RP, Combes V (2021) Circulation and cross-shelf exchanges in the Malvinas Islands shelf region. Prog Oceanogr 198:102666. https://doi.org/10.1016/j.pocean.2021.102666

Pearse JS, Bosch I (1994) Brooding in the Antarctic: Östergren had it nearly right. In: David B, Guille A, Feral J-P, Roux M-J (eds) Echinoderms through time. B. Balkema Press, pp 111–120

Pearse JS, Mcclintock JB, Bosch I (1991) Reproduction of Antarctic benthic marine invertebrates: tempos, modes, and timing. Am Zool 31(1):65–80. https://doi.org/10.1093/icb/31.1.65

Pearse JS, Mooi R, Lockhart SJ, Brandt A (2009) Brooding and species diversity in the southern ocean: selection for brooders or speciation within brooding clades? In: Krupnik I, Lang MA, Miller SE (eds) Smithsonian at the poles: contributions to international polar year science. Proceedings of the Smithsonian at the poles symposium. Smithsonian Institution Scholarly Press, pp 181–196. https://doi.org/10.5479/si.097884601X.13

Pérez AF, Boy CC, Calcagno J, Malanga G (2015) Reproduction and oxidative metabolism in the brooding sea star *Anasterias Antarctica* (Lütken, 1957). J Exp Mar Biol Ecol 463:150–157. https://doi.org/10.1016/j.jembe.2014.11.009

Pérez AF, Fraysse C, Boy CC, Epherra L, Calcagno JA (2017) Reproductive biology and energetics of the brooding sea star *Anasterias Antarctica* (Echinodermata: Asteroidea) in the Beagle Channel, Tierra del Fuego, Argentina. Rev Biol Trop 65(1–1):221–232. https://doi.org/10.15517/rbt.v65i1-1.31690

Pérez AF, Fraysse CP, Ojeda M (2024) Distribución y patrones de desarrollo de asteroideos antárticos. Rev Biol Trop 72(S1):e58868. https://doi.org/10.15517/rev.biol.trop..v72iS1.58868

Peterson RG, Whitworth T III (1989) The subantarctic and polar fronts in relation to deep water masses through the southwestern Atlantic. J Geophys Res Oceans 94:10817–10838

Piola AR, Gordon AL (1989) Intermediate waters in the southwest South Atlantic. Deep-Sea Res I Oceanogr Res Pap 36:1–16

Ponce JF, Fernández M (2014) Climatic and environmental history of Isla de los Estados, Argentina. Springer Netherlands. https://doi.org/10.1007/978-94-007-4363-2

Ponce JF, Borromei AM, Rabassa J, Martínez O (2011) Late quaternary palaeoenvironmental change in western Staten Island (54.5° S, 64° W), Fuegian Archipelago. Quat Int 233(2):89–100. https://doi.org/10.1016/j.quaint.2010.10.011

R Core Team (2025) R: a language and environment for statistical computing [Software]. R Foundation for Statistical Computing. Available: https://www.R-project.org/

Raguá-Gil JM, Gutt J, Clarke A, Arntz WE (2004) Antarctic shallow-water mega-epibenthos: shaped by circumpolar dispersion or local conditions? Mar Biol 144(5):829–839. https://doi.org/10.1007/s00227-003-1269-3

Rivadeneira PR, Brogger M, Penchaszadeh P (2017) Aboral brooding in the deep water sea star *Ctenodiscus australis* Lütken, 1871 (Asteroidea) from the Southwestern Atlantic. Deep-Sea Res I Oceanogr Res Pap 123:105–109. https://doi.org/10.1016/j.dsr.2017.03.011

Romero SI, Piola AR, Charo M, Garcia CAE (2006) Chlorophyll-a variability off Patagonia based on SeaWiFS data. J Geophys Res Oceans 111(C5):2005JC003244. https://doi.org/10.1029/2005JC003244

Salvat MB (1985) Biología de la reproducción de Anasterias Minuta Perrier (Echinodermata, Asteroidea): Especie incubadora de las costas patagónicas, PhD thesis, Universidad de Buenos Aires, Facultad de Ciencias Exactas y Naturales, Buenos Aires

Schejter L, Rimondino C, Chiesa I, Díaz de Astarloa JM, Doti B, Elías R, Escolar M, Genzano G, López-Gappa J, Tatián M, Zelaya DG, Cristobo J, Perez CD, Cordeiro RT, Bremec CS (2016) Namuncurá marine protected area: an oceanic hot spot of benthic biodiversity at Burdwood Bank, Argentina. Polar Biol 39(12):2373–2386. https://doi.org/10.1007/s00300-016-1913-2

Schejter L, Genzano G, Gaitán E, Perez CD, Bremec CS (2020a) Benthic communities in the Southwest Atlantic Ocean: conservation value of animal forests at the Burdwood Bank slope. Aquat Conserv Mar Freshwat Ecosyst 30(3):426–439. https://doi.org/10.1002/aqc.3265

Schejter L, Genzano G, Pérez CD, Acuña F, Cordeiro RTS, Silva RA, Garese A, Bremec CS (2020b) Checklist of benthic cnidaria in the SW Atlantic Ocean (54°S–56°S). Zootaxa 4878(2):201–239. https://doi.org/10.11646/zootaxa.4878.2.1

Sladen WP (1889) Report on the Asteroidea collected by HMS challenger. Zoology 30:893

Strathmann RR, Strathmann MF (1982) The relationship between adult size and brooding in marine invertebrates. Am Nat 119(1):91–101. https://doi.org/10.1086/283892

Strathmann RR, Strathmann MF, Emson RH (1984) Does limited brood capacity link adult size, brooding, and simultaneous hermaphroditism? A test with the starfish *Asterina phylactica*. Am Nat 123(6):796–818. https://doi.org/10.1086/284240

Tablado A (1982) Asteroideos Argentinos Familia Poraniidae. Comunicaciones del Museo Argentino de Ciencias Naturales "Bernardino Rivadavia". Hidrobiología 2(8):87–106

Verril AE (1914) Monograph of the shallow-water starfishes of the North Pacific Coast from the Arctic Ocean to California (C. Hart Merriam, Ed.). The Smithonian Institution, Washington, DC

Chapter 9
The Marine Molluscan Fauna of Isla de los Estados

Diego G. Zelaya and Marina Güller

Abstract Isla de los Estados, located in the Southwestern Atlantic 24 km east of Tierra del Fuego, is a Provincial Protected Area. Since the XIX century, several scientific expeditions have collected biological samples from the island; however, studies specifically focused on marine molluscs remain limited, with available data scattered across various sources. This study compiles existing information on benthic mollusc species richness from Isla de los Estados and analyzes the composition of the assemblage, along with its faunistic affinities with Patagonia and Antarctica. A total of 166 species are recorded, including 96 species of gastropods, 48 bivalves, 12 polyplacophora, 7 solenogastres, and 3 cephalopods. Most species are known from the intertidal and shallow subtidal zones (<50 m depth), with only 56 species occurring below 100 m. More than 90% of the species are shared with Patagonia and/or the Malvinas Islands, supporting the classification of the island within the Magellan Province. Notably, 57% of the species currently known from Isla de los Estados are represented by only one or two records, highlighting the incomplete state of knowledge regarding the island's molluscan fauna.

Keywords Mollusca · Invertebrate richness · Magellan region · Southwest Atlantic

D. G. Zelaya · M. Güller (✉)
Universidad de Buenos Aires, Facultad de Ciencias Exactas y Naturales, Departamento de Biodiversidad y Biología Experimental (DBBE), Buenos Aires, Argentina

Consejo Nacional de Investigaciones Científicas y Técnicas (CONICET), Buenos Aires, Argentina
e-mail: mguller@bg.fcen.uba.ar

A. Raya Rey, U. Balza (eds.), *Biodiversity and Conservation of Isla de los Estados (Chuani-sin)*, Natural and Social Sciences of Patagonia,
https://doi.org/10.1007/978-3-032-13630-5_9

9.1 Introduction

Isla de los Estados is located in the Austral Southwestern Atlantic 24 km east of Peninsula Mitre, the easternmost tip of Tierra del Fuego, from which it is separated by the Le Maire Strait (Fig. 9.1). It stretches approximately 65 km in a west-to-east direction and features a highly irregular coastline, deeply indented by numerous bays and fjords, resulting in a total shoreline of about 300 km. The island is a Provincial Natural Reserve proclaimed on its constitution in 1991. Later, in 2022, the "Área Natural Protegida Península Mitre" (Natural Protected Area) was created by *Ley Provincial* 1461 including the marine areas surrounding the Isla de los Estados, Isla de Año Nuevo and adjacent islets.

Several human arrivals and visits to Isla de los Estados have contributed to gathering information and provided valuable incidental data on the diversity of animals, and also plants, inhabiting this area. The compilations of mammals' records by Massoia and Chebez (1993) and of birds' records by Chebez and Bertonatti (1994),

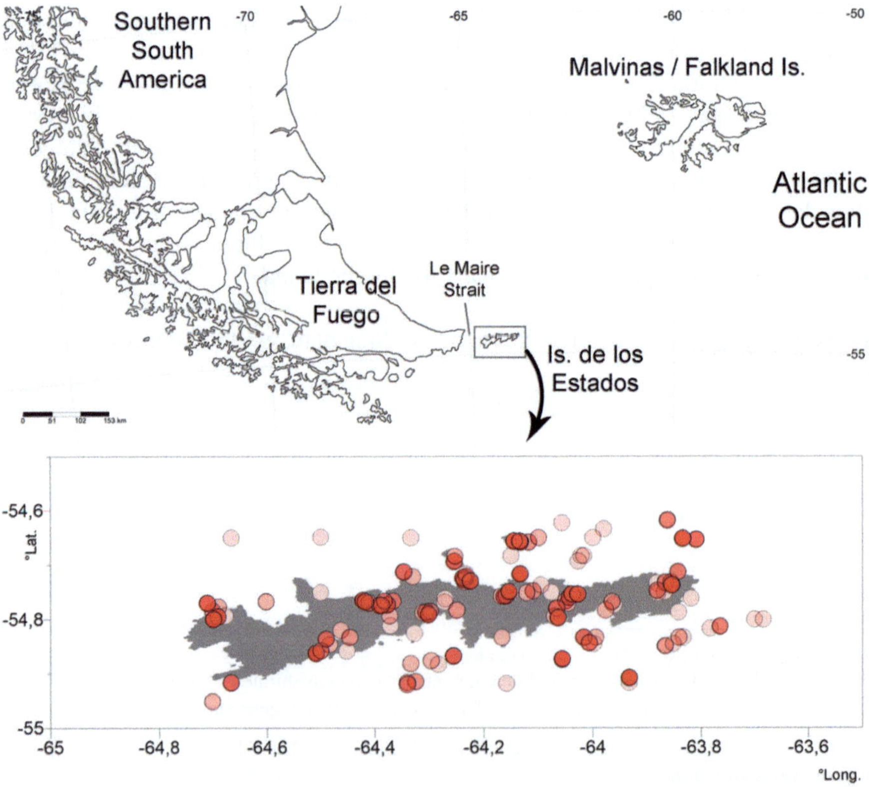

Fig. 9.1 Geographic location of the studied area and sites with records of molluscs. Fill intensity of the circles reflects the number of records in each site, with darker circles representing a higher number of species than lighter circles

are clear examples for this. However, the island invertebrate fauna, particularly marine invertebrates, has received considerably less attention, and the available information is scattered across the literature. This lack of integration hinders a comprehensive understanding of the island's invertebrate biodiversity and, consequently, limits the ability to develop effective management and conservation strategies for the area.

The first scientific reference to benthic molluscs from Isla de los Estados comes from the "Expedición Austral Argentina", an Argentine-Italian expedition carried out between 1881 and 1882, during which Decio Vinciguerra served as zoologist. The published results of the expedition include records of 18 mollusc morphospecies from Isla de los Estados (Vinciguerra 1883, 1884), although 10 of these were only reported to the genus level.

In 1902, the Swedish South Polar Expedition collected samples from two sites on the northern side of the island, one at 36 m depth and the other in the intertidal, from where 10 shelled species, and one nudibranch were reported (Strebel 1908; Odhner 1926).

Between 1932 and 1935, additional material was collected from Isla de los Estados during a series of voyages undertaken by ships of the Argentine Army. During one of these voyages aboard the A.R.A. *San Luis*, Alberto Carcelles conducted "about 80 trawls" across 21 sites, ranging from the intertidal zone to depths of 146 m (Carcelles 1935). His initial report on the collected molluscs focused on 15 large-sized species (Carcelles 1935). Carcelles later produced a series of publications (Carcelles 1942, 1944a, b, 1946a, b, 1947a, b, 1954a, b) providing further details on particular species obtained during these expeditions, along with material collected aboard the A.R.A. *Bahía Blanca* (1941) and A.R.A. *1° de Mayo* (1942). This body of material was later examined in studies by Castellanos (Castellanos 1951, 1980); and more recent contributions (e.g.: Pastorino and Zelaya 2001; Pastorino 2005a, b, 2016; Pastorino and Urteaga 2012; Liuzzi and Zelaya 2013; Güller and Zelaya 2013, 2016, 2022; Güller et al. 2015; Zelaya and Güller 2017, 2023; Urcola and Zelaya 2018, 2024; Pastorino and Griffin 2019), which frequently cited specimens from these historic expeditions. However, the precise provenance of much of this material is uncertain, as the sampling stations from these trips were not georeferenced.

In the early 1970s, the R.V. *Hero* conducted three cruises that intensively sampled Isla de los Estados and its surrounding waters. This remains the most comprehensive sampling effort undertaken in the area to date. The primary data on molluscs from these expeditions originate from Cruise 71-2 (April/May 1971) during which the island was nearly completely circumnavigated (Flint 1971). Additional data were collected during cruise 71-5 (October/November 1971) and cruise 73-2 (May 1973) (Imshaug 1972). Information on the molluscs collected during these cruises is only partially published and dispersed across multiple sources. Castellanos (1980) provided a list of gastropod, bivalve and polyplacophora species collected from 11 intertidal and 6 subtidal stations, just a small fraction of the total stations sampled during the first cruise. Dell (1990) contributed further data on a few gastropods and bivalves species. Additional studies focused on specific groups: Salvini-Plawen

(1978) on Solenogastres, and Sirenko (2006) on Polyplacophora. Several taxonomic publications have included material from these cruises, covering particular gastropods and bivalves families: Fissurellidae (McLean 1984a, b), Eatoniellidae and Rissoidae (Ponder and Worsfold 1994), Muricidae (Pastorino and Harasewych 2000; Pastorino 2005a), Naticidae (Pastorino 2005b), Scissurellidae and Anatomidae (Zelaya and Geiger 2007), Carditidae and Condylocardiidae (Güller and Zelaya 2013, 2022), Siphonariidae (Güller et al. 2015), Lasaeidae (Ituarte and Zelaya 2015), Prosiphonidae (Pastorino 2016), Cyamiidae (Urcola and Zelaya 2018) and Philobryidae (Urcola and Zelaya 2021).

In the 1980s, researchers from the Centro Austral de Investigaciones Científicas (CADIC-CONICET) conducted further collections in the region, leading to the description of two new molluscs species (Vázquez and Caldini 1989, 1990).

Between 2000 and 2003, one of the authors (D.Z.) conducted five field trips to Isla de los Estados. During these trips, D.Z. circumnavigated the island and obtained numerous mollusc samples using a 1 mm mesh net, from the intertidal to depths of 60 m. To date, only a limited number of species collected during these trips have been studied (Simone and Zelaya 2004; Zelaya and Ituarte 2004; Güller and Zelaya 2013, 2016, 2017; Urcola and Zelaya 2018; Zelaya et al. 2020; Zelaya and Güller 2023; Roche et al. 2023).

More recently, Isla de los Estados has served as a stopover and refuge for research cruises which departed from Ushuaia *en route* to Namuncurá-Banco Burdwood Marine Protected Area. These expeditions sampled 19 sites around the island from 2015 to 2017. However, molluscan data from these efforts remains limited, with information published for only a small number of species (Teso et al. 2019; Güller et al. 2020; Zelaya et al. 2020; Urcola and Zelaya 2021, 2024; Güller and Zelaya 2022, 2025; Zelaya and Güller 2023; Bonavita et al., under revision).

This contribution aims to compile the current knowledge on the diversity of benthic molluscs from Isla de los Estados and its adjacent areas. It provides an updated inventory of the known species and analyzes the composition of molluscan assemblages in the region, comparing them with those from other areas of the Argentine Sea.

9.2 Material and Methods

The study area spans from 54.6 to 55°S and 63.7 to 64.8°W. This includes not only Isla de los Estados (*s.s.*), but also its adjacent areas, notably the Islas de Año Nuevo archipelago to the north and several small islets to the south.

This contribution provides the first critical compilation of historical information on molluscs from this region. It also updates the taxonomic identity of previously reported material, based on either modern systematic revisions or the author's direct examination of specimens. This latter effort primarily concerns collections housed in the Museo de La Plata (MLP) [main repository of the material of the R.V. *Hero* reported by Castellanos (1980)] and the Museo Argentino de Ciencias Naturales "Bernardino Rivadavia" (MACN) [repository of the material collected during the trips of de A.R.A. *San Luis*, A.R.A. *Bahía Blanca* and A.R.A. *1° de Mayo* (1942); and a few lots of the R.V. *Hero*]. The Natural History Museum of Los Angeles County (LACM) and the National Museum of Natural History, Smithsonian Institution (USNM) [main repositories for the material collected during the R.V. *Hero* expeditions], were also searched for some particular lots.

9.3 Results

9.3.1 Species of Molluscs Currently Known

The currently available literature comprises 596 published records of molluscs from Isla de los Estados, representing five of the eight extant Classes of the Phylum Mollusca. The majority of these records (358) pertain to the Class Gastropoda, while records of Cephalopoda (4) and Solenogastres (9) remain scarce (Fig. 9.2a). To date, no records exist for the members of the Classes Scaphopoda, Caudofoveata and Monoplacophora in this region.

Based on these data, a total of 166 valid species of molluscs are currently recognized from Isla de los Estados, including 96 species of gastropods, 48 species of bivalves, 12 species of polyplacophora, 7 species of solenogastres, and 3 species of cephalopods (Table 9.1; Fig. 9.2b). Most species are represented by fewer than five records, with 67 known from a single record. Only seven gastropod species, one bivalve species and two polyplacophorans are represented by more than 10 records each (Fig. 9.2d).

9.3.2 *Species Doubtfully Present in the Area*

An additional four species have been cited for the studied area; however, their presence is currently considered doubtful, improbable or erroneous due to misidentifications. These cases include:

Neolepton cobbi (Cooper and Preston, 1910) reported from San Juan de Salvamento by Castellanos (1979). Later, Castellanos (1980) mentioned the same specimens as *N. concentricum* (Preston, 1912). As the original specimens have been lost (see Zelaya and Ituarte 2004), the identification cannot be confirmed, and the presence of *N. cobbi* in Isla de los Estados remains uncertain.

Siphonaria tristentis (sic) Sowerby I, 1823 recorded from the Islas Año Nuevo archipelago by Strebel (1908). According to Güller et al. (2015) the presence of this species in the southern tip of South America is considered improbable.

Calliostoma kophameli Strebel, 1905 cited from Puerto Cook by Castellanos (1980). According to MolluscaBase eds. 2025), the identity of this taxon is uncertain. The material referenced by Castellanos could not be located in the collections of the MLP or MACN to verify the identification.

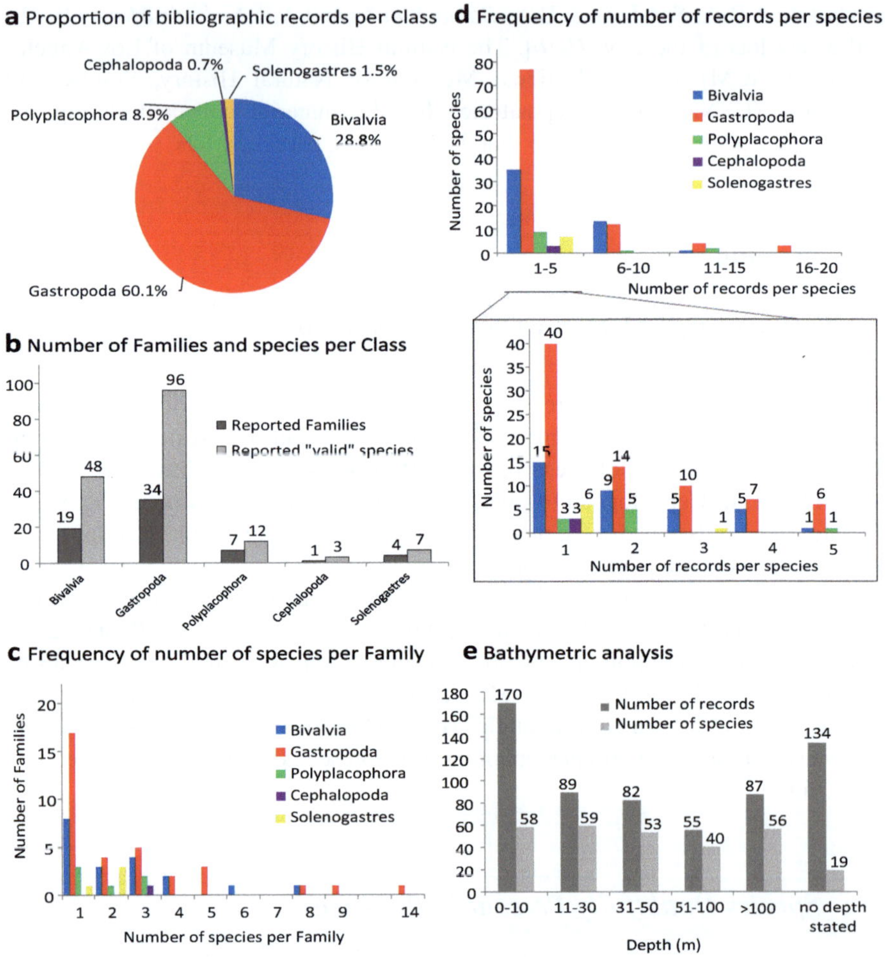

Fig. 9.2 Analysis of the currently available information on Mollusca from Isla de los Estados. (**a**) Bibliographic records per Class, (**b**) number of families and valid species per class, (**c**) number of species per family, (**d**) number of records per species, (**e**) bathymetric distribution of the species

Fissurella maxima Sowerby I, 1834 reported from Isla Zeballos, Puerto Cook and Bahía Capitán Cánepa by Castellanos (1980). As this species is distributed in central and northern Chile (Torres and Galvez Herrera 2022), its presence on Isla de los Estados is improbable.

Fissurellidea hiantula (Lamarck, 1912) reported from Puerto Presidente Roca by Castellanos (1980) is actually a subspecies of the South African species *Dendrofissurella scutellum* (Gmelin, 1791) (McLean 1984a, b). The specimens referred to by Castellanos' (1980) could not be located in the collections of the MLP or the MACN.

Table 9.1 Classes, Families and species of molluscs (arranged alphabetically) currently reported from Isla de los Estados

Family	Species	Source
Class Bivalvia		
Astartidae		
	Astarte longirostra d'Orbigny, 1846	Castellanos (1980), Dell (1990)
Carditidae		
	Carditella naviformis (Reeve, 1843)	Güller and Zelaya (2013)
	Cyclocardia compressa (Reeve, 1843)	Güller and Zelaya (2013)
	Cyclocardia thouarsii (d'Orbigny, 1846)	Güller and Zelaya (2013)
	Cyclocardia velutina (Smith, 1881)	Castellanos (1980) (under *Venericardia*), Güller and Zelaya (2013)
Condylocardiidae		
	Benthocardiella finisterra Güller and Zelaya, 2022	Güller and Zelaya (2022)
	Benthocardiella ituartei Güller and Zelaya, 2022	Güller and Zelaya (2022)
	Carditopsis flabellum (Reeve, 1843)	Güller and Zelaya (2013)
Cyamiidae		
	Cyamiocardium chuanisinense Urcola and Zelaya, 2018	Urcola and Zelaya (2018), Zelaya et al. (2020)
	Cyamiocardium dahli Soot-Ryen, 1957	Urcola and Zelaya (2018)
Gaimardiidae		
	Gaimardia mesembrina (Melvill and Standen, 1907)	Zelaya et al. (2020)
	Gaimardia trapesina (Lamarck, 1819)	Carcelles (1935) (under *Modiola*), Castellanos (1980), Güller et al. (2020)
	Kidderia minuta Dall, 1876	Zelaya et al. (2020)
	Kidderia pusilla (Gould, 1850)	Castellanos (1979, 1980)
Hiatellidae		
	Hiatella barnea (Ihering, 1907)	Zelaya and Güller (2023)
	Hiatella meridionalis (d'Orbigny, 1846)	Zelaya and Güller (2023)

(continued)

Table 9.1 (continued)

Family	Species	Source
	Hiatella umbonata Zelaya and Güller, 2023	Güller et al. (2020) (as *Hiatella* sp.O), Zelaya and Güller (2023)
Lasaeidae		
	Kellia bullata Philippi, 1845	Castellanos (1980)
	Lasaea sp.	Castellanos (1980) (as *L. miliaris* and *L. consanguinea*)
	Malvinasia sp. 2	Ituarte and Zelaya (2015)
	Mysella mabillei (Dall, 1908)	Castellanos (1980), Dell (1990)
	Waldo digitatus Zelaya and Ituarte, 2013[a]	Zelaya and Ituarte (2002) (as *Waldo parasiticus*), Zelaya and Ituarte (2013)
	Waldo paucitentaculatus Zelaya and Ituarte, 2013	Zelaya and Ituarte (2013)
Limidae		
	Acesta patagonica (Dall, 1902)	Carcelles (1947a) (under *Lima*), Castellanos (1980)
	Limea pygmaea (Philippi, 1845)	Carcelles (1947a) (under *Lima (Limatula)*), Castellanos (1980) (under *Limatula*), Dell (1990) (under *Limatula (Antarctolima)*)
Limopsidae		
	Limopsis marionensis Smith, 1885	Castellanos (1980) (as *Limopsis jousseaumei*)
Mactridae		
	Mulinia edulis (King, 1832)	Carcelles (1935), Castellanos (1980)
Mytilidae		
	Aulacomya atra (Molina, 1782)	Vinciguerra (1883, 1884) (as *Mytilus fischerianus* and *M. magellanicus*), Carcelles (1935, 1942) (as *Aulacomya magellanica*), Castellanos (1980) (as *Aulacomya ater*)
	Mytilus chilensis Hupé, 1854	Vinciguerra (1884), Carcelles (1935), Castellanos (1980)
	Perumytilus purpuratus (Lamarck, 1819)	Carcelles (1935) (under *Modiolus*), Castellanos (1980) (under *Brachydontes*)
Neilonellidae		
	Neilonella sulculata (Gould, 1852)	Castellanos (1980) (under *Nuculana*)

(continued)

Table 9.1 (continued)

Family	Species	Source
Neoleptonidae		
	Neolepton concentricum (Preston, 1912)	Zelaya and Ituarte (2004)
	Neolepton hupei Soot-Ryen, 1957	Zelaya et al. (2020)
	Pachykellya fuegiensis Güller and Zelaya, 2022	Güller and Zelaya (2022)
Pandoriidae		
	Pandora braziliensis Sowerby II, 1874	Castellanos (1980) (as *Pandora patagonica*), Güller and Zelaya (2016)
Pectinidae		
	Zygochlamys patagonica (King, 1832)	Carcelles (1935) (under *Pecten*), Castellanos (1971) (as *Chlamys patagonicus*), Castellanos (1980) (as *Chlamys patagonica* and *C. lischkei*)
Pharidae		
	Ensis macha (Molina, 1782)	Carcelles (1935), Castellanos (1980)
Philobryidae		
	Lissarca miliaris (Philippi, 1845)	Dell (1990), Güller and Zelaya (2025)
	Philobrya alata Urcola and Zelaya, 2021	Urcola and Zelaya (2021)
	Philobrya blakeana (Melvill and Standen, 1914)	Urcola and Zelaya (2024)
	Philobrya brattstromi Soot-Ryen, 1957	Urcola and Zelaya (2024)
	Philobrya delli Urcola and Zelaya, 2024	Urcola and Zelaya (2024)
	Philobrya magellanica (Stempell, 1899)	Urcola and Zelaya (2024)
	Philobrya multistriata Lamy, 1908	Urcola and Zelaya (2024)
	Verticipronus cowuti Urcola and Zelaya, 2021	Urcola and Zelaya (2021)
Sareptidae		
	Aequiyoldia sp.[b]	Dell (1990) (as *Aequiyoldia eightsi*)

(continued)

Table 9.1 (continued)

Family	Species	Source
Veneridae		
	Retrotapes exalbidus (Dillwyn, 1817)	Vinciguerra (1883) (under *Chione*), Castellanos (1980) (under *Eurhomalea*)
	Tawera elliptica (Lamarck, 1818)	Castellanos (1980) (as *Tawera gayi*)
Class Cephalopoda		
Enteroctopodidae		
	Enteroctopus megalocyathus (Gould, 1852)	Bonavita et al. under revision
	Muusoctopus eureka (Robson, 1929)	Bonavita et al. under revision
	Muusoctopus n. sp.	Bonavita et al. under revision
Class Gastropoda		
Anatomidae		
	Anatoma conica (d'Orbigny, 1841)	Zelaya and Geiger (2007)
	Anatoma euglypta (Pelseneer, 1903)	Zelaya and Geiger (2007)
Buccinoidea *incertae sedis*		
	Microdeuthria michaelseni (Strebel, 1905)	Castellanos (1980) (under *Pareuthria*)
Calliostomatidae		
	Margarella violacea (King, 1832)	Strebel (1908) (under *Photinula (Margarella)*), Castellanos (1980)
	Photinula coerulescens (King, 1832)	Carcelles (1944b), Castellanos (1980)
	Calliostoma irisans Strebel, 1905	Castellanos (1980)
	Calliostoma moebiusi Strebel, 1905	Castellanos and Fernández (1976a)
Calyptraeidae		
	Crepipatella dilatata (Lamarck, 1822)	Castellanos (1980)
	Trochita pileolus (d'Orbigny, 1841)	Pastorino and Urteaga (2012), Teso et al. (2019), Penchaszadeh et al. (2021)
	Trochita pileus (Lamarck, 1822)	Strebel (1908) (as *Calyptraea costellata*), Castellanos (1980) (as *C. costellata*); Pastorino and Urteaga (2012), Teso and Penchaszadeh (2019), Teso et al. (2019), Penchaszadeh et al. (2021)

(continued)

Table 9.1 (continued)

Family	Species	Source
Cancellaridae		
	Admete magellanica (Strebel, 1905)	Castellanos (1980)
Capulidae		
	Capulus ungaricoides (d'Orbigny, 1841)	Carcelles (1944a) (as *Capulus chilensis*)
Cingulopsidae		
	Skenella hallae Ponder and Worsfold, 1994	Ponder and Worsfold (1994)
Cominellidae		
	Pareuthria atrata (Smith, 1881)	Castellanos and Fernández (1976b) (as *Pareuthria ringei* and *P. rosea*), Castellanos (1980) (as *P. rosea*), Pastorino (2016)
	Pareuthria fuscata (Bruguière, 1789)	Castellanos and Fernández (1976b) (as *Pareuthria magellanica*), Castellanos (1980) (as *P. magellanica* and *P. plumbea*), Pastorino (2016)
	Pareuthria venustula Powell, 1951	Pastorino (2016)
Cymatiidae		
	Argobuccinum pustulosum ([Lightfoot], 1786)	Carcelles (1954a) (as *Argobuccinum argus*)
	Fusitriton magellanicus (Röding, 1798)	Carcelles (1935) (as *Argobuccinum cancellatum*), Carcelles (1954a) (under *Argobuccinum (Fusitriton)*), Castellanos (1980) (under *Argobuccinum*), Teso et al. (2019)
Dorididae		
	Doris fontainii d'Orbigny, 1836	Roche et al. (2023)
Eatoniellidae		
	Eatoniella castanea Ponder and Worsfold, 1994	Ponder and Worsfold (1994)
	Eatoniella cf. *denticula* Ponder and Worsfold, 1994	Ponder and Worsfold (1994)
	Eatoniella denticula Ponder and Worsfold, 1994	Ponder and Worsfold (1994)
	Eatoniella ebenina Ponder and Worsfold, 1994	Ponder and Worsfold (1994)
	Eatoniella glomerosa Ponder and Worsfold, 1994	Ponder and Worsfold (1994)
	Eatoniella nigra (d'Orbigny, 1841)	Ponder and Worsfold (1994)

(continued)

Table 9.1 (continued)

Family	Species	Source
	Eatoniella picea Ponder and Worsfold, 1994	Ponder and Worsfold (1994)
	Eatoniella turricula Ponder and Worsfold, 1994	Ponder and Worsfold (1994)
	Pupatonia magellanica Ponder and Worsfold, 1994	Ponder and Worsfold (1994)
Epitoniidae		
	"Cirsotrema" ctenodentatum Zelaya and Güller, 2017	Zelaya and Güller (2017)
	"Cirsotrema" magellanicum (Philippi, 1845)	Castellanos (1980) (under *Epitonium*)
	"Cirsotrema" strebeli Zelaya and Güller, 2017	Zelaya and Güller (2017)
Eubranchidae		
	Eubranchus fuegiensis Odhner, 1926	Odhner (1926)
Eulimidae		
	Pelseneeria sudamericana Pastorino and Zelaya, 2001	Pastorino and Zelaya (2001)
Fasciolariidae		
	Cavineptunea insularis (Fernández, 1977)	Fernández (1977) (under *Fasciolaria*)
Fissurellidae		
	Fissurella oriens Sowerby I, 1834	Castellanos (1980), McLean (1984a)
	Fissurella picta (Gmelin, 1791)	Carcelles (1935), Castellanos (1980); McLean (1984a)
	Fissurella radiosa Lesson, 1831	Strebel (1908) (as *Fissurella*? *darwini*), McLean (1984a)
	Fissurellidea patagonica (Strebel, 1907)	McLean (1984b)
	Puncturella conica (d'Orbigny, 1841)	Castellanos (1980)
Lepetidae		
	Iothia emarginuloides (Philippi, 1868)	Castellanos (1980) (as *Lepeta coppingeri*), Dell (1990) (as *Iothia coppingeri*)
Littorinidae		
	Laevilitorina cf. *caliginosa* (Gould, 1849)[c]	Castellanos (1980)

(continued)

Table 9.1 (continued)

Family	Species	Source
Lottiidae		
	Scurria ceciliana (d'Orbigny, 1841)	Castellanos (1980) (under *Patelloidea*)
	Scurria scurra (Lesson, 1831)	Castellanos (1980)
Mangelidae		
	Typhlodaphne strebeli Powell, 1951[d]	Carcelles (1944b) (as *Mangelia purissima*); Castellanos (1980) (as *M. purissima*)
Marginellidae		
	Volvarina warrenii (Marrat, 1876)	Carcelles (1946a) (as *Volvarina patagonica*), Carcelles and Williamson (1951) (as *V. patagonica*), Teso et al. (2019)
Mitridae		
	Paradmete crymochara (Rochebrune and Mabille, 1885)	Carcelles (1947b) (under *Mitra*)
Muricidae		
	Acanthina monodon (Pallas, 1774)	Carcelles (1935) (as *Acanthina calcar*), Carcelles (1954b) (as *Nucella (Acanthina) calcar*), Castellanos (1980)
	Trophon clenchi (Carcelles, 1953)	Pastorino (2005a)
	Trophon geversianus (Pallas, 1774)	Carcelles (1935, 1946b), Castellanos (1980), Pastorino (2005a)
	Trophon ohlini Strebel, 1904	Teso et al. (2019)
	Trophon parodizi Pastorino, 2005	Teso et al. (2019)
	Trophon plicatus ([Lightfoot], 1786)	Vinciguerra (1883) (as *Fusus (Trophon) laciniatus*), Carcelles (1946b) (as *T. laciniatus*), Pastorino (2005a)
	Xymenopsis buccineus (Lamarck, 1816)	Pastorino and Harasewych (2000)
	Xymenopsis muriciformis (King, 1832)	Castellanos (1980) (as *Xymenopsis. liratus* and *X. decolor*), Pastorino and Harasewych (2000)

(continued)

Table 9.1 (continued)

Family	Species	Source
Nacellidae		
	Nacella deaurata (Gmelin, 1791)	Strebel (1908) (as *Patinella delicatissima*), Castellanos (1980) (under *Patinigera*)
	Nacella magellanica (Gmelin, 1791)	Carcelles (1935) (under *Patinella*), Castellanos (1980) (as *Patinigera magellanica* and *P. tincta*)
	Nacella mytilina (Helbling, 1779)	Vinciguerra (1883) (under *Patella*), Carcelles (1947b), Castellanos (1980)
Naticidae		
	Euspira patagonica (Philippi, 1845)	Strebel (1908) (as *Natica soluta*), Castellanos (1980) (under *Polynices*), Dell (1990) (as *Falsilunatia soluta* in part), Pastorino (2005b)
	Falsilunatia carcellesi (Dell, 1990)	Pastorino (2005b)
	Falsilunatia soluta (Gould, 1847)	Dell (1990) (in part), Pastorino (2005b)
	Polinices cf. *uber* (Valenciennes, 1832)	Pastorino (2005b) (as *Polinices* sp., cf. *P. uber*)
	Tectonatica impervia (Philippi, 1845)	Strebel (1908) (under *Natica*), Castellanos (1980), Dell (1990), Pastorino (2005b)
Newtoniellidae		
	Eumetula (?) *pulla* (Philippi, 1845)	Carcelles (1944b) (under *Cerithiopsis*), Castellanos (1980) (under *Ataxocerithium*)
Onchididae		
	Onchidella marginata (Couthouy, 1852)	Castellanos (1980) (under *Oncidium*)
Onchidorididae		
	Acanthodoris falklandica Eliot, 1907	Roche et al. (2023)
Orbitestellidae		
	Patagorbitestella ponderi (Linse, 2002)	Simone and Zelaya (2004) (under *Orbistestella*)
Prosiphonidac		
	Antistreptus magellanicus Dall, 1902	Carcelles (1944b), Pastorino and Griffin (2019)
	Argeneuthria cerealis (Rochebrune and Mabille, 1885)	Pastorino (2016)

(continued)

Table 9.1 (continued)

Family	Species	Source
	Argeneuthria paessleri (Strebel, 1905)	Castellanos and Fernández (1976b)
	Falsimacme kobelti (Strebel, 1905)	Pastorino (2016)
Rissoidae		
	Haurakia averni (Ponder and Worsfold, 1994)	Ponder and Worsfold (1994) (under *Pusillina (Haurakia)*), Güller and Zelaya (2017)
	Haurakia cf. *averni* (Ponder and Worsfold, 1994)	Ponder and Worsfold (1994) (under *Pusillina (Haurakia)*)
	Onoba algida Ponder and Worsfold, 1994	Ponder and Worsfold (1994) (under *Onoba (?)*)
	Onoba cf. *algida* Ponder and Worsfold, 1994	Güller and Zelaya (2017)
	Onoba erugata Ponder and Worsfold, 1994	Ponder and Worsfold (1994)
	Onoba fuegoensis (Strebel, 1908)	Strebel (1908) (under *Rissoia (?Cingula)*), Ponder and Worsfold (1994), Güller and Zelaya (2017)
	Onoba georgiana (Pfeffer, 1886)	Ponder and Worsfold (1994)
	Onoba lacuniformis Ponder and Worsfold, 1994	Ponder and Worsfold (1994)
	Onoba protofimbriata Ponder and Worsfold, 1994	Ponder and Worsfold (1994)
	Onoba schythei (Philippi, 1868)	Ponder and Worsfold (1994), Güller and Zelaya (2017)
	Onoba striola Ponder and Worsfold, 1994	Ponder and Worsfold (1994)
	Onoba subincisa Ponder and Worsfold, 1994	Ponder and Worsfold (1994)
	Onoba sulcula Ponder and Worsfold, 1994	Ponder and Worsfold (1994)
	Powellisetia microlirata Ponder and Worsfold, 1994	Ponder and Worsfold (1994)
Scissurellidae		
	Scissurella clathrata Strebel, 1908	Strebel (1908), Carcelles and Williamson (1951) (under *Schizotrochus*), Zelaya and Geiger (2007)
	Scissurella petermannensis Lamy, 1910	Zelaya and Geiger (2007)

(continued)

Table 9.1 (continued)

Family	Species	Source
Siphonariidae		
	Siphonaria fuegiensis Güller, Zelaya and Ituarte, 2015	Güller et al. (2015)
	Siphonaria lateralis Gould, 1846	Castellanos (1980) (under *Kerguelenella*), Güller et al. (2015)
	Siphonaria lessonii Blainville, 1827	Carcelles (1935), Güller et al. (2015)
Skeneidae		
	Liotella crassicostata (Strebel, 1908)	Strebel (1908) (under *Cyclostrema*), Carcelles and Williamson (1951) (under *Brookula*)
Volutidae		
	Adelomelon ancilla ([Lightfoot], 1786)	Castellanos (1980)
	Argentovoluta bottai (Vázquez and Caldini, 1989)	Vázquez and Caldini (1989) (under *Miomelon*)
	Odontocymbiola magellanica (Gmelin, 1791)	Vinciguerra (1883, 1884) (under *Voluta*), Carcelles (1935) (under *Voluta*), Castellanos (1980), Teso et al. (2019)
	Odontocymbiola pescalia Clench and Turner, 1964	Vazquez and Caldini (1990) (as *Odontocymbiola rucciana* n. sp.)
	Pachycymbiola ferussacii (Donovan, 1824)	Vinciguerra (1883) (under *Voluta*)
Class Polyplacophora		
Callochitonidae		
	Callochiton puniceus (Gould, 1846)	Sirenko (2006), Teso et al. (2019)
Chaetopleuridae		
	Chaetopleura brucei Iredale, 1912[e]	Castellanos (1980)
Chitonidae		
	Tonicia calbucensis Plate, 1898	Sirenko (2006)
	Tonicia lebruni Rochebrune, 1884	Castellanos (1948), Castellanos (1951, 1956) (as *T. lebruni* and *T. rubridens*), Castellanos (1980), Carcelles and Williamson (1951) (as *T. rubridens*), Sirenko (2006)
Hemiarthridae		
	Hemiarthrum setulosum Carpenter, 1876	Castellanos (1980), Sirenko (2006)

(continued)

Table 9.1 (continued)

Family	Species	Source
Ischnochitonidae		
	Ischnochiton stramineus (Sowerby I, 1832)	Castellanos (1951, 1980) (as *Ischnochiton imitator*); Sirenko (2006); Liuzzi and Zelaya (2013)
Leptochitonidae		
	Leptochiton kerguelensis Haddon, 1886	Sirenko (2006)
	Leptochiton medinae (Plate, 1899)	Sirenko (2006), Teso et al. (2019)
	Stenosemus exaratus (Sars, 1878)[f]	Castellanos (1951, 1956) (as *Ischnochiton dorsuosus*); Sirenko (2006) (as *?Stenosemus dorsuosus*)
Mopaliidae		
	Nuttallochiton hyadesi (Rochebrune, 1889)	Sirenko (2006), Teso et al. (2019)
	Nuttallochiton martiali (Rochebrune, 1889)	Castellanos (1980), Sirenko (2006)
	Plaxiphora aurata (Spalowsky, 1795)	Carcelles (1935) (as *Plaxiphora setigera*); Castellanos (1951, 1980), Sirenko (2006)
Class Solenogastres		
Dondersiidae		
	Nematomenia ptyalosa Salvini-Plawen, 1978	Salvini-Plawen (1978)
	Nematomenia tegulata Salvini-Plawen, 1978	Salvini-Plawen (1978)
Phyllomeniidae		
	Lituiherpia spermathecata Salvini-Plawen, 1978	Salvini-Plawen (1978)
	Phyllomenia cornuadentata Salvini-Plawen, 1978	Salvini-Plawen (1978)
Pruvotinidae		
	Pruvotina megathecata Salvini-Plawen, 1978	Salvini-Plawen (1978)
	Pruvotina peniculata Salvini-Plawen, 1978	Salvini-Plawen (1978)
Solenogastres *incertae sedis*		
	Pholidoherpia lepidota Salvini-Plawen, 1978	Salvini-Plawen (1978) (under *Pholidoherpia (?)*)

(continued)

Table 9.1 (continued)

[a]The material reported as *Waldo parasiticus* by Zelaya and Ituarte (2002) was re-identified as *W. digitatus* by Zelaya and Ituarte (2013, Rosenfeld et al., 2023)

[b]There is no certainty about which species of the complex "*A. eightsii*" (see González-Wevar et al. 2019) the material found in Isla de los Estados belongs to

[c]There is no certainty about which species of the complex "*L. caliginosa*" (see Rosenfeld et al., 2023) the material found in Isla de los Estados belongs to

[d]*Typhlodaphne purissima* is known from Shag Rocks and South Georgia (Kantor et al. 2016). Zelaya (2005a) considered the reports of this species from off Argentina and Tierra del Fuego to be referable to *T. strebeli* Powell, 1951

[e]*Chaetopleura brucei* was reported from San Juan del Salvamento by Castellanos (1980). According to Sirenko (2006), this is "a very enigmatic species described from South Orkney Islands (in water that is too cold for the genus *Chaetopleura*)"; the identity of this material deserves revision

[f]Castellanos (1951, 1956) reported *Ischnochiton dorsuosus* Haddon, 1886 from Puerto Basil Hall. However, the material upon which Castellanos' records were based, was re-identified as *Stenosemus exaratus* by Liuzzi (2005)

Odostomia transluscens was erroneously reported by Castellanos and Landoni (1984) and Castellanos (1989) as collected from Isla de los Estados at "54°20′S 64°28′W". However, the actual locality is 54°20′S 65°28′W (Di Luca et al. 2021).

9.3.3 The Molluscan Assemblage of Isla de los Estados

The molluscan fauna of Isla de los Estados comprises 34 families of gastropods, 19 families of bivalves, seven families of polyplacophorans, one family of cephalopods and four families of solenogastres (Fig. 9.2b).

Among bivalves, the highest species richness is observed in the families Philobryidae and Lasaeidae, represented by eight and six species, respectively. In terms of the numbers of records, the most represented the families are Carditidae, Condylocardiidae and Mytilidae, with *Carditopsis flabellum* and *Carditella naviformis* being the most frequently recorded species (12 and 10 records, respectively). However, most bivalve families (58%) are represented by only one or two species each, and half of the bivalve species are known from just one or two records (Fig. 9.2c, d).

Among gastropods, the families Rissoidae, Eatoniellidae and Muricidae are the most species-rich, comprising 14, nine and eight species, respectively, and also account for the highest number of records (55, 66 and 47, respectively). 19 gastropod species are represented by more than five records each; however, as with bivalves, most species are known from only one or two records (Fig. 9.2c, d).

In polyplacophora, most records correspond to the families Mopaliidae and Chitonidae (17 and 16, respectively), with*Tonicia lebruni* and *Plaxiphora aurata* being the most recorded species, with 15 and 13 records, respectively. Still, eight of the 12 reported species are only known from one or two records (Fig. 9.2d).

The three species of benthic cephalopods reported from the area belong to a single Family, Enteroctopodidae, and each is represented by a single record.

The seven known species of Solenogastres in the study area belong to four families. All are represented by a single record, except for *Pruvotina megathecata* which has been reported from three localities.

Sampling in the northern side of Isla de los Estados has been more intensive than in the southern side, with 1.4 times more stations (85 vs. 60), which have yielded 2.2 times more records (367 vs. 164), and 1.3 more species (123 vs. 94 species).

From a bathymetric perspective, 54.2% of all published molluscs records come from the intertidal and shallow subtidal (less than 50 m depth), encompassing 104 species. In contrast, only 14.6% of the records originate from depths greater than 100 m, though these account for 56 species, representing 38.1% of the total fauna (Fig. 9.2e). About two thirds of the deep-water records come from sites located in the southern slope of Isla de los Estados.

At present, one bivalve species (*Benthocardiella finisterra*), one solenogastres species (*Pholidoherpia lepidota*) and six gastropod species (*Eatoniella castanea*, *Onoba algida*, *Onoba striola*, *Pelseneeria sudamericana*, *Cavineptunea insularis*, and *Argentovoluta bottai*) are known exclusively from Isla de los Estados.

9.4 Discussion

To date, exhaustive studies on the molluscan fauna of Isla de los Estados have not been conducted. This lack of comprehensive research is not due to the scarcity of samples from this area, but rather to the absence of researchers specifically dedicated to study the material collected during the various expeditions to the island. Most of the available information originates from occasional records, isolated publications or taxonomic studies on particular groups. The incomplete knowledge of the molluscan fauna of Isla de los Estados is evident in the high number of species known from only one published record, despite the fact that many of these "infrequent" species are actually common in neighboring regions. Moreover, a significant portion of the existing data lacks precise geographic (44.6% without latitude and longitude) and/or bathymetric information (22.5% without depth). The absence of georeferenced data is particularly notable for all the samples collected by Alberto Carcelles aboard the A.R.A. *San Luis*, while the lack of depth data affects records for 19 species. Even among the species for which bathymetric information is available, most records are from the intertidal zone or shallow subtidal areas (up to 50 m depth), with relatively few exceeding 100 m depth. Interestingly, 30 out of the 56 species recorded at depths greater than 100 m have not been reported from shallower waters. Despite these limitations, the compilation data presented here reveals a relatively high molluscan species richness for Isla de los Estados, with 166 species currently documented. Notably, the 48 bivalve species reported from the island represent nearly one-third of the bivalve diversity recorded for the Argentine Sea by Zelaya (2015: 169 species), and half of the species reported by Zelaya (2005b) for the Patagonian coasts. This is particularly notorious given the small size of this archipelago compared to those broader regions, suggesting significant local diversification and molluscan abundance. Similarly, the species richness of Isla de los Estados represents ca. 80% of the gastropod diversity reported by Zelaya (2006: 123 spp.), 85.7% of the diversity of polyplacophorans reported by Liuzzi (2005: 14 spp.), and 27.2% the diversity of benthic cephalopods found by Bonavita et al. (under revision) in the Atlantic coast of Patagonia.

Eight species of molluscs are currently known only from Isla de los Estados. Out of them, six species (*Benthocardiella finisterra*, *Eatoniella castanea*, *Pelseneeria sudamericana*, *Onoba algida*, *Onoba striola*, and *Pholidoherpia lepidota*) are micromolluscs described in recent years. These facts, along with the still limited knowledge of micromolluscs from the adjacent Argentine shelf, suggest that this apparent "endemism" may be an artifact of under-sampling or incomplete taxonomic assessment. In contrast, more than 90% of the bivalve species recorded from Isla de los Estados are shared with Patagonia and/or Islas Malvinas Islands. Only one species, *Limopsis marionensis*, is currently considered to be shared with Antarctic waters; though its taxonomic identity deserves revision. Similarly, the gastropod, polyplacophora, solenogastres and benthic cephalopod fauna of Isla de los Estados show strong affinities with Patagonia and/or Malvinas Islands, but exhibit very low similarity with Antarctic assemblages. Only two gastropods species (*Scissurella petermannensis* and *Anatoma euglypta*) and three Solenogastres species (*Nematomenia ptyalosa*, *N. tegulata*, and *Pruvotina peniculata*) are shared with Antarctic waters. The currently available evidence supports the inclusion of Isla de los Estados within the Magellan Region, for both shallow and deep waters. Nonetheless, it is noteworthy that some Antarctic elements appear at depths greater than 135–137 m, even though data on deep-water molluscs from the island remain scarce, with only 35 species recorded from depths exceeding 200 m.

Competing Interests This study was partially funded by a grant from Agencia I + D + i (PICTO Malvinas 2021–00017) to DZ, and a grant from the Universidad de Buenos Aires (UBACyT 20020220400045BA) to MG.

The authors are grateful to Gabriela Liuzzi and Mariela Romanelli (MACN), Diego Gutiérrez Gregoric (MLP), Lindsey Groves (LACM) and Ellen Strong (USNM) for providing access to and/or information on the collections under their care.

References

Carcelles AR (1935) Sobre algunos invertebrados marinos de la Isla de los Estados. Boletín del Centro Naval 13:1–16

Carcelles AR (1942) Nota sobre el mejillón "*Aulacomya ater*" (Chemnitz). Physis 19:180–190 + 2 pls

Carcelles AR (1944a) Las especies del género *Capulus* Montfort 1810, en aguas argentinas. Com Zool Mus Montevideo 1:1–5

Carcelles AR (1944b) Nota sobre algunos moluscos magallánicos obtenidos frente al Río de la Plata. Com Zool Mus Montevideo 1:1–11 + pl. 1

Carcelles AR (1946a) Notas sobre dos especies argentinas de Marginellidae. Notas del Mus La Plata, zool 11:51–57

Carcelles AR (1946b) Observaciones sobre algunas especies actuales y fósiles de *Trophon* de la República Argentina. Notas del Mus La Plata, zool 11:59–89

Carcelles AR (1947a) Notas sobre algunos bivalvos argentinos. Com Zool Mus Montevideo 2:1–10 + pl. 1

Carcelles AR (1947b) Notas sobre algunos gastrópodos marinos del Uruguay y la Argentina, I-IV. Com Zool Mus Montevideo 2:1–27 + pls. 1–7

Carcelles AR (1954a) Especies sudamericanas de *Argobuccinum* Brugière 1792. Com Mus Argent Cienc Nat Bernardino Rivadavia, zool 2:243–254

Carcelles AR (1954b) Observaciones sobre los géneros *Nucella*, *Chorus* y *Concholepas*. Com Mus Argent Cienc Nat Bernardino Rivadavia, zool 2:255–273 + pls. 1–5

Carcelles AR, Williamson SI (1951) Catálogo de los moluscos marinos de la Provincia Magallánica. Rev Inst Nac Invest Cienc Nat, Ciencias Zoológicas 2:225–383

Castellanos ZJA de (1979) Micromoluscos poco conocidos del sur argentino–chileno. Neotropica 25:133–140

Castellanos ZJA de (1980) Moluscos litorales de Isla de los Estados. Obra del Centenario del Mus de La Plata 6:53–73

Chebez JC, Bertonatti CC (1994) La avifauna de la Isla de los Estados, Islas de Año Nuevo y mar circundante (Tierra del Fuego, Argentina). L.O.L.A., Buenos Aires

Castellanos ZJA de (1948) Sobre seis especies de poliplacóforos argentinos. Notas del Mus de La Plata, zool 13:192–202

Castellanos ZJA de (1951) Poliplacóforos del Museo Argentino de Ciencias Naturales. Com Inst Nac Invest Cienc Nat anexo al Mus Argent Cienc Nat Bernardino Rivadavia 1:1–31

Castellanos ZJA de (1956) Catálogo de los poliplacóforos argentinos y de aguas vecinas al Estrecho de Magallanes. Rev Mus La Plata (n ser). S Zool 6:465–486

Castellanos ZJA de (1971) Los *Chlamys* más comunes del Mar Argentino. Neotropica 17:55–66

Castellanos ZJA de, Fernández DE (1976a) Los géneros *Calliostoma* y *Neocalliostoma* del Mar Argentino con especial referencia al área subantártica. Rev Mus La Plata (n ser), secc Zool 12:135–156

Castellanos ZJA de, Landoni NA (1984) Nueva contribución al conocimiento de los micromoluscos de la plataforma continental Argentina. Rev Mus La Plata (n ser), Secc Zool 13: 291–299.

Castellanos ZJA de (1989) Novedades sobre micromoluscos subantárticos (Mollusca, Gastropoda). Neotropica 36:89–92

Castellanos ZJA de, Fernández DE (1976b) Las especies del género *Paraeuthria* Strebel, 1905 del Mar Argentino (Mollusca Buccinulidae). Neotropica 22:9–14

Dell RK (1990) Antarctic Mollusca with special reference to the fauna of the Ross Sea. Bull R Soc NZ 27:1–311

Di Luca J, Güller M, Zelaya DG (2021) Pyramidellidae (Gastropoda: Heterobranchia) from the end of the world. Malacologia 63:225–242

Fernández D (1977) Nueva especie de *Fasciolaria* de Argentina (Moll. Gastr.). Neotropica 23:27–29

Flint OS Jr (1971) R/V Hero Cruise 71-2 to Isla de los Estados. Antarct J US 6:263–265

González-Wevar CA, Gérard K, Rosenfeld S et al (2019) Cryptic speciation in Southern Ocean *Aequiyoldia eightsii* (Jay, 1839): Mio-Pliocene trans-Drake Passage separation and diversification. Prog Oceanogr 174:44–54. https://doi.org/10.1016/j.pocean.2018.09.004

Güller M, Zelaya DG (2013) The families Carditidae and Condylocardiidae in the Magellan and Perú-Chile provinces (Bivalvia: Carditoidea). Zootaxa 3682:201–239. https://doi.org/10.11646/zootaxa.3682.2.1

Güller M, Zelaya DG (2016) Unravelling the identity of *Pandora* species (Bivalvia: Pandoridae) from southern South America. J Molluscan Stud 82:440–448. https://doi.org/10.1093/mollus/eyw008

Güller M, Zelaya DG (2017) New insights into the diversity of rissoids from sub-antarctic and antarctic waters (Gastropoda: Rissooidea). Polar Biol 40:1923–1937. https://doi.org/10.1007/s00300-017-2108-1

Güller M, Zelaya DG (2022) The smallest marine bivalves from the end of the world (Tierra del Fuego, Isla de los Estados and Burdwood Bank). Polar Biol 45:777–787. https://doi.org/10.1007/s00300-022-03018-3

Güller M, Zelaya DG (2025) "*Lissarca miliaris*" (Bivalvia: Philobryidae): a species able to cross the Antarctic convergence or a species complex? Polar Biol 48:18. https://doi.org/10.1007/s00300-024-03337-7

Güller M, Zelaya DG, Ituarte C (2015) How many *Siphonaria* species (Gastropoda: Euthyneura) live in southern South America? J Molluscan Stud 82:80–96. https://doi.org/10.1093/mollus/eyv036

Güller M, Puccinelli E, Zelaya DG (2020) The Antarctic circumpolar current as a dispersive agent in the Southern Ocean: evidence from bivalves. Mar Biol 167:143. https://doi.org/10.1007/s00227-020-03746-2

Imshaug HA (1972) R/V Hero cruise 71-5 to Isla de los Estados. Antarct J US 7:42–44

Ituarte CF, Zelaya DG (2015) A redescription of *Malvinasia* Cooper & Preston, 1910 (Bivalvia: Galeommatoidea), with the description of a new species. J Molluscan Stud 81:455–465. https://doi.org/10.1093/mollus/eyv027

Kantor YI, Harasewych MG, Puillandre N (2016) A critical review of Antarctic Conoidea (Neogastropoda). Moll Res 36:153–206. https://doi.org/10.1080/13235818.2015.1128523

Liuzzi MG (2005) Biogeografía y sistemática de los poliplacóforos de Tierra del Fuego e Islas del Arco de Scotia. Universidad de Buenos Aires, Tesis de Licenciatura

Liuzzi MG, Zelaya DG (2013) Egg-hull ultrastructure of *Ischnochiton stramineus* (Sowerby, 1832), a south American brooding chiton (Chitonina: Ischnochitonidae). J Molluscan Stud 79:372–377. https://doi.org/10.1093/mollus/eyt029

Massoia E, Chebez JC (1993) Mamíferos silvestres del archipiélago fueguino. L.O.L.A., Buenos Aires

McLean JH (1984a) Systematics of *Fissurella* in the Peruvian and Magellanic faunal provinces (Gastropoda: Prosobranchia). Contrib Sci (Los Angel Calif) 354:1–70

McLean JH (1984b) Shell reduction and loss in fissurellids: a review of genera and species in the *Fissurellidea* group. Am Malacol Bul 2:21–34

MolluscaBase (2025) Accessed through: World Register of Marine Species at: https://www.marinespecies.org/

Odhner NH (1926) Die Opistobranchien. Further Zoological Results of the Swedish Antarctic Expedition 1901–1903, 2:1–100, pls 1–3

Pastorino G (2005a) A revision of the genus *Trophon* Montfort, 1810 (Gastropoda: Muricidae) from southern South America. The Nautilus 119:55–82

Pastorino G (2005b) Recent Naticidae (Mollusca: Gastropoda) from the Patagonian coast. The Veliger 47:225–258

Pastorino G (2016) Revision of the genera *Pareuthria* Strebel, 1905, *Glypteuthria* Strebel, 1905 and *Meteuthria* Thiele, 1912 (Gastropoda: Buccinulidae) with the description of three new genera and two new species from southwestern Atlantic waters. Zootaxa 4179:301–344

Pastorino G, Griffin M (2019) Gastropods of the genus *Antistreptus* as examples of persistent molluscan lineages in the Neogene of the southwestern Atlantic. J Paleontol 93:916–924. https://doi.org/10.1017/jpa.2019.8

Pastorino RSG, Harasewych MG (2000) A revision of the patagonian genus *Xymenopsis*. The Nautilus 114:38–58

Pastorino G, Urteaga D (2012) A taxonomic revision of the genus *Trochita* Schumacher, 1817 (Gastropoda: Calyptraeidae) from the southwestern Atlantic. The Nautilus 126:68–78. https://www.biodiversitylibrary.org/page/50437262

Pastorino G, Zelaya D (2001) New species of the eulimid genus *Pelseneeria* (Mollusca: Gastropoda) from Staten Island, Argentina. The Veliger 44:310–314

Penchaszadeh P, Arrighetti F, Aldea C et al (2021) Reproductive pattern of *Trochita pileolus* (d'Orbigny, 1841) (Gastropoda: Calyptraeidae) and different types of intracapsular development in *Trochita* species of South America. Polar Biol 45:59–69. https://doi.org/10.1007/s00300-021-02972-8

Ponder WF, Worsfold TM (1994) A review of the rissoiform gastropods of southwestern South America (Mollusca: Gastropoda). Contrib Sci (Los Angel Calif) 445:1–63. https://doi.org/10.5962/p.208081

Powell AWB (1951) Antarctic and Subantarctic Mollusca: Pelecypoda and Gastropoda. Discov Rep 26:47–196, pls 5–10. http://biodiversitylibrary.org/page/5632211

Roche A, Cetra N, Núñez V, Darrigran G (2023) Catálogo de las especies de Nudibranchia (Gastropoda: Heterobranchia) de la Colección de Moluscos del Museo de La Plata, Argentina. Acta Zoológica Lilloana 67:159–175. https://doi.org/10.30550/j.azl/2023.67.1/2023-04-13

Rosenfeld S, Segovia NI, Maturana CS, Aldea C, Saucède T, Brickle P, Spencer HG, Poulin E, González-Wevar CA (2023) A revision of the higher latitude periwinkle species *Laevilitorina caliginosa* sensu lato. Zool Jour Lin Soc 202:zlad171. https://doi.org/10.1093/zoolinnean/zlad171

Salvini-Plawen L von (1978) Antarktische und subantarktische Solenogastres (eine Monographie: 1898–1974). Zoologica(Stuttgart) 128:1–305

Simone LRL, Zelaya DG (2004) A new *Orbitestella* (Gastropoda: Heterobranchia: Orbitestellidae) from Tierra del Fuego, Argentina. The Nautilus 118:160–166

Sirenko BI (2006) Report on the present state of our knowledge with regard to the chitons (Mollusca: Polyplacophora) of the Magellan Strait and Falkland Islands. Venus 65:81–89

Strebel H (1908) Die Gastropoden. In: Wissenschaftliche Ergebnisse der Schwedischen Südpolar-Expedition 1901–1903 unter Leitung von Dr. Otto Nordenskjöld, vol 6. Lithographisches Institut des Generalstabs, Stockholm, pp 1–111, 6 pls. https://www.biodiversitylibrary.org/page/6227397

Teso V, Penchaszadeh P (2019) Development of the gastropod *Trochita pileus* (Calyptraeidae) in the sub-Antarctic southwestern Atlantic. Polar Biol 42:171–178. https://doi.org/10.1007/s00300-018-2412-4

Teso V, Urteaga D, Pastorino G (2019) Assemblages of certain benthic molluscs along the southwestern Atlantic: from subtidal to deep sea. BMC Ecol 19:49. https://doi.org/10.1186/s12898-019-0263-7

Torres FI, Galvez Herrera O (2022) La familia Fissurellidae (Mollusca, Gastropoda) en las costas de Chile: catálogo descriptivo y revisión taxonómica. Publicación Ocasional del Museo Nacional de Historia Natural, Chile 75:5–88. https://publicaciones.mnhn.gob.cl/668/w3-propertyname-1354.html

Urcola MR, Zelaya DG (2018) The genus *Cyamiocardium* Soot-Ryen, 1951 (Bivalvia: Cyamiidae) in sub-Antarctic and Antarctic waters. Polar Biol 41:1157–1174. https://doi.org/10.1007/s00300-018-2275-8

Urcola MR, Zelaya DG (2021) New species of Philobryidae (Bivalvia: Pteriomorphia) from Patagonian and Antarctic waters. Zool Anz 291:23–33. https://doi.org/10.1016/j.jcz.2021.02.001

Urcola MR, Zelaya DG (2024) The genus *Philobrya* J.G. Cooper, 1867 (Bivalvia: Philobryidae) in Patagonia and adjacent Antarctic waters. Zootaxa 5437:151–192. https://doi.org/10.11646/ZOOTAXA.5437.2.1

Vázquez JA, Caldini CH (1989) Nuevo subgénero y nueva especie de Volutidae (moll. Gastropoda) para aguas subantárticas de la Argentina. Com Mus Argent Cienc Nat Bernardino Rivadavia, zool 2:169–174

Vazquez JA, Caldini CH (1990) *Odontocymbiola rucciana* sp. nov. (Moll. Gastropoda: Volutidae) del Canal Beagle, nuevos aportes a diagnosis de subfamilia Odontocymbiolinae Clench y Turner, 1964 y nuevas observaciones sobre *Odontocymbiola pescalia* Clench y Turner, 1964. Com Mus Argent Cienc Nat Bernardino Rivadavia, hidrobiología 2:176–192

Vinciguerra D (1883) Apuntes zoológicos sobre la Isla de los Estados. In: Bove G (ed) Expedición Austral Argentina. Informes Preliminares. Instituto Geográfico Argentino, Buenos Aires, pp 55–68

Vinciguerra D (1884) Sulla fauna dell'America australe. BSGI 1884:785–811

Zelaya DG (2005a) Systematics and zoogeography of marine gastropod molluscs of South Georgia. Spixiana 28:109–139

Zelaya DG (2005b) The bivalves from the scotia arc islands: species richness and faunistic affinities. Sci Mar 69(Suppl 2):113–122

Zelaya DG, Geiger DL (2007) Species of Scissurellidae and Anatomidae from sub-Antarctic and Antarctic waters (Gastropoda: Vetigastropoda). Malacologia 49:393–443

Zelaya DG, Güller M (2017) Undercover speciation of wentletraps (Caenogastropoda: Epitoniidae) in the southwestern Atlantic. Zootaxa 4286:41–69. https://doi.org/10.11646/ZOOTAXA.4286.1.2

Zelaya DG, Güller M (2023) Decrypting the feared genus *Hiatella* (Bivalvia): south American species. Zool J Linnean Soc 199:882–905. https://doi.org/10.1093/zoolinnean/zlad064

Zelaya DG, Ituarte C (2002) The identity of *Waldo parasiticus* (Dall, 1876) and description of *Waldo trapezialis* new species (Bivalvia Galeommatoidea). The Nautilus 116:109–117. https://www.biodiversitylibrary.org/part/95155

Zelaya DG, Ituarte C (2004) The genus *Neolepton* Monterosato, 1875 in southern South America (Bivalvia: Neoleptonidae). J Molluscan Stud 70:123–137. https://doi.org/10.1093/mollus/70.2.123

Zelaya DG (2006) Biodiversidad y zoogeografía de gasterópodos y bivalvos magallánicos y antárticos. PhD thesis, Universidad de Buenos Aires, Facultad de Ciencias Exactas y Naturales, Buenos Aires

Zelaya DG, Ituarte C (2013) Two new species of *Waldo* Nicol, 1966 from sub-Antarctic waters (Bivalvia: Galeommatoidea). Mar Biol Res 9:776–784. https://doi.org/10.1080/17451000.2013.765584

Zelaya DG (2015) Marine Bivalves from the Argentine Coast and Continental Shelf: Species Diversity and Assessment of the Historical Knowledge. Am Malacol Bull 33:245–262

Zelaya DG, Güller M, Ituarte C (2020) Filling a blank in bivalve taxonomy: an integrative analysis of Cyamioidea (Mollusca: Bivalvia). Zool J Linnean Soc 190:558–591. https://doi.org/10.1093/zoolinnean/zlz144

Chapter 10
Uncovering the Terrestrial and Aquatic Invertebrates of Isla de los Estados, Argentina

María Eugenia Lopez, Francisco J. Sola, Carina I. Argañaraz, and Amira Salom

Abstract Invertebrates play crucial roles in ecosystems and represent a significant proportion of global biodiversity. Particularly, in freshwater ecosystems, aquatic invertebrates play a key role in ecosystem structure and function. Many taxa, such as Odonata, Plecoptera, Ephemeroptera, have life cycles that connect aquatic and terrestrial environments. Southern Patagonia is home to invertebrate groups with Gondwanan distribution, as well as relictual species, while the sub-Antarctic region is particularly notable for the high number of endemic aquatic species. However, invertebrates have historically been under-studied, with only sporadic collections dating back to the seventeenth century. Modern efforts have considered expanding our understanding of Southern Patagonian invertebrates, adding hundreds of species

M. E. Lopez (✉)
Consejo Nacional de Investigaciones Científicas y Técnicas (CONICET), Ushuaia, Tierra del Fuego, Argentina

Dirección Regional Patagonia Austral, Administración de Parques Nacionales (APN), Ushuaia, Tierra del Fuego, Argentina

Universidad Nacional de Tierra del Fuego (UNTDF), Ushuaia, Tierra del Fuego, Argentina

Centro Austral de Investigaciones Científicas (CADIC), Ushuaia, Tierra del Fuego, Argentina
e-mail: melopez@untdf.edu.ar

F. J. Sola
Consejo Nacional de Investigaciones Científicas y Técnicas (CONICET), Ushuaia, Tierra del Fuego, Argentina

Universidad Nacional de Tierra del Fuego (UNTDF), Ushuaia, Tierra del Fuego, Argentina

C. I. Argañaraz
Center of Excellence in Telesensing of Environment and Model Prediction of Severe events (CETEMPS), University of L'Aquila, L'Aquila, Italy

A. Salom
Consejo Nacional de Investigaciones Científicas y Técnicas (CONICET), Ushuaia, Tierra del Fuego, Argentina

Centro Austral de Investigaciones Científicas (CADIC), Ushuaia, Tierra del Fuego, Argentina

A. Raya Rey, U. Balza (eds.), *Biodiversity and Conservation of Isla de los Estados (Chuani-sin)*, Natural and Social Sciences of Patagonia,
https://doi.org/10.1007/978-3-032-13630-5_10

and morphospecies, especially of Diptera, Coleoptera, and Arachnida. Despite these advances, Isla de los Estados has remained largely unexplored due to its hazardous coasts and lack of a permanent human population. This chapter provides a comprehensive overview of the aquatic and terrestrial invertebrate fauna on Isla de los Estados, by synthesizing current knowledge and incorporating new findings from recent sampling efforts.

Keywords Macroinvertebrate biodiversity · Sub- Antarctic region · Terrestrial ecosystems · Inland waters

10.1 Introduction

Arthropoda is one of the best-known invertebrate phyla. Today, arthropods play essential roles in nearly all terrestrial and freshwater ecosystems, contributing to key processes such as pollination, pest control, decomposition, nutrient cycling, and serving as a primary food resource for numerous animal taxa (Maggi et al. 2024). Further, their ecological importance, combined with sensitivity to environmental change, relatively short generation times, and specific habitat requirements, makes them useful bioindicators. Changes in arthropod community composition, species abundance, or physiological traits can offer early warnings of habitat degradation, climate change, pollution, or the effectiveness of restoration efforts (Fenoglio et al. 2020; Arenhardt et al. 2021; Farfán-Beltrán et al. 2022). Arthropods are traditionally divided into two major monophyletic groups: Chelicerata (including sea spiders, horseshoe crabs, spiders, scorpions, and their relatives) and Mandibulata (myriapods, insects, and crustaceans) (Giribet and Edgecombe 2019). Their evolutionary history dates back approximately 521 million years ago, coinciding with the appearance of the earliest terrestrial flora. This long co-evolutionary history suggests deep ecological interactions within terrestrial ecosystems (Misof et al. 2014; Giribet and Edgecombe 2019).

Many insects and arachnids in the Southern Hemisphere exhibit what are known as 'Gondwanan distributions': lineages that originated and began diversifying when the continent of Gondwana was still connected. As a result, distant regions such as South America, South Africa, Australia, and New Zealand may share taxonomic groups that would be unlikely to disperse across today's vast ocean distances (Derkarabetian et al. 2021; Galli and Rellini 2020; Harms et al. 2024).

Insects form the biological foundation of terrestrial ecosystems. They cycle nutrients, pollinate plants, disperse seeds, maintain soil structure and fertility, control populations of other organisms, and provide key food resources (Majer 1987). Yet, knowledge and conservation of Southern Hemisphere insect faunas lags substantially behind those of Europe and North America, where the relatively low-diverse but well-documented faunas benefit from a long tradition of natural history and public engagement which enables conservation needs to be assessed and addressed (New and Samways 2014). Within the Southern Hemisphere, where each region (South America, South Africa, Australia, and New Zealand) is presumed to have distinctive insect faunas and high levels of endemism, the temperate South America's southernmost reaches, together with the isolated Sub-Antartctic islands recognized for their "exceptionally high conservation value" (Chown et al. 2008), insects diversity remains especially under-studied (New and Samways 2014). This represents a critical gap in our understanding of global biodiversity.

Arachnids, on the other hand, are generally underrepresented in Arctic, Antarctic and sub-Antarctic regions, likely due to the region's harsh climatic conditions which limit food availability. Spiders (orders Mygalomorphae and Araneomorphae) and harvestmen (Opiliones) occur in southern Argentina and Chile, but with far lower diversity compared with the Northern Hemisphere (Pugh 2004; Szymkowiak et al. 2007). Mygalomorphs are typically large-bodied spiders including more than 3000 species. In contrast, Araneomorphae form a highly diversified group with over 47,000 described species, exhibiting a wide range of life strategies, behavioral traits, and morphological characteristics (Kulkarni et al. 2023). Together these infraorders account for 99% of all spider species on Earth (Dimitrov and Hormiga 2021).

Spiders are primarily predators, with insects constituting their main prey. As such, they play a critical role in pest control in agricultural systems (Armendano and González 2011) and in managing populations of insect vectors (Lawania et al. 2013). Additionally, spiders begin to disperse shortly after hatching, exhibiting two main dispersal strategies: locomotion (walking) and 'ballooning' (aerial dispersal) (Bonte et al. 2003; Bell et al. 2005). Ballooning spiders can move over hundreds or even thousands of kilometers, records range from 900 to 3000 km (Bell et al. 2005). Spiders have even been found aboard ships more than 300 km from the nearest coastline, having "surfed the air". Due to this high vagility, spiders are often among the first colonizers following natural disturbances, such as volcanic eruptions on newly formed islands (Szymkowiak et al. 2007). Additionally, some studies suggest that spiders with higher ballooning capacity tend to be generalists with broader ecological tolerances, whereas habitat specialists typically exhibit lower vagility (Bonte et al. 2003). Therefore, studying the few arachnids present in the Subantarctic region can increase our understanding of dispersal mechanisms, ecological roles and adaptations to these cold, wind-swept landscapes (Beron 2018).

Of particular interest are the complex invertebrate communities hosted in freshwater bodies (i.e. rivers, streams, lakes, ponds, etc). These communities, mainly composed of arthropods, mollusks and worms, link aquatic and terrestrial environments and trophic webs, as many species have complex life cycles occurring in both the aquatic and terrestrial environments. For instance, the adults of the orders Ephemeroptera, Plecoptera, Trichoptera, Odonata and Diptera are terrestrial, while their larval stage is aquatic. In contrast, several Coleoptera families spend their entire life cycle in water, with winged adults capable of flight and dispersal to other locations (Hanson et al. 2010). In the temperate and Sub-Antarctic ecosystems of the Southern Hemisphere life cycles consisting of one or two generations per year are commonly observed due to a low annual temperature range (Hanson et al. 2010; Contador et al. 2014).

Aquatic macroinvertebrates play a key role in the ecological network, functioning as a food source for top predators such as birds and fish, and as consumers of lower trophic levels like plankton (Wallace and Webster 1996; Lopez et al. 2021). Additionally, these communities are widely regarded as a highly effective indicator of ecosystem status. Its use in detecting the effects of land use change has been extensively documented (Kenney et al. 2009). Despite their critical role in maintaining freshwater ecosystem functions and their substantial contribution to global biodiversity, aquatic macroinvertebrate communities are often underrepresented in conservation assessments and overlooked in aquatic management strategies (Collier et al. 2016; Cardoso et al. 2020; Samways et al. 2020). Despite some recent efforts, Austral Patagonia freshwater arthropod fauna also remains poorly known.

Islands serve as natural laboratories for studying biodiversity and evolutionary processes, often hosting unique faunal assemblages (Leclère and Gerhold 2024). Isla de los Estados, a remote island within the Tierra del Fuego archipelago, remains virtually unexplored with respect to its arthropod fauna. Its geographic isolation and harsh environmental conditions have limited both biological exploration and research, and anthropogenic disturbance. Therefore, understanding the local arthropod fauna of Isla de los Estados is not merely an academic endeavor, it is a critical component of conservation planning.

10.1.1 A History of Entomological Exploration in the Tierra del Fuego Archipelago

Early exploration voyages in this region were primarily focused on navigation and discovery. However, they often included naturalists who conducted preliminary observations and collected biological specimens. One of the earliest expeditions to Tierra del Fuego was that of Captain James Cook (1768–1771) (Cook 1999). Accompanied by botanists Joseph Banks and Daniel Solander, the expedition gathered extensive plant collections from previously uncharted regions, including areas within Tierra del Fuego. A pivotal moment in the scientific exploration of the region came with the voyage of the HMS Beagle (1831–1836), with naturalist Charles Darwin. Darwin dedicated considerable time in Tierra del Fuego, meticulously documenting his observations and collecting specimens across a diverse wide range of taxa, including insects (Trifilo 1960).

The early assessment of Fuegian insect Fauna being 'poor' by Charles Darwin underscores the need for more comprehensive and systematic studies to accurately assess the true diversity of insects in this region. The nineteenth and twentieth centuries saw a series of additional scientific expeditions to Tierra del Fuego, some of which included entomological studies. One notable example is the Swedish Magellanic Expedition (1907–1909), under the leadership of Carl Skottsberg. While this expedition primarily focused on botanical, geographical, and geological inquiries within Patagonia and Tierra del Fuego (Skottsberg 1911), it also collected biological specimens, including insects, that are likely preserved in Swedish natural history collections.

Despite the broader scientific interest in Tierra del Fuego archipelago over the centuries, Isla de los Estados has received relatively little attention with respect to its aquatic and terrestrial invertebrate biodiversity. This notable absence in the historical literature points to a significant gap in our scientific understanding and highlights the unexplored nature of the invertebrate fauna inhabiting this remote island.

10.1.2 Isla de los Estados: A Unique Ecosystem at the Edge of the World

Isla de los Estados, is an Argentine island located approximately 30 kilometers off the eastern extremity of Tierra del Fuego in the south-Atlantic Ocean. Its subantarctic climate is marked by persistently cold temperatures, frequent precipitation averaging around 2000 mm annually, and strong, constant winds (Ponce and Fernández 2014). The island's landscape is predominantly covered by dense, low-lying subantarctic evergreen forests and a diverse array of peatland communities (Salom et al., Chap. 4, this volume). The forests are primarily dominated by *Nothofagus betuloides* (southern beech), a key plant species in the sub-Antarctic region, and by *Drimys winteri* (winter's bark). Notably, Isla de los Estados is the only region in Argentina where Magellanic moorland occurs (Arana et al. 2021). Relatively isolated from major human disturbance, with access mostly limited to guided tours and a naval station, Isla de los Estados holds considerable ecological significance as part of the Tierra del Fuego archipelago, an area renowned for its distinctive biodiversity and regarded as one of the planet's last remaining true wilderness areas (Savoretti et al. 2021). The island's harsh climatic conditions and the dominance of *Nothofagus* vegetation and moorlands likely play a crucial role in shaping the composition and diversity of its arthropod community, favoring species with adaptations to cold, wet, and windy environments, as well as species with specific ecological ties to southern beech forests.

The colonization of arthropods on Isla de los Estados likely occurred through different mechanisms over geological time. During the Last Glacial Maximum, the island was connected to the Mitre Peninsula (on the Isla Grande of Tierra del Fuego), which itself was linked to the South American mainland (Ponce and Fernández, Chap. 2, this volume). This land bridge may have facilitated initial colonization. Following the retreat of glaciers around 18,000 years ago and the subsequent formation of the Le Maire Strait (Ponce and Fernández 2014), new colonization pathways emerged, including aerial dispersal (ballooning), bird-mediated transport, oceanic rafting (Szymkowiak et al. 2007), and more recently, human introduction (Pugh 2004). In insects, highly vagile species, such as butterflies, tend to colonize new areas more easily, leading to homogenization, while less vagile species, like ants, experience higher rates of species loss and heterogenization.

10.1.3 Scope and Objectives

This chapter provides a comprehensive overview of the terrestrial and aquatic invertebrate diversity found on Isla de los Estados. It synthesizes current knowledge of these communities and incorporates new findings from recent sampling efforts conducted on the island (2018–2022). Where possible, the species recorded on Isla de los Estados are compared with those from nearby islands (Tierra del Fuego Main Island, the Malvinas Islands, and/or South Georgia). Cladistic biogeographic studies (Morrone 1993; Morrone et al. 1994; Soares and de Carvalho 2005) have shown that the Magellanic Tundra province shares a close phylogenetic relationship with the Magellanic Forest, Valdivian Forest, and Malvinas Islands provinces. Despite differences in their geological histories, these regions exhibit notable similarities in their arthropod fauna (Pugh 2004; Arana et al. 2021).

10.2 Sampling and Analysis Methodology

10.2.1 Terrestrial Ecosystem

In a study conducted on Franklin Bay, Isla de los Estados, we collected arthropod samples using 36 pitfall traps during the austral summers of 2018, 2021 and 2022. The objective was to establish baseline knowledge of the island's invertebrate community. This information was critical for assessing the impact of introduced ungulates through ongoing closures experiments (Salom 2025). Traps were deployed in two characteristic environments of the island: the Subantarctic Evergreen Forest and the Maritime Tussock Grasslands (Fig. 10.1). Further details on these environments can be found in Ponce and Fernández (2014) and in Salom et al. (Chap. 4, this volume).

A total of 1439 terrestrial arthropods were collected, representing two main classes: Insecta and Arachnida (primarily spiders, order Araneae). These groups were analyzed separately. Specimens were identified to genus and species when possible; otherwise, they were assigned to morphospecies.

10.2.2 Inland Waters

We conducted a study in six shallow lakes on San Juan del Salvamento (Isla de los Estados) and Observatorio island (archipelago of the Año Nuevo Islands) in 2022 and 2024 (Fig. 10.2a, b). We perform a qualitative analysis of the aquatic macroinvertebrate community. Samples were collected using a hand net with a D-shaped frame, and each sample consisted of multiple sweeps performed along the edge of the shallow lakes. Macroinvertebrates were identified to the lowest possible

Fig. 10.1 (**a**) Maritime Tussock (*Poa flabellata*) grasslands. (**b**) *Nothofagus betuloides* forest. (**c**) *Eristalis croceimaculata* syrphid fly. (**d**) *Aegorhinus vitulus* weevil. (**e**) *Ceroglossus suturalis* ground beetle. (**f**) *Erichius femoralis* lucanid beetle. (Credits: Joaquín Cano (iNaturalist) (**c**), wp-polzin (iNaturalist) (**f**))

taxonomic level using regional keys and recent publications (Domínguez et al. 2006; Michat et al. 2008; Domínguez and Fernández 2009; Stark et al. 2009; Libonatti et al. 2011; Hamada et al. 2018).

10.2.3 Previous Records from Isla de los Estados and the Surrounding Islands

The taxa identified within the aquatic and terrestrial communities will be presented in tabular form. The Isla de los Estados column displays the individuals recorded after analysis of the samples taken, as described above. The other columns present a comparison with previous records from surrounding islands: Isla Grande de Tierra del Fuego, Malvinas Islands and South Georgia. This information is crucial for two reasons: Firstly, it facilitates the identification of novel records; Secondly, it enables the identification of gaps in information concerning these communities.

Fig. 10.2 Aquatic macroinvertebrate community. (**a**, **b**) Shallow lakes of Observatorio island. (**c**) *Hyalella* spp. (**d**) *Chilina sp*. (**e**, **f**) *Lancetes angusticollis* larva and adult. (**g**) Limnephilidae larva (**h**) *Monocosmoecus hyadesi* larva with case. (**i**) *Austrocosmoecus hirsutus* larva. (**j**) *Meridialaris chiloeensis* larva. (**k**) *Edwardsina* sp. larva (**l**) *Notoperla* sp. Larva. (**m**) Chironomidae larva. (**n**) *Rheochorema magellanicum*. (Credits: Samanta Subires (**a**, **b**), María Eugenia Lopez (**c**–**i**, **k**–**n**), Eduardo Domínguez (**j**))

10.3 Results

10.3.1 Terrestrial Insects of Isla de los Estados

We collected 1146 insect specimens, from which we identified 104 distinct morphospecies. Specimens were examined and identified to species or morphospecies level using appropriate taxonomic keys. Records of insect species reported from Isla de los Estados are extremely scarce, with only 25 species listed in GBIF.org and a few articles (GBIF.org a; Posadas 2002, 2012; Roig-Juñent et al. 2024) (Table 10.1, Fig. 10.1). Among the 104 morphospecies collected, we recognized 22 Coleoptera, 67 Diptera, and 15 Hymenoptera. The most prominent families identified were Carabidae, Curculionidae, Limoniidae, Ephrydidae, and Sphaeroceridae.

Carabidae or ground beetles are a highly diverse family found in nearly all terrestrial habitats worldwide. Most ground beetles are predatory generalist feeders. Curculionidae, or weevils, represent the herbivorous beetles, some of which specialize on living woody tissue. Within Diptera, Limoniidae (crane flies) and Ephrydidae (shore flies) are associated with humid environments with rich soils or high availability of organic matter, usually in the form of algae.

The diversity cited by the literature focused primarily on Coleoptera, with Diptera and Hymenoptera remaining underrepresented in terms of available information (Munari 2015). This reflects a global trend (Bickel et al. 2009; Raguso 2020; Haas et al. 2021; Theron et al. 2023; Zhang et al. 2025), driven in part by geographic research gaps, taxonomic biases, and limited resources and expertise.

Knowing that the 25 previously cited species (Table 10.1) are included in our collection, our sampling adds at least 79 novel morphospecies. Most of these are likely to correspond to species already present in Tierra del Fuego, and possibly the Malvinas Islands, as suggested by the region's biogeographic patterns (Spinelli et al. 2006; Morrone 2000; Posadas 2012; Roig-Juñent et al. 2024).

Despite regional studies offering some insights, the reviewed literature reveals a striking lack of research specifically targeting the terrestrial insect fauna of Isla de los Estados. This points to a significant knowledge gap in understanding the island's biodiversity. Therefore, our results significantly improve local knowledge of terrestrial insects, especially for the underrepresented Hymenoptera and Diptera diversity. Future research should prioritize systematic surveys of the island's insect fauna, employing a range of sampling techniques to better capture its diversity. Long-term monitoring will also be critical to detect changes in insect diversity and abundance over time, particularly in light of environmental changes and the known impacts of introduced species, such as red deer (*Cervus elaphus*) (Rodriguez-Cabal et al. 2019; Zhu et al. 2023). Additionally, the increasing risk of species introductions through ecotourism and human activity at the naval station in Parry Port warrant attention.

Table 10.1 List of insect species previously recorded for Isla de los Estados (from literature and GBIF)

Family	Species	IDLE	Isla Grande	S. Georgia	Malvinas	References
Curculionidae	*Aegorhinus vitulus* (J.C.Fabricius 1775)		X			GBIF.org
	Telurus caudiculatus (Morrone and Anderson 199)		X			GBIF.org, Posadas (2012)
	Germaniellus rugipennis					Posadas (2012)
	Falklandius antarcticus (Stierlin 1903)	X	X		X	Posadas (2002)
Cerambycidae	*Microplophorus magellanicus* (Blanchard 1851)		X			GBIF.org
Carabidae	*Kenodactylus audouini* (Guérin-Méneville 1830)	X	X		X	GBIF.org
	Cascellius gravesii (Curtis 1839)	X	X		X	GBIF.org
	Ceroglossus suturalis (Fabricius 1775)	X	X			GBIF.org
	Migadops latus (Guérin-Méneville 1841)		X		X	GBIF.org
	Trechisibus antarcticus (Dejean 1831)		X			GBIF.org
	Metius malachiticus (Dejean 1828)	X	X			Roig-Juñent et al. (2024)
	Omalodera discoidalis (Blanchard 1842)		X			Roig-Juñent et al. (2024)
	Abropus carnifex (Fabricius 1775)		X			Roig-Juñent et al. (2024)
	Nothocascellius aeneoniger (G.R.Waterhouse 1840)		X			Roig-Juñent et al. (2024)
	Nothotrechisibus hornensis (Fairmaire)		X			Roig-Juñent et al. (2024)
	Trechisibus darwinii (Jeannel 1927)		X		X	Roig-Juñent et al. (2024)
	Pseudomigadops darwinii (G.R.Waterhouse 1842)		X			GBIF.org, Roig-Juñent et al. (2024)
	Merizodus soledadinus (Guérin-Méneville 1830)		X	X		GBIF.org
Lucanidae	*Erichius femoralis* (Guérin-Méneville 1839)	X		X		Paulsen (2006)
Aenictopecheidae	*Gamostolus subantarcticus* (Berg 1883)	X	X			GBIF.org
Hydrobiosidae	*Rheochorema magellanicum* (Flint 1974)		X			GBIF.org
Philopteridae	*Acutifrons australis* (Guimaraes 1942)					GBIF.org
Ephydridae	*Scatophila angusta* (Becker 1896)	X				GBIF.org
Helcomyzidae	*Paractora rufipes* (Macquart 1844)		X			GBIF.org
Syrphidae	*Eristalis croceimaculata* (Jacobs 1900)		X		X	GBIF.org

Column IDLE indicates species confirmed from collected specimens from our study. Isla Grande, South Georgia and Malvinas columns indicate the presence of species at those locations (from literature and GBIF)

10.3.2 Diversity of Arachnids in Isla de los Estados

In total, 293 individuals were collected, of which 83% were adults and 17% juveniles, representing 23 arachnid species (21 spiders and two opilionids). Only one specimen could not be identified to genus or species level and remains classified as a morphospecies (Linyphiidae msp 1). Six species were highly abundant (>10 individuals each), while five were represented by only a single individual (singletons).

The sex ratio was 0.65:1 (female:male), indicating higher male activity in midsummer, likely related to mating behavior constrained by harsh seasonal conditions. In contrast, Usher (1983) reported a female-biased sex ratio in the Malvinas Islands (3:1) and a balanced ratio in South Georgia Island (1:1). Another noteworthy result is the high proportion of adults in our samples. Typically, adults represent less than 50% of the individuals in similar studies (Cardoso et al. 2008), though this pattern has also been observed in previous studies in Tierra del Fuego (Argañaraz et al. 2020, 2023). As Pugh (2004) noted, areas below 45° S are characterized by low arachnid diversity. Beyond species richness, other ecological differences may exist compared to temperate and tropical regions, and further research is needed to identify factors shaping these patterns.

We compiled 39 arachnids records for Isla de los Estados, summarized in Table 10.2. Most arachnid species present on Isla de los Estados are also found on Isla Grande (Tierra del Fuego) (37 species); 13 species are shared with the Malvinas Islands; and only three (two linyphiid species and one anyphaenid) with South Georgia. This distribution pattern reflects both the island's proximity to the mainland (<30 km), which facilitates colonization via ballooning, and historical land connections, which could explain the presence of opilionids (which do not disperse aerially). Additionally, some species may have been introduced by humans, as has occurred in the Malvinas Islands (Lavery and Snazell 2023) and South Georgia Islands (Pugh 2004). Nevertheless, most species appear closely related to South American taxa, and some are endemic to the region.

The most diverse families were Anyphaenidae (Fig. 10.3a) and Linyphiidae, each with ten species. Anyphaenids included all South American genera, some monospecific, such as *Axyracrus elegans* and *Selknamia minima* (Ramírez 2003). Linyphiidae is one of the most diverse spider families globally (640 genera and 4941 species, World Spider Catalog 2025), it was represented on the island by the South American genera. These fall into two groups: (1) widespread genera with many species (*Laminacauda*, *Dubiaranea*, and *Neomaso*), and (2) genera restricted to southern latitudes, such as *Notiomaso*, which includes nine species, several described from the Malvinas Islands and South Georgia. We also recorded *Notholepthyphantes* and *Rhabdogyna* (each with two species described) and *Intecymbium* with a single species known from Chile and Argentina.

Other genera restricted to Southern Argentina and Chile included *Emmenomma*, *Anisacate*, and *Rubrius* (Macrobunidae family), *Cybaeolus* (Hahniidae), and *Mecysmauchenius* (Mecysmaucheniidae) (Fig. 10.3b), also present in the Malvinas Islands. Additional continental restricted genera were *Sheranapis* (Anapidae),

Table 10.2 List of arachnid species present on Isla de los Estados, Argentina

Class	Order	Family	Species	IDLE	Isla Grande	Malvinas	S. Georgia	References
Arachnida	Araneae	Anapidae	*Normplatnicka* (Rix and Harvey 2010)					GBIF.org
			Sheranapis (Platnick and Forster 1989)					GBIF.org
		Anyphaenidae	*Araiya coccinea* (Simon 1884)	X	X			Ramírez (2003), GBIF.org
			Axyracrus elegans (Simon 1884)	X	X			Ramírez (2003), GBIF.org
			Coptoprepes sp.[a]	X	X			Ramírez (2003), Barone et al. (2016)
			Monapia vittata (Simon 1884)		X			Ramírez (2003), GBIF.org
			Negayan excepta (Tullgren 1901)	X	X	X		Lopardo (2005), Lavery (2017)
			Negayan paduana (Karsch 1880)		X	X	X	Ramírez (2003), Lavery (2017), GBIF.org
			Oxysoma saccatum (Tullgren 1902)		X			Ramírez (2003), GBIF.org
			Philisca hyadesi (Simon 1884)		X			Ramírez (2003), GBIF.org
			Selknamia minima (Ramírez 2003)		X			Ramírez (2003)
			Tomopisthes horrendous (Nicolet 1849)		X			Ramírez (2003)
		Hahniidae	*Cybaeolus* (Simon 1884)[a]		X	X		Lavery (2017), Burgo et al. (2025), GBIF.org
			Hahnia sp.[a]	X	X	X		Pugh (2004), Lavery (2017)
			Intihuatana antarctica (Simon 1902)		X			Dupérré and Harms (2018)
		Linyphiidae	*Dubiaranea* (Mello-Leitão 1943)[a]		X			Argañaraz et al. (2020), GBIF.org
			Intecymbium antarcticum (Simon 1895)		X			Miller (2007), Argañaraz et al. (2023)
			Laminacauda plagiata (Tullgren 1901)	X	X			Miller (2007), Lavery and Snazell (2013)

Arachnida	Araneae	Linyphiidae	*Neomaso* cf. *minimus*	X	X			Argañaraz et al. (2023)
			Neomaso patagonicus (Tullgren 1901)	X	X			Argañaraz et al. (2023), GBIF.org
			Neomaso sp.[a]	X	X	X	X	Millidge (1985), Pugh (2004), GBIF.org
			Notholepthyphantes cf. *erythrocerus*[a]	X	X			Argañaraz et al. (2020)
			Notiomaso cf. *exonychus*	X	X	X		Argañaraz et al. (2023)
		Lycosidae	*Lycosa australis* (Simon 1884)	X	X			Argañaraz et al. (2020), GBIF.org
		Macrobunidae	*Anisacate fuegianum* (Simon 1884)	X	X	X		Lavery (2017), GBIF.org
			Emmenomma oculatum (Simon 1884)	X	X	X		Almeida-Silva et al. (2015), Lavery (2017)
			Rubrius antarcticus (Karsch 1880)	X	X	X		Roth (1967), Lavery (2017) (cited the genus *Rubrius*), GBIF.org
			Rubrius castaneifrons (Simon 1884)		X	X		Roth (1967), Lavery (2017) (cited the genus *Rubrius*), GBIF.org
		Mecysmaucheniidae	*Mecysmauchenioides nordenskjoldi* (Tullgren 1901)		X			Grismado and Lopardo (2003), GBIF.org
			Mecysmauchenius segmentatus (Simon 1884)	X	X	X		Grismado and Lopardo (2003), Lavery (2017), GBIF.org
		Synotaxidae	*Tekellina araucana* (Marusik, Eskov and Ramírez 2022)		X	X		Ramírez et al. (2022), GBIF.org
		Tetragnathidae	*Diphya* (Nicolet 1849)[a]		X			Simon (1896), GBIF.org
		Theridiidae	*Seikirkiella luisi* (Levi 1967)		X			GBIF.org
		Thomisidae	*Coenypha ditissima* (Nicolet 1849)	X	X			GBIF.org
	Opiliones	Caddidae	*Acropsopilio* sp.[a]	X	X			Maury et al. (1996), Argañaraz et al. (2023)
		Neopilionidae	*Thrasychirus* cf. *dentichelis*	X	X			Acosta and Pérez-González (2023), GBIF.org

Cross indicates the presence of the species in the present survey (IDLE) and on other nearby islands (Isla Grande-Tierra del Fuego, Malvinas Island and South Georgia), references confirm the presence at each of those location

[a]Indicates the presence of the genus and/or species of the genus

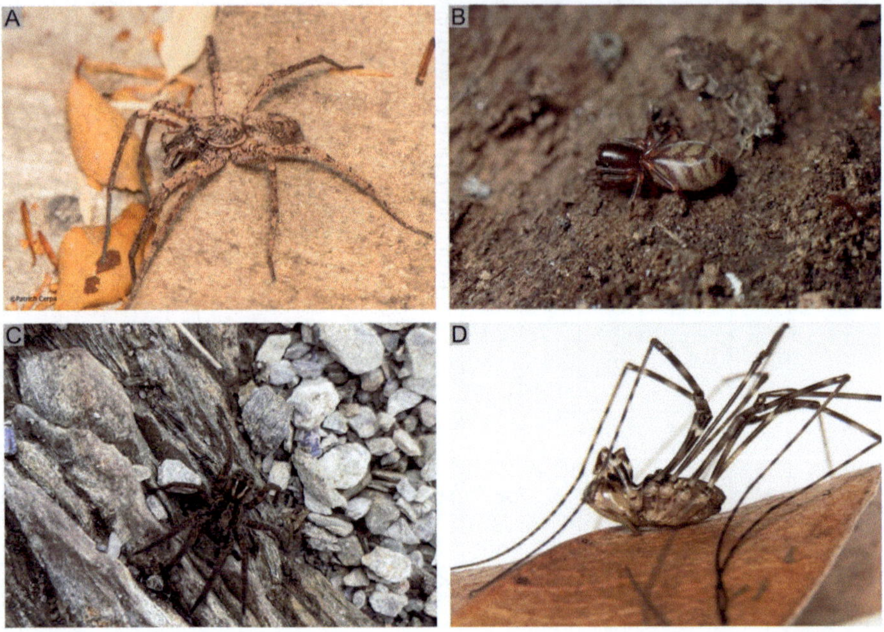

Fig. 10.3 (**a**) *Tomopisthes horrendus* (Nicolet, 1849). (**b**) *Mecysmauchenius segmentatus* Simon, 1884. (**c**) *Lycosa australis* Simon, 1884. (**d**) *Thrasychirus* sp. (Credits: Patrich Cerpa (iNaturalist) (**a**), Marisol Toledo Araneda (iNaturalist) (**b**), André Terwei (iNaturalist) (**c**), Abel Perez-Gonzalez (Instagram: @abelperez.g) (**d**))

Intihuatana (Hahniidae), *Selkirkiella* (Theridiidae), *Coenypha* (Thomisidae), and *Mecysmauchenioides* (Mecysmaucheniidae). We also recorded genera with broader distribution, such as *Lycosa* (Lycosidae) (Fig. 10.3c), *Tekellina* (Synotaxidae), *Episinus* (Theridiidae), *Dyphina* (Tetragnathidae), *Hahnia* (Hahniidae), and *Normplatnicka* (Anapidae) (World Spider Catalog 2025).

Among opilionids we found species from two families. *Acropsopilio* sp. (possibly *A. chilensis*) from the family Caddidae which includes very small harvestmen (1–2 mm) and shows a disjunct distribution across southern Chile and Argentina (sub-Antarctic), the Yungas, and southern Brazil (Paranaense and Cerrado Forest) (Maury et al. 1996). The Neopilionidae family, with representatives in South America, Africa, Australia, New Caledonia, and New Zealand, includes the genus *Thrasychirus* (Fig. 10.3d) which is restricted to South America (Giribet et al. 2021). These distributions reflect historical biogeographic processes. In the past, forests in Argentina and Chile formed a continuous corridor between north and south (Maury et al. 1996), and during the Gondwana period, continents were in close proximity (Giribet et al. 2021).

Additionally, southern Patagonia hosts several relict spider families with Gondwanan distribution that have not been recorded on Isla de los Estados, such as Austrochilidae, Hexathelidae, Malkaridae, and Orsolobidae (Grismado et al. 2014).

Given the ecological similarity and shared geological history, it is plausible these groups also occur on the island. Many species, especially those from the more diverse families, are only a few millimeters long, requiring intensive sampling to adequately survey this inconspicuous fauna and better understand their richness and ecological requirements.

10.3.3 Diversity of Aquatic Macroinvertebrates on Isla de los Estados

First, we describe the taxa found during the survey, then supplement the list with records from the GBIF.org database and studies from Isla Grande de Tierra del Fuego, Malvinas Islands and South Georgia (GBIF.org b, c; Elgueta and Jezek 2014; Anderson et al. 2018; Zagarola et al. 2017).

For the aquatic macroinvertebrate community, we recorded Crustacea, Gastropoda and Insecta (Table 10.3). Amphipods (Crustacea) are abundant in freshwater environments and inhabit various microhabitats such as leaf litter, under rocks and aquatic vegetation. They feed on particulate organic matter found in the substrate. Adult sizes range from 7 to 12 mm. We recorded individuals of the genus *Hyalella* (Fig. 10.2c), which is widely distributed in South America and exhibits high diversity and complexity due to low morphological differentiation among at least 50 described species (Rincón et al. 2016; Rogers et al. 2020).

The gastropods found belong to the genus *Chilina*, the only genus in the family Chilinidae (Fig. 10.2d). This family is endemic to South America and inhabits a range of freshwater environments with clean, oxygenated water and variable temperatures (Cuezzo 2009). These organisms are scrapers. Their shells often feature characteristic dark zigzag bands (Gutierrez Gregoric 2020). Like other gastropods, they play an important role in the diet of both aquatic and terrestrial predators (Cuezzo et al. 2020).

Insects represented the group with the highest diversity, including the orders Trichoptera, Coleoptera and Diptera. Larvae of Trichoptera (caddisfly) are common in freshwater habitats, with some building shelters attached to the substrate and others creating portable cases using silk and surrounding material. These insects are sensitive to environmental changes and are typically found in clean, oxygenated waters. Adults are terrestrial and resemble moths with long antennae (Rincón et al. 2016). The species *Austrocosmoecus hirsutus* and *Monocosmoecus hyadesi*, belonging to the family Limnephilidae, are shredders that feed on leaf litter (Fig. 10.2g–j). Their portable shelters are built from local materials such as leaf fragments, sand grains, pebbles, twigs, or even empty mollusk shells (Pes et al. 2018). In contrast, *Rheochorema magellanicum* (Fig. 10.2n), a member of the family Hydrobiosidae, is a free-living predator that does not construct larval refuges (Angrisano and Sganga 2009). This species is the sole one documented thus far for Isla de los Estados (GBIF.org).

Table 10.3 List of aquatic macroinvertebrates present on Isla de los Estados and Observatorio island (IDLE & O)

Class	Order	Family	Specie	IDLE & O	Isla Grande	S. Georgia	Malvinas	References
Crustacea	Amphipoda	Hyaleliiade	*Hyalella* spp. (Smith 1874)	X	X		X	GBIF.org
Mollusca	Gastropoda	Chilinidae	*Chilina* sp. (Gray 1828)	X	X		X	GBIF.org
Insecta	Ephemeroptera	Beatidae	*Andesiops torrens* (Lugo-Ortiz and McCafferty 1999)		X			Zagarola et al. (2017), GBIF.org
		Leptophlebiidae	*Meridialaris chiloeensis* (Demoulin 1955)		X			GBIF.org
		Nesameletidae	*Metamonius anceps* (Eaton 1883)		X			Zagarola et al. (2017), GBIF.org
	Plecoptera	Gripopterygidae	*Aubertoperla kuscheli* (Illies 1963)		X			GBIF.org
			Notoperla fuegiana (Enderlein 1905)		X			GBIF.org
			Notoperla archiplatae (Illies 1958)		X			GBIF.org
			Rithroperla rossi (Froehlich 1960)		X			Zagarola et al. (2017), GBIF.org
	Trichoptera	Limnephilidae	*Austrocosmoecus hirsutus* (Schmid 1955)	X	X			GBIF.org
			Monocosmoecus hyadesi (Mabille 1888)	X	X			Zagarola et al. (2017), GBIF.org
		Hydrobiosidae	*Rheochorema magellanicum* (Flint 1974)	X	X			Zagarola et al. (2017), GBIF.org
	Coleoptera	Dytiscidae	*Lancetes angusticollis* (Curtis 1839)	X	X	X		GBIF.org
		Elmidae	*Luchoelmis* sp. (Spangler and Staines 2004)		X			GBIF.org
	Diptera	Chironomidae		X	X	X	X	Zagarola et al. (2017), GBIF.org
		Simulidae	*Gigantodax rufescens* (Edwards 1931)		X			GBIF.org
		Blephariceridae	*Edwardsina* sp. (Edwards 1929)		X			GBIF.org
		Muscidae			X			GBIF.org
		Tipulidae			X			GBIF.org
		Psychodidae		X	X			Elgueta and Jezek (2014)

Cross indicates the presence of the taxa in the present survey (IDLE&O) and previous records for the nearby islands (Isla Grande de Tierra del Fuego, Malvinas Island and South Georgia)

Coleoptera, the most diverse orders globally, includes over 350,000 described species (Rincón et al. 2016). The family Dytiscidae includes aquatic beetles known as diving beetles due to their habit of surfacing periodically to replenish their air supply. All stages of their life cycle, except the pupal stage, are aquatic. Both larvae and adults are predators that feed on insects, crustaceans, leeches and even small fish. Adults are strong swimmers thanks to their streamlined bodies and retain their ability to fly, allowing them to disperse in search of new habitats (Michat et al. 2023). In this study, both larvae and adults of *Lancetes angusticollis* were recorded; this species has also been documented in Isla Grande Tierra del Fuego and South Georgia (Fig. 10.2e–f).

From Diptera, we recorded larvae of two families: Chironomidae and Psychodidae. Of the 11 Chironomidae subfamilies described worldwide, nine are found in South America (Rincón et al. 2016). This family includes representatives of various functional feeding groups, and their diets can vary depending on environmental conditions. Chironomids (Fig. 10.2m) are among the most important biomass producers in the aquatic ecosystems they inhabit, owing to their reproductive output (Paggi et al. 2023). Psychodidae are small, hairy flies, and some genera have aquatic immature stages adapted to habitats such as phytotelmata, mudflies, pond margins and shallow streams. However, descriptions of larval stages for aquatic genera remain scarce, making records relatively rare (Cordeiro and Wagner 2018).

Finally, in this section, we would like to highlight some taxa commonly found in continental water bodies on Isla Grande de Tierra del Fuego and its surroundings, but not yet recorded on Isla de los Estados, likely due to limited sampling effort.

Mayflies (Ephemeroptera) and stoneflies (Plecoptera) are two insect groups that play important roles in freshwater ecology. The nymphs of mayflies occur in nearly all types of inland aquatic environments. These nymphs display a variety of feeding strategies, including shredding, scraping and, in some cases, predation. Their widespread distribution and sensitivity to environmental changes make them reliable indicators of water quality. Adult mayflies do not feed and have a short lifespan, dedicated primarily to mating, oviposition, and dispersal (Domínguez et al. 2006). Species commonly recorded in freshwater ecosystems of Isla Grande de Tierra del Fuego include *Metamonius anceps*, *Meridialaris chiloeensis* (Fig. 10.2j) and *Andesiops torrens* (Table 10.2) (Zagarola et al. 2017; GBIF.org). Stonefly larvae typically inhabit cold, shaded rivers and streams with high dissolved oxygen levels, although some species can develop in cold, oligotrophic lakes. These larvae are highly sensitive to environmental changes, such as increased sedimentation or organic pollution that can alter oxygen concentration (Fig. 10.2l). Together with Ephemeroptera and Trichoptera, they are often used in water quality indices such as the EPT index. However, the effectiveness of such tools depends on adequate taxonomic knowledge of the local macroinvertebrate community (Pessacq and Duarte Simoes 2023).

Two additional families of dipterans are noteworthy due to their specialized morphology and ecological niches. Larvae of the family Blephaceridae possess a series of sucking discs on their abdomen that allow them to adhere to substrates in fast-flowing, oxygen-rich streams. They scrape epilithon from rocks for food. The genus

Edwardsina sp. (Fig. 10.2k), which includes 32 species, 11 of which are distributed from central to southern Chile and Argentina, is a notable example (Gil-Azevedo 2018).

The Simulidae (blackflies) are a cosmopolitan family of aquatic dipterans. Their larvae attach to submerged substrate using a posterior circlet located at the distal end of the abdomen. They possess labral fans on the head, which they use to filter suspended particles from the water column (Hamada and do Nascimento 2018). *Gigantodax rufescens* is one such species documented in the Argentinian part of the Isla Grande de Tierra del Fuego.

10.4 Conclusions

The arthropod fauna of Isla de los Estados remains a largely unexplored frontier of biodiversity knowledge. Although insights can be inferred from studies conducted across the broader Tierra del Fuego archipelago and on other sub-Antarctic islands, the specific richness, composition, and ecological roles of arthropods on this remote island are still poorly understood. Therefore, our findings substantially enhance the understanding of the island's biodiversity, especially for invertebrates. The island's unique geographic and ecological context marked by isolation, harsh climate, and distinctive vegetation, suggests the presence of fascinating and perhaps unique assemblage of species, including endemics adapted to its environmental conditions. There is a pressing need for further research to document and understand this hidden biodiversity. Comprehensive surveys, taxonomic studies, and ecological investigations are essential to uncover the diversity of Isla de los Estados's arthropod fauna and to generate the baseline data necessary for informed effective conservation of its aquatic and terrestrial ecosystems.

Competing Interest On behalf of all authors, the corresponding author declares no conflict of interest.

This work was made possible thanks to the financial support to the PI Andrea Raya Rey from Agencia Nacional de Promoción Científica y Tecnológica (PICT 2017 N° 897), the Antarctic Research Trust and the Wildlife Conservation Society Argentina, We thank the Argentine Army, the General Direction of Protected Areas and Biodiversity of Tierra del Fuego Province Government and the National Park Administration for their institutional support. Special thanks to the crew of the ship ARA Granville, and to Henk Boersma and Micky Porco Fisher from the Sarah Vorwek, Puerto Beagle vessels, respectively, for safely transporting us to and from the island. We also thank Ulises Balza, Guillermo Retamar, Bruno Riviotti, Romina Mansilla, Rodrigo Iturraspe, Guillermo Gargiulo, Fernando Encinas, Emanuel Mendieta, Samanta Dodino, Joaquín Cano and Samanta Subires for their assistance with logistics and/or fieldwork. We are also grateful to the Agroforestry Resources Laboratory (CADIC) for their support in providing space and resources for arachnid storage and identification, and to Alexandra Elbakyan for facilitating access to scientific literature.

References

Acosta LE, Pérez-González A (2023) Opiliones. In: Claps LE, Roig-Juñent SA, Morrone JJ (eds) Biodiversidad de Artrópodos Argentinos volumen 5. Editorial INSUE UNT, San Miguel de Tucumán, pp 83–105

Almeida-Silva LM, Griswold CE, Brescovit AD (2015) Revision of Emmenomma Simon (Amaurobiidae, Macrobuninae). Zootaxa 3931(3):387–400. https://doi.org/10.11646/zootaxa.3931.3.3

Anderson CB, Johnson M, Lopez ME (2018) Establishing habitat-specific indicator species in Tierra del Fuego with freshwater macroinvertebrates. N Z J Mar Freshw Res 52(1):145–154

Angrisano EB, Sganga JV (2009) Trichoptera. In: Domínguez E, Fernández H (eds) Macroinvertebrados bentónicos sudamericanos. Sistemática y biología. Fundación Miguel Lillo, Tucumán, pp 255–307

Arana MD, Natale ES, Ferretti NE et al (2021) Esquema biogeográfico de la República Argentina

Arenhardt TCP, Vitorino MD, Martins SV (2021) Insecta and collembola as bioindicators of ecological restoration in the Ombrophilous dense Forest in southern Brazil. Floresta e Ambiente 28(4):e20210008

Argañaraz CI, Benitez J, Soler R et al (2023) Urbanization effect of homogenization on ground-dwelling arachnids' diversity in natural forest and peatland remnants. J Insect Conserv 27:283–294. https://doi.org/10.1007/s10841-022-00453-9

Argañaraz CI, Martínez Pastur GJ, Ramírez MJ et al (2020) Ground-dwelling spiders and understory vascular plants on Fuegian austral forests: community responses to variable retention management and their association to natural ecosystems. For Ecol Manag 474:118375. https://doi.org/10.1016/j.foreco.2020.118375

Armendano A, González A (2011) Efecto de las arañas (Arachnida: Araneae) como depredadoras de insectos plaga en cultivos de alfalfa (Medicago sativa) (Fabaceae) en Argentina. Rev Biol Trop 59(4):1651–1662

Barone ML, Werenkraut V, Ramírez MJ (2016) New species and phylogenetic relationships of the spider genus Coptoprepes using morphological and sequence data (Araneae: Anyphaenidae). Zootaxa 4175(5):436–448. https://doi.org/10.11646/zootaxa.4175.5.2

Bell JR, Bohan DA, Shaw EM et al (2005) Ballooning dispersal using silk: world fauna, phylogenies, genetics and models. Bull Entomol Res 95:69–114. https://doi.org/10.1079/BER2004350

Beron P (2018) Regional Arachnogeography. In: Zoogeography of Arachnida. Monographiae Biologicae, vol 94. Springer, Cham. https://doi.org/10.1007/978-3-319-74418-6_7

Bickel D, Pape T, Meier R (eds) (2009) Diptera diversity: status, challenges and tools. Brill, Leiden

Bonte D, Vandenbroecke N, Lens L et al (2003) Low propensity for aerial dispersal in specialist spiders from fragmented landscapes. Proc R Soc B 270:1601–1607. https://doi.org/10.1098/rspb.2003.2432

Burgo A, Catley K, Grismado CJ, Dupérré N et al (2025) Systematics of the colour-polymorphic spider genus *Cybaeolus*, with comments on the phylogeny of the family Hahniidae (Araneae). Zool J Linnean Soc 203:zlae165. https://doi.org/10.1093/zoolinnean/zlae165

Cardoso P, Barton PS, Birkhofer K et al (2020) Scientists' warning to humanity on insect extinctions. Biol Conserv 242:108426

Cardoso P, Scharff N, Gaspar C et al (2008) Rapid biodiversity assessment of spiders (Araneae) using semi-quantitative sampling: a case study in a Mediterranean forest. Insect Conserv Diver 1:71–84. https://doi.org/10.1111/j.1752-4598.2007.00008.x

Chown SL, Lee JE, Shaw JD (2008) Conservation of Southern Ocean Islands: invertebrates as exemplars. Journal of Insect Conservation, 12(3-4), 277–291. https://doi.org/10.1007/s10841-008-9151-8

Collier KJ, Probert PK, Jeffries M (2016) Conservation of aquatic invertebrates: concerns, challenges and conundrums. Aquat Conserv Mar Freshwat Ecosyst 26(5):817–837

Contador T, Kennedy J, Ojeda J et al (2014) Ciclos de vida de insectos dulceacuícolas y cambio climático global en la ecorregión subantártica de Magallanes: investigaciones ecológicas a largo plazo en el Parque Etnobotánico Omora, Reserva de Biosfera Cabo de Hornos (55 S). Bosque (Valdivia) 35(3):429–437

Cook J (1999) The voyages of captain cook. Wordsworth Editions, London

Cordeiro DP, Wagner R (2018) Family Psychodidae. In: Thorp and Covich's freshwater invertebrates. Academic Press, Boston, pp 765–770

Cuezzo MG, Gregoric DEG, Pointier JP et al (2020) Phylum Mollusca. In: Thorp and Covich's freshwater invertebrates. Academic Press, Boston, pp 261–430

Cuezzo MG (2009) Mollusca Gastropoda. In: Domínguez E, Fernández H (eds) Macroinvertebrados bentónicos sudamericanos. Sistemática y biología. Fundación Miguel Lillo, Tucumán, pp 595–629

Derkarabetian S, Baker CM, Giribet G (2021) Complex patterns of Gondwanan biogeography revealed in a dispersal-limited arachnid. J Biogeogr 48(6):1336–1352

Dimitrov D, Hormiga G (2021) Spider diversification through space and time. Annu Rev Entomol 66:225–241. https://doi.org/10.1146/annurev-ento-061520-083414

Domínguez E, Fernández HR (2009) Macroinvertebrados bentónicos sudamericanos: sistemática y biología. Fundación Miguel Lillo, Tucumán, p 656

Domínguez E, Molineri C, Pescador ML et al (2006) Ephemeroptera of South America. In: Adis J, Arias JR, Rueda-Delgado GK, Wantzen M (eds) Aquatic Biodiversity in Latin America (ABLA), vol 2. Pensoft, Sofia-Moscow, p 646

Dupérré N, Harms D (2018) Raising the dead: rediscovery and redescription of some lost spider types (Araneae) described by Eugène Simon. Evol Syst 2:1–20. https://doi.org/10.3897/evolsyst.2.24122

Elgueta M, Jezek J (2014) Nuevos registros de Psychodidae (Diptera), con una lista de especies citadas para Chile. In: Anales del Instituto de la Patagonia, vol 42. Universidad de Magallanes, Punta Arenas, pp 71–84

Farfán-Beltrán ME, Chávez-Pesqueira M, Hernández-Cumplido J, Cano-Santana Z (2022) A quick evaluation of ecological restoration based on arthropod communities and trophic guilds in an urban ecological preserve in Mexico City. Rev Chil Hist Nat 95:1–13

Fenoglio MS, Rossetti MR, Videla M (2020) Negative effects of urbanization on terrestrial arthropod communities: a meta-analysis. Glob Ecol Biogeogr 29:1412–1429

Galli L, Rellini I (2020) The geographic distribution of Protura (Arthropoda: Hexapoda): a review. Biogeographia 35:51–69. GBIF.org. 2025, April 9

GBIF.org – Global Biodiversity Information Facility. Occurrence Download. https://doi.org/10.15468/dl.pz25nu

GBIF.org a – Global Biodiversity Information Facility. Occurrence Download. https://doi.org/10.15468/dl.jxqg5k

GBIF.org b – Global Biodiversity Information Facility. Occurrence Download. https://doi.org/10.15468/dl.fts95q

GBIF.org c – Global Biodiversity Information Facility. Occurrence Download. https://doi.org/10.15468/dl.ecpf29

Gil-Azevedo LH (2018) Family Blephariceridae. In: Thorp and Covich's freshwater invertebrates. Academic Press, pp 759–763

Giribet G, Edgecombe GD (2019) The phylogeny and evolutionary history of arthropods. Curr Biol 29(12):592–602. https://doi.org/10.1016/j.cub.2019.04.057

Giribet G, Sheridan K, Baker CM et al (2021) A molecular phylogeny of the circum-Antarctic opiliones family Neopilionidae. Invertebr Syst 35(8):827–849. https://doi.org/10.1071/IS21012

Grismado CJ, Lopardo L (2003) Nuevos datos sobre la distribución geográfica de las familias australes de arañas Malkaridae y Mecysmaucheniidae (Arachnida: Araneae), con la descripción de la hembra de *Mecysmauchenius thayerae* Forster Platnick. Revista Ibérica de Aracnología 8:37–43

Grismado C, Ramírez MJ, Izquierdo MA (2014) Araneae: taxonomía, diversidad y clave de identificación de familias de la Argentina. In: Roig-Juñent S, Claps LE, Morrone JJ (eds) Biodiversidad de Artrópodos Argentinos, vol 3. Editorial INSUE – UNT, Entre Ríos, pp 55–94

Gutierrez Gregoric DE (2020) Chilinoidea, Chilinidae. In: Thorp and Covich's freshwater invertebrates. Academic Press, Boston, pp 295–301

Haas M, Baur H, Schweizer T et al (2021) Tiny wasps, huge diversity – a review of German Pteromalidae with new generic and species records (Hymenoptera: Chalcidoidea). Biodivers Data J 9:e77092. https://doi.org/10.3897/BDJ.9.e77092

Hamada N, do Nascimento JMC (2018) Family Simuliidae. In: Thorp and Covich's freshwater invertebrates. Academic Press, Boston, pp 701–718

Hamada N, Thorp JH, Rogers DC (2018) Keys to neotropical Hexapoda. In: Thorp and Covich's Freshwater Invertebrates, vol III. Academic, Boston

Hanson P, Springer M, Ramirez A (2010) Capítulo 1: Introducción a los grupos de macroinvertebrados acuáticos. Rev Biol Trop 58:3–37

Harms D, Harvey MS, Roberts JD et al (2024) Tectonically driven climate change and the spread of temperate biomes: insights from dragon pseudoscorpions (Pseudotyrannochthoniidae), a globally distributed arachnid lineage. J Biogeogr 51(6):1032–1048

Kenney MA, Sutton-Grier AE, Smith RF et al (2009) Benthic macroinvertebrates as indicators of water quality: the intersection of science and policy. Terr Arthropod Rev 2(2):99

Kulkarni S, Wood HM, Hormiga G (2023) Advances in the reconstruction of the spider tree of life: a roadmap for spider systematics and comparative studies. Cladistics 39:479–532. https://doi.org/10.1111/cla.12557

Lavery AH (2017) Annotated checklist of the spiders, harvestmen, and pseudoscorpions of The Falkland Islands and South Georgia. Arachnology 17(5):210–228. https://doi.org/10.13156/arac.2017.17.5.210

Lavery AH, Snazell RG (2013) The spiders of The Falkland Islands 1: Erigoninae (Araneae, Linyphiidae). Arachnology 16(2):37–60. https://doi.org/10.13156/arac.2013.16.2.37

Lavery AH, Snazell RG (2023) The spider of the Falkland Island 2: non-erigonine Linyphiidae. Arachnology 19(4):681–698

Lawania KK, Trigunayat K, Trigunayat MM (2013) Spider in mosquito control. Int J Environ Eng Manag 4(4):331–338

Leclère T, Gerhold P (2024) Phylogenetic diversity of plant and insect communities on islands. Ecol Evol 14(12):e70660. https://doi.org/10.1002/ece3.70660

Libonatti ML, Michat MC, Torres PL (2011) Clave para los adultos de las subfamilias, tribus y géneros de Dytiscidae de la Argentina (Coleoptera: Adephaga). Revista de la Sociedad Entomológica Argentina 70(3–4):317–336

Lopardo L (2005) Phylogenetic revision of the spider genus Negayan (Araneae, Anyphaenidae, Amaurobioidinae). Zool Scr 34(3):245–277. https://doi.org/10.1111/j.1463-6409.2005.00194.x

Lopez ME, Saad J, Izaguirre I et al (2021) Aquatic macroinvertebrates from littoral vegetated lakes in arid Patagonia: the influence of fish introduction on community structure. Aquat Conserv Mar Freshwat Ecosyst 31(7):1873–1885

Maggi T, Pardo L, Chroil R (2021) Insect pollinators: a key to ecosystem resilience and food security. Pollinators 1:35–50

Majer J (1987) The conservation and study of invertebrates in remnants of native vegetation. In: Saunders D, Burbidge A, Hopkins A (eds) Nature conservation: the role of remnants of native vegetation. Surrey Beatty & Sons, Chipping Norton, pp 333–335

Maury EA, Pinto da Rocha R, Morrone JJ (1996) Distribution of *Acropsopilio chilensis* Silvestri, 1904 in southern South America (Opiliones, Palpatores, Caddidae). Biogeographica 72(3):127–132

Michat MC, Archangelsky M, Bachmann AO (2008) Claves genéricas para la identificación de las larvas de Dytiscidae de la Argentina (Coleoptera: Adephaga). Revista de la Sociedad Entomológica Argentina 67(3–4):17–36

Michat MC, Archangelsky A, Benetti CJ (2023) Dytiscidae. In: Biodiversidad de artrópodos argentinos. Sociedad Entomologica Argentina, Mendoza, pp 137–158

Miller JA (2007) Review of erigonine spider genera in the Neotropics (Araneae: Linyphiidae, Erigoninae). Zool J Linnean Soc 149(1):1–263. https://doi.org/10.1111/j.1096-3642.2007.00233.x

Millidge AF (1985) Some Linyphiid spiders from South America (Araneae, Linyphiidae). Am Mus Novit 2836:1–78

Misof B, Liu S, Meusemann K et al (2014) Phylogenomics resolves the timing and pattern of insect evolution. Science 346(6210):763–767

Morrone JJ (1993) Revisión sistemática de un nuevo género de Rhytirrhinini (Coleoptera: Curculionidae), con un análisis biogeográfico del dominio Subantártico. Boletın de la Sociedad de Biologıa de Concepcion 64:121–145

Morrone J (2000) Biogeographic delimitation of the Subantarctic subregion and its provinces. Revista Del Museo Argentino de Ciencias Naturales, Nueva Serie 2(1):1–15

Morrone JJ, Roig-Juñent S, Crisci JV (1994) Cladistic biogeography of terrestrial Subantarctic beetles (Insecta: Coleoptera) from southern South America. Natl Geogr Res Explor 10:104–115

Munari L (2015) Is the diversity of the beach flies adequately known? Some notes on the state of the art of current knowledge (Diptera: Canacidae). Boll Soc Entomol Ital 147(3):99–111

New TR, Samways MJ (2014) Insect conservation in the southern temperate zones: an overview. Austral Entomol 53(1):26–31

Paggi AC, Donato MH, Siri A (2023) Chironomidae. In: Biodiversidad de artrópodos argentinos (pp. 248–285–158). Sociedad Entomologica Argentina, Mendoza

Paulsen MJ (2006) Monographic revision of the southern south American stag beetles of the genera Pycnosiphorus Solier and Erichius Maes (Coleoptera: Lucanidae: Lucaninae), and molecular systematics of Lucanidae. University of Nebraska-Lincoln

Pes AM, Holzenthal RW, Sganga JV et al (2018) Order trichoptera. In: Thorp and Covich's freshwater invertebrates. Academic Press, Boston, pp 237–324

Pessacq P, Duarte Simoes TV (2023) Plecoptera. In: Biodiversidad de artrópodos argentinos (pp. 248-83-105). Sociedad Entomológica Argentina, Mendoza

Ponce JF, Fernández M (2014) Climatic and environmental history of Isla de los Estados, Argentina. Springer, Netherlands. https://doi.org/10.1007/978-94-007-4363-2_6

Posadas PE (2002) Los curculionoidea (Insecta: coleoptera) de la Tierra del Fuego e Islas Malvinas. Thesis. https://doi.org/10.35537/10915/4618

Posadas P (2012) Species composition and geographic distribution of Fuegian Curculionidae (Coleoptera: Curculionoidea). Zootaxa 3303(1):1–36. https://doi.org/10.11646/ZOOTAXA.3303.1.1

Pugh PJA (2004) Biogeography of spiders (Araneae: Arachnida) on the islands of the Southern Ocean. J Nat Hist 38(12):1461–1487. https://doi.org/10.1080/0022293031000155403

Raguso RA (2020) Don't forget the flies: dipteran diversity and its consequences for floral ecology and evolution. Appl Entomol Zool 55(1):1–7

Ramírez MJ (2003) The spider subfamily Amaurobioidinae (Araneae, Anyphaenidae): a phylogenetic revision at the generic level. Bull Am Mus Nat Hist 277:1–262

Ramírez MJ, Magalhaes ILF, Pizarro-Araya J et al (2022) A new species of the spider genus Tekellina Levi, 1957 from Chile, with a broadened definition of the family Synotaxidae (Arachnida, Araneae). Zool Anz 301:76–90. https://doi.org/10.1016/j.jcz.2022.08.005

Rincón J, Merchán D, Rojas D et al (2016) Los Macroinvertebrados acuáticos de los ríos del Parque Nacional Cajas, 1st edn. Don Bosco, Ecua-dor

Rodriguez-Cabal MA, Barrios-Garcia MN, Greyson-Gaito CJ et al (2019) Non-native ungulates indirectly impact foliar arthropods but not soil function. Biol Invasions 21:3077–3084

Rogers DC, Fredes NA, Martínez PA et al (2020) Phylum Arthropoda: introduction and Arachnida. In: Thorp and Covich's freshwater invertebrates. Academic Press, Boston, pp 523–559

Roig-Juñent S, Griotti M, Miras D et al (2024) Geadephaga beetles (Insecta: Coleoptera) from the sub-Antarctic Tierra del Fuego archipelago: diversity, distribution and taxonomic key. Polar Biol 47(12):1441–1456. https://doi.org/10.21203/rs.3.rs-3850609/v1

Roth VD (1967) A review of the south American spiders of the family Agelenidae (Arachnida, Araneae). Bull Am Mus Nat Hist 134(5):297–346

Salom A (2025) Herbívoros introducidos en Isla de los Estados: estado poblacional e impactos producidos por el ciervo colorado (*Cervus elaphus*) y la cabra (*Capra hircus*). Dissertation, Universidad de Buenos Aires

Samways MJ, Barton PS, Birkhofer K et al (2020) Solutions for humanity on how to conserve insects. Biol Conserv 242:108427

Savoretti A, Coronato A, Roig C et al (2021) Isla Grande de Tierra del Fuego and Isla de los Estados Peatlands. Springer, Cham, pp 155–171. https://doi.org/10.1007/978-3-030-60683-1_9

Simon E (1896) Arachnides recueillis à la Terre-de-feu par M. Carlos Backhausen. (2e Mémoire). Anales del Museo Nacional de Buenos Aires 5(ser. 2, t. 2):141–145

Skottsberg C (1911) The wilds of Patagonia: a narrative of the Swedish expedition to Patagonia, Tierra del Fuego and The Falkland Islands in 1907–1909. E. Arnold, London

Soares EDG, De Carvalho CJB (2005) Biogeography of Palpibracus (Diptera: Muscidae): an integrative study using panbiogeography, parsimony analysis of endemicity, and component analysis. Regionalización biogeográfica en Iberoamérica y tópicos afines, pp 485–494

Spinelli G, Marino P, Posadas P (2006) The Patagonian species of the genus Atrichopogon Kieffer, with a biogeographic analysis based on Forcipomyiinae (Diptera: Ceratopogonidae). Insect Syst Evol 37(3):301–324

Stark BP, Froehlich C, Zuniga MC (2009) South American stoneflies (Plecoptera). In: Adis J, Arias JR, Golovatch S, Wantzen KM, Rueda-Delgado G (eds) Aquatic Biodiversity of Latin America (ABLA), vol 5. Pensoft, Sofia-Moscow, p 154

Szymkowiak P, Górski G, Bajerlein D (2007) Passive dispersal in arachnids. Biol Lett 44(2):75–101

Theron GL, Anderson B, Cozien RJ et al (2023) We don't know the half of it: morphological and molecular evidence reveal dramatic underestimation of diversity in a key pollinator group (Nemestrinidae). Invertebr Syst 37(1):1–13

Trifilo SS (1960) Darwin and the second beagle expedition in Tierra del Fuego. Pac Hist Rev 29(3):221–229. https://doi.org/10.2307/3636161

Usher MB (1983) Spiders from Beauchene Island, Falkland Island, South Atlantic. J Zool (Lond) 200:571–582

Wallace JB, Webster JR (1996) The role of macroinvertebrates in stream ecosystem function. Annu Rev Entomol 41(1):115–139

World Spider Catalog (2025) World Spider Catalog. Version 26. Natural History Museum Bern. http://wsc.nmbe.ch, https://doi.org/10.24436/2

Zagarola JPA, Martínez Pastur GJ, Lopez ME et al (2017) Assessing the effects of urbanization on streams in Tierra del Fuego. Ecol Austral 27(1):045–054

Zhang YM, Bossert S, Spasojevic T (2025) Evolving perspectives in Hymenoptera systematics: bridging fossils and genomes across time. Syst Entomol 50(1):1–31

Zhu Y, Veen GF, Heinen R et al (2023) Large mammalian herbivores affect arthropod food webs via changes in vegetation characteristics and microclimate. J Ecol 111(9):2077–2089

Chapter 11
Fish Diversity in Isla de los Estados and Surrounding Areas

Javier H. Rojo, Daniel O. Bruno, Facundo Llompart, Tomás Chalde, and Sergio M. Delpiani

Abstract Knowledge about fish diversity in Isla de los Estados is relatively limited and outdated compared to the well-documented biodiversity of the nearby Isla Grande de Tierra del Fuego. This lack of data may lead to an underestimation of the number of species inhabiting this remote region of the southern Atlantic Ocean. This chapter reviews the fish species found in freshwater, coastal, and surrounding marine areas of Isla de los Estados. Research on the island's freshwater fish diversity is scarce, with one native species identified to date. In contrast, 16 species have been recorded in coastal marine zones, some of which were described more than 120 years ago. Additionally, 25 species have been reported in the surrounding marine areas, with several species obtained from commercial fishing vessels. This study highlights the urgent need for new sampling efforts to produce a comprehensive checklist of fish biodiversity. An updated inventory, along with basic biological information, such as local distribution, reproduction patterns and interactions with other taxonomic groups, is essential for the development of effective conservation and management policies.

J. H. Rojo (✉)
Centro Austral de Investigaciones Científicas (CADIC-CONICET),
Ushuaia, Tierra del Fuego, Argentina

D. O. Bruno · F. Llompart · S. M. Delpiani
Centro Austral de Investigaciones Científicas (CADIC-CONICET),
Ushuaia, Tierra del Fuego, Argentina

Instituto de Ciencias Polares, Ambiente y Recursos Naturales, Universidad Nacional de Tierra del Fuego, Antártida e Islas del Atlántico Sur (ICPA–UNTDF),
Ushuaia, Tierra del Fuego, Argentina

T. Chalde
Centro Austral de Investigaciones Científicas (CADIC-CONICET),
Ushuaia, Tierra del Fuego, Argentina

Universidad Tecnológica Nacional, Facultad Regional Tierra del Fuego,
Ushuaia, Tierra del Fuego, Argentina

A. Raya Rey, U. Balza (eds.), *Biodiversity and Conservation of Isla de los Estados (Chuani-sin)*, Natural and Social Sciences of Patagonia,
https://doi.org/10.1007/978-3-032-13630-5_11

Keywords Southern Atlantic ocean · Chondrichthyans · Teleosts · Tierra del Fuego

11.1 Introduction

Isla de los Estados (IDLE) is located at the southern tip of South America, approximately 240 km east of the city of Ushuaia and 30 km from the eastern shore of Isla Grande de Tierra del Fuego (IGTDF), separated by the Le Maire Strait (Fig. 11.1). IDLE is characterized by a cold, insular, and windy climate, which makes access difficult. The only permanent settlement on the island is an Argentine Army post. These adverse conditions have likely hindered scientific exploration efforts.

The island is surrounded by deep channels where permanent oceanographic fronts are generated due to the area's hydrology and topography (Acha et al. 2004). From the west, the region receives modified subantarctic waters (salinities between 26 and 32) formed by the mixing of full salinity Pacific Ocean waters with fresh water inputs from melting ice, snow, and continental runoff along the Beagle Channel (Balestrini et al. 1998; Giesecke et al. 2021). These waters flow in a west-to-east direction, driven by sea level differences between the Pacific and Atlantic

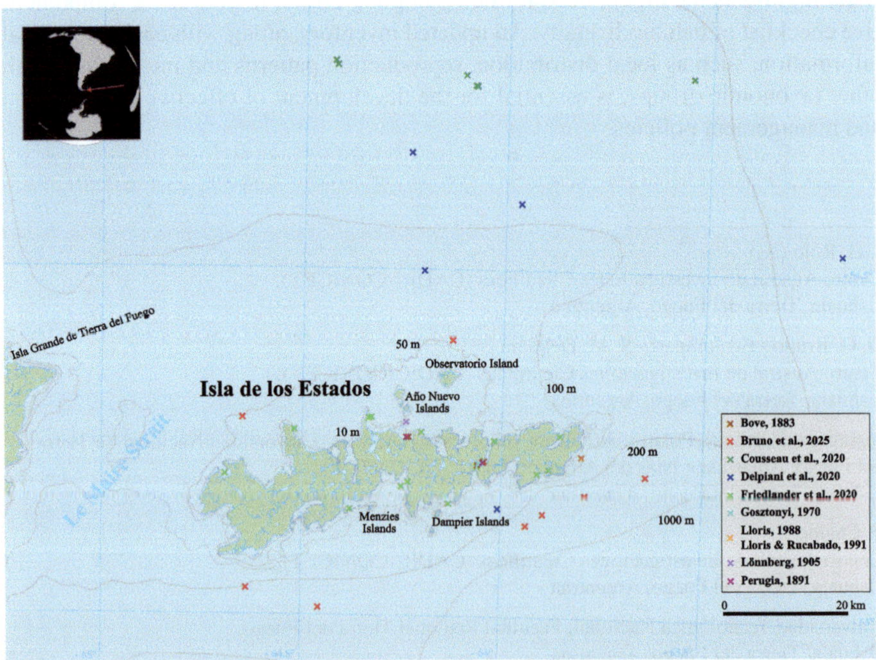

Fig. 11.1 Map of Isla de los Estados with the location of the fish sampling sites. Grey lines represent isobaths at 10, 50, 100, 200 and 1000 m

oceans and by prevailing westerly winds (Cucco et al. 2022; Martín et al. 2023). In turn, these waters mix with saltier waters (<33.9) from the Cape Horn Current, which enters the continental shelf through the Le Maire Strait (Acha et al. 2004) and the western channel of the Burdwood Bank (Matano et al. 2019). This results in a complex current system that circulates around the island (Piola and Matano 2001). Its 470 km-long coastline resembles that of eastern IGTDF and is characterized by an irregular margin with numerous fjords and bays (Ponce and Fernández 2014). The subtidal zone depicts extensive kelp forests dominated by macroalgae such as *Macrocystis pyrifera* and *Lessonia* spp., with a rocky substrate intermixed with sand and/or silt (Dayton 1985; Friedlander et al. 2020; Hüne et al. 2021). Sea temperatures in the area ranged between 5 and 7 °C (Dudley and Crow 1983).

The earliest documentation of fish species in IDLE dates back to a report from the Italian expedition to Antarctica at the end of the nineteenth century (Bove 1883). Between that time and the early twentieth century, only two additional records, derived from European South American expeditions, mention fish species in the area (Perugia 1891; Lönnberg 1905). Since then, very few studies have focused on freshwater and coastal ichthyofauna of IDLE (Gosztonyi 1970; Lloris 1988; Nardi et al. 2019; Friedlander et al. 2020). Additionally, several oceanographic surveys and commercial fishing operations have documented fish species in the surrounding marine areas of IDLE at depths ranging from 40 to 250 m (Fig. 11.1).

There are approximately 290 permanent freshwater bodies on the island, over 80% of which have a surface smaller than 0.1 km^2 (Fig. 11.1). In addition, around 280 creeks and small streams are distributed across the island. However, records of freshwater fish are limited to a few streams discharging directly into the sea, which likely leads to an underestimation of the island's freshwater fish diversity.

From an ichthyological perspective, IDLE falls within the Atlantic sector of the Magellanic Province. This biogeographic region spans the Patagonian shelf from Península Valdés (43°S) to Cape Horn (55°S), extending northward along the continental shelf (Balech and Ehrlich 2008; Cousseau et al. 2020). Excluding the Pacific sector, 254 fish species have been recorded in the Magellanic Province (Cousseau et al. 2020). Among bony fish, a high level of endemism is observed in some taxa well adapted to temperate-cold environments such as zoarcids, notothenioids, and liparids (Balech and Ehrlich 2008; Cousseau et al. 2020). For chondrichthyans, only 17 species have been documented, represented by a significant number of batoids from the families Arhynchobatidae and Rajidae, as well as two sharks and one holocephalan (Sabadin et al. 2020). However, climate change, fishing pressure, and biological invasion by alien species are altering fish community composition across southern Patagonia through temperature change, acidification, hypoxia or competition (Galván et al. 2022).

Drawing from the existing literature, this chapter provides a comprehensive review of the documented fish diversity in the freshwater, coastal, and surrounding marine environments of IDLE. It also includes available information on ichthyoplankton in nearby marine areas. This work serves as the foundation for future efforts to enhance knowledge and develop conservation strategies for the island's fish species.

11.2 Fish Species of Isla de los Estados

As in IGTDF, the number of freshwater fish species in IDLE is very low (Table 11.1, Fig. 11.2). Over the last 140 years, only one native species was recorded in the island's freshwater bodies, the puyen *Galaxias maculatus* (Jenyns 1842). It was found in both landlocked and diadromous forms (Bove 1883; Perugia 1891; Gosztonyi 1970; Ferriz and Gómez 2015). However, the presence of additional native species cannot be ruled out. To date, no exotic species, such as salmonids, have been documented either through conventional sampling methods (e.g., electrofishing or trawl nets) or environmental DNA analysis (Nardi et al. 2019).

In contrast, the marine ichthyofauna of IDLE is characterized by species adapted to the extreme conditions of southern waters and occupying a wide range of depths and ecological niches (Table 11.1, Fig. 11.2). The dominant group is the nototheniids (family Nototheniidae), known for their adaptation to low temperature (Balushkin 2000). These benthic fish, with robust bodies and large pectoral fins, inhabit coastal and deep waters and play a key role in southern marine food webs as predators of benthic invertebrates and small fish. Noteworthy species include *Paranotothenia magellanica* (Forster 1801), *Patagonotothen brevicauda* (Lönnberg 1905),

Table 11.1 Fish species documented in Isla de los Estados and surrounding marine areas

Ambient	Order	Family	Species	IUCN conservation status
Freshwater	Galaxiiformes	Galaxiidae	*Galaxias maculatus*	Least concern
Marine coastal	Myxiniformes	Myxinidae	*Myxine australis*	Least concern
	Ophidiiformes	Ophidiidae	*Genypterus blacodes*	Vulnerable
	Perciformes	Bovichtidae	*Cottoperca trigloides*	Least concern
		Harpagiferidae	*Harpagifer bispinis*	Least concern
		Nototheniidae	*Paranotothenia magellanica*	Not evaluated
			Patagonotothen brevicauda	Not evaluated
			Patagonotothen cornucola	Least concern
			Patagonotothen longipes	Not evaluated
			Patagonotothen sima	Not evaluated
			Patagonotothen squamiceps	Not evaluated
			Patagonotothen tessellata	Data deficient
		Zoarcidae	*Austrolycus depressiceps*	Not evaluated
			Crossostomus chilensis	Data deficient
			Maynea puncta	Not evaluated
			Phucocoetes latitans	Least concern
			Pogonolycus marinae	Not evaluated

(continued)

Table 11.1 (continued)

Ambient	Order	Family	Species	IUCN conservation status
Marine surrounding areas	Carcharhiniformes	Scyliorhinidae	*Schroederichthys bivius*	Least concern
	Clupeiformes	Clupeidae	*Sprattus fuegensis*	Least concern
	Gadiformes	Gadidae	*Micromesistius australis*	Not evaluated
		Moridae	*Salilota australis*	Data deficient
		Merlucciidae	*Macruronus magellanicus*	Not evaluated
			Merluccius australis	Not evaluated
	Lamniformes	Lamnidae	*Lamna nasus*	Vulnerable
	Myctophiformes	Myctophidae	*Gymnoscopelus braueri*	Least concern
	Myxiniformes	Myxinidae	*Myxine affinis*	Least concern
	Perciformes	Agonidae	*Agonopsis chiloensis*	Not evaluated
		Nototheniidae	*Dissostichus eleginoides*	Not evaluated
			Eleginops maclovinus	Not evaluated
			Patagonotothen elegans	Not evaluated
			Patagonotothen jordani	Not evaluated
			Patagonotothen ramsayi	Data deficient
			Patagonotothen wiltoni	Not evaluated
		Sebastidae	*Sebastes oculatus*	Least concern
	Rajiformes	Arhynchobatidae	*Bathyraja albomaculata*	Vulnerable
			Bathyraja brachyurops	Near threatened
			Bathyraja cousseauae[a]	Least concern
			Bathyraja macloviana	Near threatened
			Bathyraja magellanica	Least concern
			Sympterygia bonapartii	Near threatened
			Psammobatis normani	Least concern
			Psammobatis rudis[a]	Least concern
	Scombriformes	Centrolophidae	*Icichthys australis*	Not evaluated
			Seriolella porosa	Not evaluated
		Stromateidae	*Stromateus brasiliensis*	Not evaluated
	Squaliformes	Squalidae	*Squalus acanthias*	Vulnerable

[a]Including egg capsules

P. cornucola (Richardson 1844), *P. longipes* (Steindachner 1876), *P. sima* (Richardson 1845), *P. squamiceps* (Peters 1877), and *P. tessellata* (Richardson 1845) (Bove 1883; Gosztonyi 1970; Friedlander et al. 2020).

The zoarcids (family Zoarcidae) are also well represented. These fish have elongated, slender bodies adapted to a bottom-dwelling lifestyle, often residing in crevices and burrows. This family occurs across a broad depth range, from shallow to abyssal waters (Anderson 1990), and preys mainly on small invertebrates.

Fig. 11.2 Some of the most representative freshwater and marine coastal fish species documented in IDLE

Representative species include *Austrolycus depressiceps* Regan, (1913), *Crossostomus chilensis* (Regan 1913), *Maynea puncta* (Jenyns 1842), *Phucocoetes latitans* Jenyns (1842) and *Pogonolycus marinae* (Lloris 1988) (Bove 1883; Gosztonyi 1970; Lloris 1988; Lloris and Rucabado 1991; Friedlander et al. 2020).

Ophidiids (family Ophidiidae) are deep-sea predators recognised for their elongated bodies and sharp teeth (Nielsen 1999), feeding actively on fish and squid. The only species recorded in the area is *Genypterus blacodes* (Forster 1801) (Bove 1883; Lönnberg 1905).

The bovichtids (family Bovichthyidae) are fish characterized by robust bodies, large heads, and well-developed pectoral fins adapted for manoeuvrability in complex benthic environments. They primarily consume crustaceans and small fish. *Cottoperca trigloides* (Forster 1801) is the only bovichtid reported in IDLE (Lattuca et al. 2020; Friedlander et al. 2020).

Hagfish (family Myxinidae) are jawless fish known for producing copious amounts of mucus. They burrow into the seabed and scavenge on dead or dying animals (Fernholm 1998). *Myxine australis* Jenyns (1842) has been reported in IDLE (Bove 1883).

In the intertidal and coastal zones of the fjords, the thornfish, *Harpagifer bispinis* (Forster 1801), is found (Lönnberg 1905; Gosztonyi 1970; Friedlander et al. 2020). This species belongs to the Harpagiferidae family, mainly distributed in sub-Antarctic and Antarctic regions inhabiting shallow waters associated with rocky substrates and soft bottoms. They are small fish with elongated and depressed bodies, robust heads, and distinctive cephalic spines. These sedentary fish have antifreeze glycoproteins in their blood an adaptation reminiscent of its Antarctic ancestor (Hüne and Vega 2016; Llompart et al. 2020).

11.3 Marine Surrounding Area of Isla de los Estados

In the marine surrounding areas of IDLE —delimited based on ichthyofaunal similarities, topography, and substrate type, following the geographical framework proposed by Cousseau et al. (2020)— numerous fish species have been documented through oceanographic surveys. These surveys include expeditions aboard Buque Oceanográfico ARA "Puerto Deseado", Guardacosta SB-15 "Tango" and Guardacosta GC-189 "Prefecto García", Buque de Investigación Pesquera "Capitán Oca Balda", Buque de Investigación Pesquera "Doctor Eduardo L. Holmberg", and Buque de Investigación Pesquera Oceanográfica "Víctor Angelescu", and by commercial fishing vessels. These efforts have recorded a variety of species, including fish eggs (from chondrichthyans and teleosts) and larvae (Figs. 11.3 and 11.4).

Agonids (family Agonidae) are bottom-dwelling fish with bodies covered in bony plates that inhabit rocky and sandy substrates and feed on small invertebrates (Shiko and Mecklenburg 2004). *Agonopsis chiloensis* (Jenyns 1842) is the only agonid species recorded in the area (Ehrlich et al. 1999).

The hakes (Order Gadiformes) are benthopelagic fish distributed worldwide, many of which are of high commercial value (Cohen et al. 1990). Documented species include *Macruronus magellanicus* Lönnberg, 1907, *Merluccius australis* (Hutton 1872), *Micromesistius australis* (Norman, 1937), and *Salilota australis* (Günther 1878) (Delpiani et al. 2020).

Rockfish or rockcods (family Sebastidae) are also present. These species range from intertidal zones to deep waters and primarily feed on small fish and

Fig. 11.3 Hagfish and some of the teleost fish species documented in the marine surrounding areas of IDLE

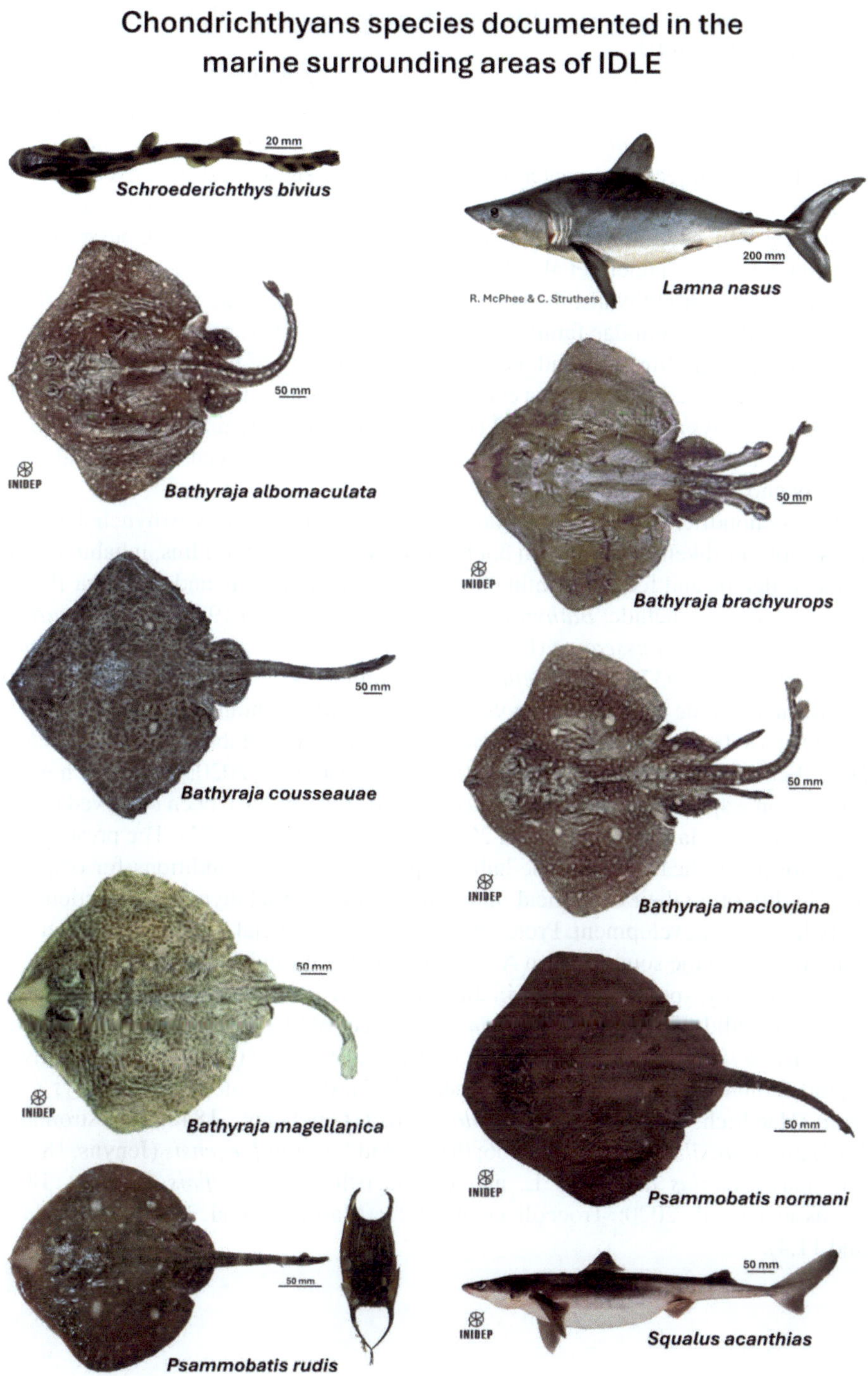

Fig. 11.4 Some of the chondrichthyans documented in the marine surrounding areas of IDLE

crustaceans (Marcinkevicius 2019). Only *Sebastes oculatus* Valenciennes, 1833 has been reported from IDLE.

Within the nototheniids, two species are particularly relevant to fisheries: the Patagonian toothfish *Dissostichus eleginoides* Smitt, 1898 and the Longtail Southern cod *Patagonotothen ramsayi* (Regan 1913). *D. eleginoides* is highly valued commercially and subject to regulated fisheries due to its susceptibility to overexploitation (Troccoli et al. 2023; Wöhler et al. 2024), while *P. ramsayi* plays a central ecological and economic role in Patagonian coastal ecosystems (Cousseau and Perrotta 2004; Riccialdelli et al. 2020).

Among chondrichthyans, the porbeagle shark, *Lamna nasus* (Bonnaterre 1788), a member of the Lamnidae family, has been identified. This species is known for its active and pelagic lifestyle, and a diet composed mainly of bony fish and cephalopods (Compagno 2008; Belleggia et al. 2021; Chiaramonte et al. 2024). The catshark *Schroederichthys bivius* (Müller and Henle 1838) from the family Scyliorhinidae, is also found. This small benthic species preys on crustaceans and small fish (Compagno 1984).

Other chondrichthyans reported are the skates from the family Arhynchobatidae. These bottom-dwelling rays, with flat bodies and broad pectoral fins, inhabit a wide range of depths and feed on benthic invertebrates (McEachran and di Sciara 1995). Reported species include: *Bathyraja albomaculata* (Norman 1937), *B. brachyurops* (Fowler 1910), *B. cousseauae* (Díaz de Astarloa and Mabragaña 2004), *B. macloviana* (Norman 1937), *B. magellanica* (Philippi 1902), *Sympterygia bonapartii* Müller and Henle 1841, *Psammobatis normani* McEachran, 1983 and *P. rudis* (Günther 1870) (Ruocco et al. 2006; Belleggia et al. 2008; Mabragaña 2011; Scenna 2011; Scenna and Díaz De Astarloa 2014; Cousseau et al. 2020; Matusevich et al. 2023). Some species, such as *B. cousseauae* and *P. rudis*, have been observed to use the area for egg-laying (Mabragaña 2011; Matusevich et al. 2023). The presence of egg capsules indicates that these habitats provide suitable conditions for oviposition, highlighting their ecological importance for chondrichthyan reproduction and early life-stage development. Protecting these areas is crucial for skates biodiversity conservation in the southwestern Atlantic Ocean (Matusevich et al. 2023).

Additionally, species recorded in the marine surrounding areas of IDLE include: the nototheniids *Eleginops maclovinus* (Cuvier 1830), *Patagonotothen elegans* (Günther 1878), *P. jordani* (Thompson 1916), and *P. wiltoni* (Regan 1913); the myctophid *Gymnoscopelus braueri* (Lönnberg 1905); the centrolophids *Icichthys australis* Haedrich, 1966, and *Seriolella porosa* Guichenot, 1848; the stromatid *Stromateus brasiliensis* Fowler, 1906; the clupeid *Sprattus fuegensis* (Jenyns, 1842); the shark *Squalus acanthias* L. and the hagfish *Myxine affinis* Günther, 1870 (Cousseau et al. 2020; Troccoli et al. 2023; Martínez et al. 2023) (Figs. 11.3 and 11.4).

11.4 Ichthyoplankton in the Marine Areas Surrounding Isla de los Estados

The oceanographic fronts present in the surrounding marine areas of IDLE promote eddy-induced upwellings, enhance nutrient availability, and create retention mechanisms for planktonic organisms, which in turn provide feeding opportunities for fish larvae (Acha et al. 2018). Previous studies have highlighted the marine zone surrounding IDLE as an important nursery area for fish (Ehrlich et al. 1999; Bruno et al. 2025). Ehrlich et al. (1999) proposed that the southern coast of Tierra del Fuego Island functions as a major spawning ground, with currents transporting planktonic eggs and larvae, especially those of *S. fuegensis* and *S. australis*, towards IDLE. Bruno et al. (2025) subsequently documented abundant eggs of both *S. fuegensis* and *S. australis* around the island, particularly during spring.

At least nine fish taxa in the larval stage have been recorded in the area (Fig. 11.5). In general, larval abundance and diversity peak between spring and summer,

Fig. 11.5 Larval stage fish taxa identified in the marine area around Isla de los Estados

correlating with optimal conditions, namely water temperature for early larvae and salinity for late larvae (Bruno et al. 2025). Late larvae are most abundant in salinity values around 34, while early larvae are associated with temperatures between 7 and 9 °C. These conditions reflect the influence of upwelling branches of the Antarctic Circumpolar Current that, in conjunction with waters from the Cape Horn Current, ascends the shelf around IDLE. This process, combined with tidal currents and vertical mixing (Matano et al. 2019), supplies surface layers with deep-water nutrients that stimulate phytoplankton productivity (Guinder et al. 2020; Malits et al. 2023). This high primary productivity supports a rich community of herbivorous zooplankton, particularly nauplii and copepods (Spinelli et al. 2020), while retention processes enhance prey-predator encounters (Matano et al. 2019).

Among the nine larval taxa identified, three dominate throughout most of the year: *S. fuegensis*, *Patagonotothen* spp. and myctophids (family Myctophidae). Molecular analyses have identified at least two nototheniids species (*P. squamiceps* and *P. wiltoni*) and one species of myctophid (*G. braueri*) (Bruno et al. 2025). These taxa are ecologically significant: *S. fuegensis* and nototheniids (mainly *P. ramsayi*) are key trophic regulators (Riccialdelli et al. 2020), while myctophids have been reported as prey items for commercially valuable fish (Cousseau and Perrotta 2004), seabirds (Raya Rey et al. 2007) and marine mammals (Dellabianca et al. 2023). Consequently, the marine area around IDLE should be considered of high priority in the monitoring of nearby Marine Protected Areas, Banco Burdwood-Namuncurá I and II, and Yaganes, given its role as a feeding and migratory area for seabirds (Dodino et al. 2024) and marine mammals (Dellabianca et al. 2023) and as a spawning and nursery area for fish (Bruno et al. 2025).

11.5 Conservation Status

The conservation status of fish species in IDLE remains largely unknown, underscoring the need for new ichthyological surveys in this remote area. Of the 46 fish species documented in IDLE, 52% have a "Data Deficient" or "Not Evaluated" status, 32% are "Least Concern", and about 15% are "Vulnerable" or "Near Threatened" (Table 11.1). However, sustainable management is essential even for species listed as "Data Deficient" or "Not Evaluated", as there is limited or no biological, ecological, or distributional information available for accurate assessment under IUCN standards. Therefore, special attention should be given to these species until more comprehensive information becomes available and their conservation status can be evaluated appropriately (IUCN 2025). This is particularly relevant for exploited species such as *D. eleginoides*, which is currently listed as "Not Evaluated" (Wöhler et al. 2024).

Even less is known about the conservation status of freshwater fish in IDLE. To date, only the small puyen, *G. maculatus* has been recorded. It is thus essential to determine not only whether other native species are present, but also if introduced species have already colonized the island. The presence of introduced species is highly likely, as many freshwater systems in IGTDF host established populations of various anadromous salmonids (Chalde and Llompart 2021), which may act as potential sources of propagules.

11.6 Conclusions

Only a few studies have recorded fish species from both freshwater and marine environments in IDLE. This area is characterized by a low richness of freshwater and similar marine fish composition to those found in IGTDF. All species recorded in IDLE have also been documented in IGTDF. However, this does not preclude the possibility that additional species already described in IGTDF may be found in IDLE's freshwater bodies or marine coastal zones through further research.

Freshwater fish diversity in IDLE is extremely low, with only one species recorded: the native *G. maculatus*. This species is widespread in freshwater environments of IGTDF and also inhabits northern Patagonia, the Malvinas Islands, Chile, Australia, and New Zealand (Cussac et al. 2020). However, many water bodies remain unexplored in IDLE. Thus, other native species characteristic of IGTDF, such as *Aplochiton zebra* (Jenyns, 1842) *A. taeniatus* (Jenyns, 1842), or *Geotria macrostoma* (Burmeister 1868), may be discovered with further research (McDowall et al. 2005; Veyñ et al. 2025). During the Last Glacial Maximum, around 24,000 years ago, IDLE and IGTDF were connected, as the sea level was approximately 140 m lower and the Le Maire Strait was dry (Ponce and Fernández 2014). This historical connection may have enabled the dispersal of freshwater species to new habitats, shaping the current ichthyofauna of IDLE. To date, no exotic species have been identified in IDLE. Nonetheless, it is likely that species of the family Salmonidae are already established on the island, given the widespread presence of anadromous species such as *Oncorhynchus mykiss* (Walbaum 1792), *O. tshawytscha* (Walbaum 1792), or *O. kisutch* (Walbaum 1792) across IGTDF (Chalde et al. 2023).

In IDLE's marine coastal areas, nototheniids are the most abundant family, with *P. squamiceps* being the most frequently recorded species (Friedlander et al. 2020). Most marine coastal fish species documented exhibit benthic habits and have a strong association with kelp forests (Friedlander et al. 2020; Hüne et al. 2021, Kaminsky et al., Chap. 5, this volume). The southern coast of IDLE is subject to higher wave exposure and features a steeper depth gradient than the northern coast (Dayton 1985; Ponce and Rabassa 2012), which could lead to differences in fish assemblages. However, Friedlander et al. (2020) found no significant differences between the ichthyofauna from northern and southern marine coasts, possibly due to methodological limitations, as their survey was based primarily on scuba diving in kelp forest habitats.

In the coming years, additional sampling surveys are expected to be conducted across IDLE to expand knowledge of both freshwater and marine coastal ichthyofauna. These efforts will include conventional sampling methods, such as electrofishing, trawling, and scuba diving, and will be complemented with molecular techniques like eDNA analysis to maximize species detection. Oceanographic surveys in waters surrounding IDLE are also expected to focus on fish in the surrounding areas of IDLE. Furthermore, integrating diet studies of birds and mammals may assist in identifying the fish species on the island (Raya Rey and Schiavini 2005; Raya Rey et al. 2007). An updated fish species list and newly gathered information are essential for improving our understanding of conservation status and for implementing effective management plans for Isla de los Estados and its surrounding marine ecosystems.

Competing Interests The authors have no competing interests to declare.

References

Acha EM, Mianzan HW, Guerrero RA, Favero M, Bava J (2004) Marine fronts at the continental shelves of austral South America. J Mar Syst 44(1–2):83–105. https://doi.org/10.1016/j.jmarsys.2003.09.005

Acha EM, Ehrlich MD, Muelbert JH, Pájaro M, Bruno D, Machinandiarena L, Cadaveira M (2018) Ichthyoplankton associated to the frontal regions of the southwestern Atlantic. In: Hoffmeyer MS, Sabatini ME, Brandini FP, Calliari DL, Santinelli NH (eds) Plankton ecology of the southwestern Atlantic. Springer, Cham, pp 219–246. https://doi.org/10.1007/978-3-319-77869-3_11

Anderson ME (1990) Studies on the Zoarcidae (Teleostei: Perciformes) of the southern hemisphere. V. Two new genera and three new species. Proc Calif Acad Sci 47(7):273–299

Balech E, Ehrlich MD (2008) Esquema biogeográfico del Mar Argentino. Revista de Investigación y Desarrollo Pesquero 19:45–75

Balestrini CF, Manzella G, Lovrich G (1998) Simulación de corrientes en el canal Beagle y bahía Ushuaia, mediante un modelo bidimensional, Informe técnico Nº 98. Servicio de Hidrografía Naval, Buenos Aires, pp 1–58

Balushkin AV (2000) Species composition, systematics and phylogeny of the suborder Notothenioidei. J Ichthyol 40:S212–S231

Belleggia M, Mabragaña E, Figueroa DE, Scenna LB, Barbini SA, Díaz De Astarloa JM (2008) Food habits of the broad nose skate, *Bathyraja brachyurops* (Chondrichthyes, Rajidae), in the south-West Atlantic. Sci Mar 72(4):701–710. https://doi.org/10.3989/scimar.2008.72n4701

Belleggia M, Colonello J, Cortés F, Figueroa DE (2021) Eating catch of the day: the diet of porbeagle shark *Lamna nasus* (Bonnaterre 1788) based on stomach content analysis, and the interaction with trawl fisheries in the South-Western Atlantic (52° S–56° S). J Fish Biol 99(5):1591–1601. https://doi.org/10.1111/jfb.14864

Bonnaterre PJ (1788) Tableau encyclopédique et méthodique des trois regnes de la nature. Ichthyologie. Chez Panckoucke, Paris

Bove G (1883) Patagonia, Terra del Fuoco, mari Australi; rapporto. Tipografía del Regio Instituto Sordo-Muti, Genova

Bruno DO, Riccialdelli L, Becker Y, Martin J, Victorio MF, Acha EM, Fernández DA (2025) Ichthyoplankton temporal and spatial patterns in sub-Antarctic coastal and oceanic waters of the southwestern Atlantic. Estuar Coast Shelf Sci 318:109260. https://doi.org/10.1016/j.ecss.2025.109260

Burmeister H (1868) Actas Soc. Paleontol. B. Aires 1, 1st edn.

Chalde T, Llompart FM (2021) Fish distribution in a southern Patagonian river invaded by Chinook salmon. Hydrobiologia 848:3489–3501. https://doi.org/10.1007/s10750-021-04603-z

Chalde T, Nardi CF, Reibel J, Quinn TP (2023) Detection of juvenile coho salmon (*Oncorhynchus kisutch*) indicates widespread successful reproduction and different life-history strategies in the extreme south of Patagonia. Biol Invasions 25:3443–3452. https://doi.org/10.1007/s10530-023-03116-1

Chiaramonte GE, Alarcos AJ, Vazquez DM, Bovcon N, Delpiani SM, Chávez LN, Dellacasa RF, Mabragaña E (2024) Recurrent juvenile stranding of the porbeagle shark *Lamna nasus* along the south West Atlantic coast. Revista del Museo Argentino de Ciencias Naturales ns 26(2):171–184

Cohen DM, Inada T, Iwamoto T, Scialabba N (eds) (1990) FAO species catalogue. 10: Gadiform fishes of the world: (Order Gadiforms). An annotated and illustrated catalogue of cods, hakes, grenadiers and other gadiform fishes known to date, vol 10. FAO Fisheries Synopsis, Rome

Compagno LJV (1984) Sharks of the world. An annotated and illustrated catalogue of shark species known to date. Part 1. Hexanchiformes to Lamniformes, vol 4. FAO Fisheries Synopsis, Rome

Compagno LJV (2008) Pelagic elasmobranch diversity. In: Camhi MD, Pikitch EK, Babcock EA (eds) Sharks of the open ocean, 1st edn. Wiley, Chichester, pp 14–23. https://doi.org/10.1002/9781444302516.ch2

Cousseau MB, Perrotta RG (2004) Peces marinos de Argentina: Biología, distribuci6n, pesca, 3rd edn. Publicaciones Especiales INIDEP, Buenos Aires

Cousseau MB, Pequeño G, Mabragaña E, Lucifora LO, Martínez P, Giussi A (2020) The Magellanic Province and its fish fauna (South America): several provinces or one? J Biogeogr 47(1):220–234. https://doi.org/10.1111/jbi.13735

Cucco A, Martín J, Quattrocchi G, Fenco H, Umgiesser G, Fernández DA (2022) Water circulation and transport time scales in the Beagle Channel, southernmost tip of South America. J Mar Sci Eng 10(7):941. https://doi.org/10.3390/jmse10070941

Cussac VE, Barrantes ME, Boy CC, Górski K, Habit E, Lattuca ME, Rojo JH (2020) New insights into the distribution, physiology and life histories of south American galaxiid fishes, and potential threats to this unique fauna. Diversity 12(5):178–193. https://doi.org/10.3390/d12050178

Cuvier G (1830) Histoire naturelle des poissons. Tome cinquième. Livre cinquième. Des Sciénoïdes. Livrecinquième. Des Sciénoïdes. v. 5: i-xxviii + 1-499 + 4 pp., Pls. 100-140. [Cuvier authored volume. i-xx + 1-374 inStrasbourg edition.]

Dayton PK (1985) The structure and regulation of some south American kelp communities. Ecol Monogr 55(4):447–468. https://doi.org/10.2307/2937131

Dellabianca NA, Torres MA, Ordoñez C, Fioramonti N, Raya Rey A (2023) Marine protected areas in the southern south-West Atlantic: insights from marine top predator communities. Aquat Conserv Mar Freshwat Ecosyst 33(5):472–487. https://doi.org/10.1002/aqc.3935

Delpiani SM, Bruno DO, Vazquez DM, Llompart F, Delpiani GE, Fernández DA, Rosso JJ, Mabragaña E, Díaz De Astarloa JM (2020) Structure and distribution of fish assemblages at Burdwood Bank, the first sub-Antarctic marine protected area "Namuncurá" in Argentina (southwestern Atlantic Ocean). Polar Biol 43(11):1783–1793. https://doi.org/10.1007/s00300-020-02744-w

Díaz De Astarloa JM, Mabragaña E (2004) Bathyraja cousseauae sp.n.: A new softnose skate from theSouthwestern Atlantic (Rajiformes, Rajidae). Copeia 2004(2):326–335. https://doi.org/10.1643/CI 03 121R2

Dodino S, Balza U, Riccialdelli L, Polito MJ, Pütz K, Raya Rey A (2024) Pre-molt dispersal and use of marine protected areas by southern rockhopper penguins (*Eudyptes chrysocome*) at the southernmost oceanic regions of South America. Prog Oceanogr 229:103369. https://doi.org/10.1016/j.pocean.2024.103369

Dudley TR, Crow GE (1983) A contribution to the flora and vegetation of Isla de los Estados (Staten Island). American Geophysical Union, Tierra del Fuego. https://doi.org/10.1029/AR037

Ehrlich MD, Sánchez RP, De Ciechomski JD, Machinandiarena L, Pájaro M (1999) Ichthyoplankton composition, distribution and abundance on the southern Patagonian shelf and adjacent waters. INIDEP Documento Científico 5:37–65

Fernholm B (1998) Hagfish systematics. In: Jørgensen JM, Lomholt JP, Weber RE, Malte H (eds) The biology of hagfishes. Springer, Netherlands, pp 33–44. https://doi.org/10.1007/978-94-011-5834-3_3

Ferriz RA, Gómez SE (2015) Polimorfismo en poblaciones diadrómicas y lacustrinas de *Galaxias maculatus*, Argentina, Sudamérica. Bioikos 29(2):19–25

Forster JR (1801) In: Bloch ME, Schneider JG (eds) M.E. Blochii, Systema Ichthyologiae iconibus cx illustratum.Post obitum auctoris opus inchoatum absolvit, correxit, interpolavit Jo. Gottlob Schneider, Saxo. Berolini. SumtibusAuctoris Impressum et Bibliopolio Sanderiano Commissum

Fowler HW (1906) New, rare or little known scombroids, No. 3. Proceedings of the Academy of Natural Sciences ofPhiladelphia 58:114–122

Fowler HW (1910) Notes on batoid fishes. Proceedings of the Academy of Natural Sciences of Philadelphia62:468–475

Friedlander AM, Ballesteros E, Bell TW, Caselle JE, Campagna C, Goodell W, Hüne M, Muñoz A, Salinas-de-León P, Sala E, Dayton PK (2020) Kelp forests at the end of the earth: 45 years later. PLoS One 15(3):e0229259. https://doi.org/10.1371/journal.pone.0229259

Galván DE, Bovcon ND, Cochia PD, González RA, Lattuca ME, Reinaldo MO, Rincón-Díaz MP, Romero MA, Vanella FA, Venerus LA, Svendsen GM (2022) Changes in the specific and biogeographic composition of coastal fish assemblages in Patagonia, driven by climate change, fishing, and invasion by alien species. In: Helbling EW, Narvarte MA, González RA, Villafañe VE (eds) Global change in Atlantic coastal Patagonian ecosystems. Springer International Publishing, pp 205–231. https://doi.org/10.1007/978-3-030-86676-1_9

Giesecke R, Martín J, Piñones A, Höfer J, Garcés-Vargas J, Flores-Melo X, Alarcón E, Durrieu De Madron X, Bourrin F, González HE (2021) General hydrography of the Beagle Channel, a subantarctic interoceanic passage at the southern tip of South America. Front Mar Sci 8:621822. https://doi.org/10.3389/fmars.2021.621822

Gosztonyi AE (1970) Los peces de la expedición científica a la Isla de los Estados, Argentina (noviembre-diciembre de 1967). Physis 30(80):173–180

Guichenot A (1848) Fauna Chilena. Peces. In: Gay C (ed) Historia fisica y politica de Chile. Zoología. v. 2. Paris Santiago. [Ictiologia plates 1-11 are in Atlas, v. 2; of 2 sets of plates at CAS, 1 set has some figures in color. Platespublished in 1854. Sometimes author given as Valenciennes.]

Guinder VA, Malits A, Ferronato C, Krock B, Garzón-Cardona J, Martínez A (2020) Microbial plankton configuration in the epipelagic realm from the Beagle Channel to the Burdwood Bank, a marine protected area in sub-Antarctic waters. PLoS One 15(5):e0233156. https://doi.org/10.1371/journal.ponc.0233156

Haedrich RL (1966) The stromateoid fish genus Icichthys: notes and a new species. Videnskabelige Meddelelserfra Dansk Naturhistorisk Forening, Kjøbenhavn 59(129):199–213

Hutton FW (1872) Fishes of New Zealand. Catalogue with diagnoses of the species. Colonial Museum andGeological Survey Department Wellington 1-93 + 95-133:1–12

Günther ACLG (1870) Catalogue of the fishes in the British Museum. British Museum (Natural History), Departmentof Zoology, London

Günther ACLG (1878) Preliminary notices of deep-sea fishes collected during the voyage of the H. M.S."Challenger." Annals and Magazine of Natural History (Series 5) 2(5):179–187

Hüne M, Vega R (2016) Feeding habits in two sympatric species of Notothenioidei, *Patagonotothen cornucola* and *Harpagifer bispinis*, in the Chilean Patagonian channels and fjords. Polar Biol 39(12):2253–2262. https://doi.org/10.1007/s00300-016-1892-3

Hüne M, Friedlander AM, Ballesteros E, Caselle JE, Sala E (2021) Assemblage structure and spatial diversity patterns of kelp forest-associated fishes in southern Patagonia. PLoS One 16(9):e0257662. https://doi.org/10.1371/journal.pone.0257662

IUCN (2025) The IUCN red list of threatened species. Version 2025-1. https://www.iucnredlist.org

Jenyns L (1842) Part IV, Fish. In: Darwin C (ed) The zoology of the voyage of H.M.S. Beagle, under the commandof Captain Fitzroy, R.N. during the years 1832 to 1836. Smith, Elder Co., London, United Kingdom, pp 1–215

Lattuca ME, Llompart FM, Avigliano E, Renzi M, Leva ID, Boy CC, Vanella FA, Barrantes ME, Fernández DA, de Albuquerque CQ (2020) First insights into the growth and population structure of *Cottoperca trigloides* (Perciformes, Bovichtidae) from the southwestern Atlantic Ocean. Front Mar Sci 7:421. https://doi.org/10.3389/fmars.2020.00421

Llompart FM, Fernández DA, Aureliano D, La Mesa M (2020) Life-history traits of the Magellan plunderfish *Harpagifer bispinis* (Forster, 1801) in the Beagle Channel (Tierra del Fuego, South America). Polar Biol 43(10):1643–1654. https://doi.org/10.1007/s00300-020-02735-x

Lloris D (1988) *Haushia marinae* gen n., sp. N. (Osteichthyes, Zoarcidae), capturada en la isla de los Estados (Argentina). Miscelánea Zoológica 12:245–250

Lloris D, Rucabado J (1991) Ictiofauna del Canal Beagle (Tierra del Fuego), aspectos ecológicos y análisis biogeográfico, vol 8. Instituto Español de Oceanografía, Madrid

Lönnberg E (1905) The fishes of the Swedish south polar expedition. Schwedische Südpolar-Expedition 1901–1903 5(6):1–72

Lönnberg E (1907) Fische. (Ergebnisse der Hamburger Magalhaensischen Sammelreise. Lfg. 8. No. 6.). L Friederichsen & Co, Hamburg :1–16. https://doi.org/10.5962/bhl.title.63772

Mabragaña E (2011) Las rayas del género *Psammobatis* de la plataforma Argentina: Biología y ecología. In: Contribuciones sobre biología, pesca y comercialización de tiburones en la Argentina: Aportes para la elaboración del Plan de Acción Nacional (Primera edición). Consejo Federal Pesquero, Buenos Aires, pp 135–151

Malits A, Ibarbalz FM, Martín J, Flombaum P (2023) Higher biotic than abiotic natural variability of the plankton ecosystem revealed by a time series along a subantarctic transect. J Mar Syst 238:103843. https://doi.org/10.1016/j.jmarsys.2022.103843

Marcinkevicius M (2019) Biología reproductiva y alimentación del escrófalo, Sebastes oculatus Valenciennes 1833 (Pisces, Sebastidae) en el área central del Golfo San Jorge (Argentina). PhD thesis, Universidad Nacional de La Plata

Martín J, Alonso G, Dragani W, Meyerjürgens J, Giesecke R, Cucco A, Fenco H (2023) General circulation and tidal wave propagation along the Beagle Channel. J Mar Syst 103889:1–15. https://doi.org/10.1016/j.jmarsys.2023.103889

Martínez PA, Wöhler OC, Troccoli GH, Di Marco EJ (2023) Efectos del establecimiento de las áreas marinas protegidas en la pesquería Argentina de la merluza negra (*Dissostichus eleginoides*). Marine Fish Sci 36(3):267–288. https://doi.org/10.47193/mafis.3632023010906

Matano RP, Palma ED, Combes V (2019) The Burdwood Bank circulation. J Geophys Res Oceans 124(10):6904–6926. https://doi.org/10.1029/2019JC015001

Matusevich F, Vazquez DM, Gabbanelli V, Díaz De Astarloa JM, Mabragaña E (2023) Chondrichthyans from the southern tip of South America with emphasis on the marine protected area Namuncurá-Burdwood Bank: exploring egg nursery grounds. Polar Biol 46(4):357–372. https://doi.org/10.1007/s00300-023-03128-6

McDowall RM, Allibone RM, Chadderton WL (2005) Falkland Islands freshwater fishes. A natural history. Falklands Conservation, London

McEachran JD (1983) Results of the research cruises of FRV "Walther Herwig" to South America. LXI. Revision of the South American skate genus Psammobatis Günther, 1870 (Elasmobranchii: Rajiformes, Rajidae). Archiv für Fischereiwissenschaft 34(1):23–80

McEachran JD, di Sciara N (1995) FAO species identification guide. Sharks and rays of the world. Part 2. Batoid fishes, vol 4. FAO Fisheries Synopsis, Rome

Müller J, Henle J (1838) Systematische Beschreibung der Plagiostomen. Veit und comp, Berlin

Müller J, Henle J (1841) Systematische Beschreibung der Plagiostomen. Veit und comp, Berlin

Nardi CF, Fernández DA, Vanella FA, Chalde T (2019) The expansion of exotic Chinook salmon (*Oncorhynchus tshawytscha*) in the extreme south of Patagonia: an environmental DNA approach. Biol Invasions 21(4):1415–1425. https://doi.org/10.1007/s10530-018-1908-8

Nielsen JG (ed) (1999) Ophidiiform fishes of the world (order Ophidiiformes): an annotated and illustrated catalogue of pearlfishes, cusk-eels, brotulas, and other ophidiiform fishes known to date, vol 18. FAO Fisheries Synopsis, Rome

Norman JR (1937) Coast fishes. Part II. The Patagonian region. Discovery reports 16:1–150. https://doi.org/10.5962/bhl.part.29326

Perugia A (1891) Appunti sopra alcuni pesci Sud-Americani. In: En *Annali del Museo Civico di Storia Naturale di Genova*, vol X. Tipografía del R. Instituto Sordo-Muti, Genova

Peters W (1877) Übersicht der während der von 1874 bis 1876 unter dem Commando des Hrn. Capitän z. S.Freiherrn von Schleinitz ausgeführten Reise S. M. S. Gazelle gesammelten und von der Kaiserlichen Admiralitätder Königlichen Akademie der Wissenschaften übersandten Fische. Monatsberichte der Königlichen PreussischenAkademie der Wissenschaften zu Berlin 1876 :831–854

Philippi RA (1902) Descripcion de cinco nuevas especies chilenas: del órden de los plagióstomos. Anales de laUniversidad de Chile 109:303–315

Piola AR, Matano RP (2001) Brazil and Falklands (Malvinas) currents. In: Encyclopedia of ocean sciences. Elsevier, pp 340–349. https://doi.org/10.1006/rwos.2001.0358

Ponce JF, Fernández M (2014) Climatic and environmental history of Isla de los Estados, Argentina. Springer, Netherlands. https://doi.org/10.1007/978-94-007-4363-2

Ponce JF, Rabassa J (2012) Geomorfología glaciar de la Isla de los Estados, Tierra del Fuego, Argentina. Rev Soc Geol Esp 25(1–2):67–84

Raya Rey A, Schiavini A (2005) Inter-annual variation in the diet of female southern rockhopper penguin (*Eudyptes chrysocome chrysocome*) at Tierra del Fuego. Polar Biol 28(2):132–141. https://doi.org/10.1007/s00300-004-0668-3

Raya Rey A, Trathan P, Schiavini A (2007) Inter-annual variation in provisioning behaviour of southern rockhopper penguins *Eudyptes chrysocome chrysocome* at Staten Island, Argentina. Ibis 149(4):826–835. https://doi.org/10.1111/j.1474-919X.2007.00718.x

Regan CT (1913) The Antarctic fishes of the Scottish National Antarctic Expedition. Transactions of the RoyalSociety of Edinburgh 49(2):229–292

Riccialdelli L, Becker Y, Fioramonti N, Torres M, Bruno D, Raya Rey A, Fernández D (2020) Trophic structure of southern marine ecosystems: a comparative isotopic analysis from the Beagle Channel to the oceanic Burdwood Bank area under a wasp-waist assumption. Mar Ecol Prog Ser 655:1–27. https://doi.org/10.3354/meps13524

Richardson J (1844) Ichthyology of the voyage of H. M. S. Erebus, Terror, under the command of Captain SirJames Clark Ross, R. N., F. R. S. In: Richardson J, Gray JE (eds) The zoology of the voyage of H. M. S. Erebus,Terror, under the command of Captain Sir J. C. Ross, R. N., F. R. S., during the years 1839 to 1843. E. W. E. W.Janson, London, United Kingdom

Richardson J (1845) Ichthyology of the voyage of H. M. S. Erebus, Terror, under the command of Captain SirJames Clark Ross, R. N., F. R. S. In: Richardson J, Gray JE (eds) The zoology of the voyage of H. M. S. Erebus,Terror, under the command of Captain Sir J. C. Ross, R. N., F. R. S., during the years 1839 to 1843. E. W. E. W.Janson, London, United Kingdom

Ruocco NL, Lucifora LO, Díaz De Astarloa JM, Wöhler O (2006) Reproductive biology and abundance of the white-dotted skate, *Bathyraja albomaculata*, in the Southwest Atlantic. ICES J Mar Sci 63(1):105–116. https://doi.org/10.1016/j.icesjms.2005.08.007

Sabadin DE, Lucifora LO, Barbini SA, Figueroa DE, Kittlein M (2020) Towards regionalization of the chondrichthyan fauna of the Southwest Atlantic: a spatial framework for conservation planning. ICES J Mar Sci 77(5):1893–1905. https://doi.org/10.1093/icesjms/fsaa064

Scenna LB (2011) Biología y ecología reproductiva de las especies del género Bathyraja (Elasmobranchii, Rajidae) en la plataforma continental Argentina. PhD thesis, Universidad Nacional de Mar del Plata

Scenna LB, Díaz De Astarloa JM (2014) Reproductive biology of the Magellan skate, *Bathyraja magellanica* (Chondrichthyes, Rajidae), in the South-Western Atlantic. Mar Freshw Res 65(9):766. https://doi.org/10.1071/MF13144

Shiko BA, Mecklenburg CW (2004) Family Agonidae Swainson 1839—poachers. Cali Acad Sci, Annotated Checklists of Fishes 30:27

Smitt FA (1898) Poissons de l'expédition scientifique à la Terre de Feu. II. Bihang till Kongl. Svenska vetenskapsakademienshandlingar. Stockholm 24(4):1–80

Spinelli ML, Malits A, García Alonso VA, Martín J, Capitanio FL (2020) Spatial gradients of spring zooplankton assemblages at the open ocean sub-Antarctic Namuncurá marine protected area/Burdwood Bank, SW Atlantic Ocean. J Mar Syst 210:103398. https://doi.org/10.1016/j.jmarsys.2020.103398

Steindachner F (1876) Ichthyologische Beiträge (III). Bemerkungen über Serranus nebulifer und S. clathratus sp.Gird. Sitzungsberichte der Kaiserlichen Akademie der Wissenschaften Mathematisch-NaturwissenschaftlicheClasse 72(1):29–96

Thompson WF (1916) Fishes collected by the United States Bureau of Fisheries steamer Albatross during 1888,between Montevideo, Uruguay, and Tome, Chile, on the voyage through the Straits of Magellan. Proceedings ofthe United States National Museum 50(2133):401–476. https://doi.org/10.5479/si.00963801.50-2133.401

Troccoli GH, Martínez PA, Di Marco EJ, Waessle JA, Wöhler OC (2023) Patrones migratorios de la merluza negra (*Dissostichus eleginoides*) en el Océano Atlántico Sudoccidental. Marine Fish Sci 36(3):245–265. https://doi.org/10.47193/mafis.3632023010907

Valenciennes A (1833) Histoire naturelle des poissons. Tome neuvième. Suite du livre neuvième. DesScombéroïdes. v. 9: i-xxix + 3 pp. + 1-512, Pls. 246-279. [Cuvier authored pp. 1-198, 330-359, 372-427;Valenciennes the balance. i-xxiv + 1-379 in Strasbourg edition. Pp. 429-512 contains additions and corrections forvols. 2 through 5.]

Veyñ M, Chalde T, Nardi C (2025) Optimisation of an environmental DNA-based approach for assessing freshwater fish biodiversity in southernmost South America. Aquat Conserv Mar Freshwat Ecosyst 35(3):e70112. https://doi.org/10.1002/aqc.70112

Walbaum JJ (1792) Petri Artedi sueci genera Piscium inquibus systema totum ichthyologiae proponitur cumclassibus, ordinibus, generum characteribus, specierum diffentiis, observationibus plumiris. In: Ichthyologiae parsIII. Rose, Grypeswaldiae

Wöhler OC, Martínez PA, Troccoli GH, Di Marco EJ (2024) Patagonian toothfish (*Dissostichus eleginoides*) stocks in south American waters and its implications for fishery management. Marine Fish Sci 38(1):109–144. https://doi.org/10.47193/mafis.3812025010110

Chapter 12
The Birds of Chuani-sin Isla de los Estados Archipelago

Ulises Balza, Joaquín Cano, Samanta Dodino, Nicolás A. Lois, Klemens Pütz, and Andrea Raya Rey

Abstract Birds are arguably the best-known taxonomic group of Chuani-sin–Isla de los Estados. Early explorers collected specimens that were later designated as holotypes for several species. The area has been recognized as an Important Bird Area, primarily due to the significance of its seabird populations—both those breeding on the island and those foraging in its surrounding waters.

In this chapter, we update the status of bird species on Isla de los Estados based on 30 years of continuous fieldwork, which includes tracking studies, stable isotope analyses, periodic population estimates of target species, and a decade of camera trap monitoring. We also review historical and current knowledge of the island's avifauna, derived from peer-reviewed literature as well as grey sources, and provide a state-of-the-art synthesis of ecological research conducted at the site.

U. Balza (✉)
Instituto de Ciencias Polares, Ambiente y Recursos Naturales, Universidad Nacional de Tierra del Fuego, Ushuaia, Tierra del Fuego, Argentina

J. Cano
Centro Austral de Investigaciones Científicas (CADIC-CONICET), Ushuaia, Tierra del Fuego, Argentina

S. Dodino
Centro Austral de Investigaciones Científicas (CADIC-CONICET), Ushuaia, Tierra del Fuego, Argentina

Instituto de Desarrollo Económico e Innovación, Universidad Nacional de Tierra del Fuego, Ushuaia, Tierra del Fuego, Argentina

Instituto de Ciencias Polares, Ambiente y Recursos Naturales, Universidad Nacional de Tierra del Fuego, Ushuaia, Tierra del Fuego, Argentina

N. A. Lois
Instituto de Ecología Genética y Evolución de Buenos Aires (IEGEBA-CONICET), Buenos Aires, Argentina

Departamento de Ciencias de la Atmosfera y de los Océanos, Facultad de Ciencias Exactas y Naturales, Universidad de Buenos Aires (DCAO-FCEyN-UBA), Buenos Aires, Argentina

A. Raya Rey, U. Balza (eds.), *Biodiversity and Conservation of Isla de los Estados (Chuani-sin)*, Natural and Social Sciences of Patagonia,
https://doi.org/10.1007/978-3-032-13630-5_12

We highlight the current status of key species and present novel data, including new species records, breeding events, and the discovery of previously unknown colonies. Seventy nine bird species are regularly recorded on Isla de los Estados, with about half being breeders. The archipelago supports globally significant populations of two penguin species and one raptor, and it is believed to be an important area for other globally threatened species. Additionally, it serves as a critical feeding area for several pelagic seabirds.

Although knowledge remains biased toward seabirds and grassland birds, this review provides essential baseline information and identifies key knowledge gaps, supporting future research and conservation in one of the most important bird conservation areas of the Southern Ocean.

Keywords Eudyptes chrysocome · Phalcoboenus australis · Andean condor · Tussock grasslands · Spheniscus magellanicus

12.1 Introduction

Birds are likely the best-known taxonomic group of the Chuani-sin–Isla de los Estados archipelago. In addition to early natural history records, the ecology of some bird populations on the island has been studied in detail over the past three decades. Isla de los Estados is recognized as both an Important Bird Area and a Key Biodiversity Area (BirdLife International 2025), primarily due to the regional and global significance of seabird breeding populations. Species qualifying under this assessment include the Striated Caracara (*Phalcoboenus australis*), the Magellanic Penguin (*Spheniscus magellanicus*), the Southern Rockhopper Penguin (*Eudyptes chrysocome*), and the Imperial Shag (*Leucocarbo atriceps*).

Early explorers settled on the island and collected bird specimens that would later serve as holotypes. For example, numerous specimens were collected on Año Nuevo Island (present-day Observatorio Island) by James Cook's crew in 1775, which became type specimens for their respective species. These included the Striated Caracara, the Southern Giant Petrel (*Macronectes giganteus*), the Magellanic Penguin, and the Upland Goose (*Chloephaga picta*) (Lysaght 1959).

Later, Argentine scientists such as Alberto Castellanos collected birds for national museum collections (Castellanos 1935, 1937). This was followed by expeditions from the Smithsonian Institution, which collected 322 specimens and documented 51 bird species. Their surveys were also the first to record seabird colonies across

K. Pütz
Antarctic Research Trust, Bremervörde, Germany

A. Raya Rey
Centro Austral de Investigaciones Científicas (CADIC-CONICET), Ushuaia, Tierra del Fuego, Argentina

Instituto de Ciencias Polares, Ambiente y Recursos Naturales, Universidad Nacional de Tierra del Fuego, Ushuaia, Tierra del Fuego, Argentina

Wildlife Conservation Society, Representación Argentina, Buenos Aires, Argentina the archipelago, adding a spatial dimension to avifaunal knowledge (Flint 1971; Imshaug 1972). Although subsequent works like Kühnemann (1976) were not focused on birds, they included incidental avian observations that were later incorporated into broader assessments, such as those of Chebez and Bertonatti (1994), who synthesized existing bird records for the region.

Juan Carlos Chebez (1962–2011), a renowned Argentine naturalist, spent extended periods on the island observing birds. Although he recorded fewer species than earlier expeditions, he referenced Kühnemann as an authority and reported several species not observed before or since, including some conspicuous species such as the Magellanic Woodpecker (*Campephilus magellanicus*), the Ruddy-headed Goose (*Chloephaga rubidiceps*), and the Chimango Caracara (*Phalcoboenus chimango*), among others. Similarly, other bird species have remained in checklists based on historical records from more than a century ago, despite never being recorded again. For example, the Black-necked Swan (*Cygnus melancoryphus*) was listed by Argentine journalist Roberto Payró, who cited a conversation with a coast guard officer stationed on the island (Payró 1898). Three other species, the Grey-breasted Seedsnipe (*Thinocorus orbignyianus*), the Eared Dove (*Zenaida auriculata*) and the Double-Collared Seedeater (*Sporophila caerulescens*) were recorded during the Argentine-Italian "Expedición Austral Argentina" in the 1880s but have not been documented since (Salvadori 1900).

Beginning in 1995, research visits by scientists from Argentina's National Research Council (CONICET) became more frequent (Parera et al. 1997; Schiavini et al. 1999), initiating long-term ecological monitoring and conservation studies. Despite these efforts, published information on the avifauna of Isla de los Estados and Año Nuevo islands remains scarce, outdated, and largely confined to grey literature. In this chapter, we summarize and update knowledge on the birds of the archipelago, based on nearly 30 years of fieldwork, mainly focused on seabirds and coastal environments. Our goal is to clarify the current species composition by including only those species confirmed through long-term monitoring or repeated records across different sources, while excluding isolated observations or vagrant individuals. We also review the ecological status of key species and highlight knowledge gaps and research priorities for the future.

12.2 Bird Richness and Characteristics

While the maritime tussock grasslands have been the primary focus of avian research, the forests of the archipelago remain comparatively understudied. Tussock grasslands are known to support high bird diversity and prey abundance, as well as provide thermal protection for breeding birds on other Subantarctic islands (Contador et al. 2020; Smith and Prince 1985). As a result, both bird biomass and species richness are higher in the coastal areas of Isla de los Estados, despite these habitats covering only 2.3% of the island's total area (Benitez et al. 2019; Salom et al., Chap 4, this volume).

To date, 79 bird species are regularly recorded on Isla de los Estados and in its surrounding waters, with 41 species confirmed or suspected as breeders (Table 12.1).

Table 12.1 Table of regularly recorded species on Isla de lo Estados and its surrounding waters

Order	Family	Scientific name	Common name	IUCN status	Breeding	Observations
Sphenisciformes	*Spheniscidae*	*Aptenodytes patagonicus*	*King penguin*	*LC*	*1*	*Large colony extirpated in the nineteenth century; scattered new breeding events observed from 2004.*
		Eudyptes chrysolophus	*Macaroni penguin*	*NT*	*1*	*New breeding species; up to two individual pairs incubating among rockhopper penguins.*
		Spheniscus magellanicus	*Magellanic penguin*	*LC*	*1*	
		Eudyptes chrysocome	*Southern rockhopper penguin*	*VU*	*1*	
Procellariiformes	*Diomedeidae*	*Diomedea exulans*	*Snowy albatross*	*VU*	*0*	
		Diomedea epomophora	*Southern royal albatross*	*VU*	*0*	
		Diomedea chrysostoma	*Gray-headed albatross*	*EN*	*0*	
		Thalassarche melanophris	*Black-browed albatross*	*LC*	*0*	
	Procellariidae	*Pachyptila belcheri*	*Slender-billed prion*	*LC*	*0*	
		Macronectes giganteus	*Southern giant petrel*	*LC*	*1*	
		Macronectes halli	*Northern giant petrel*	*LC*	*0*	*New recorded species. Feeding on water and on land from Magellanic penguin carcasses.*
		Fulmarus glacialoides	*Southern fulmar*	*LC*	*0*	
		Halobaena caerulea	*Blue petrel*	*LC*	*0*	
		Pelecanoides magellani	*Magellanic diving-petrel*	*LC*	*0*	
		Pelecanoides urinatrix	*Common diving-petrel*	*LC*	*0*	
		Ardenna gravis	*Great shearwater*	*LC*	*0*	
		Ardenna grisea	*Sooty shearwater*	*NT*	*1*	*New breeding species.*
		Daption capense	*Pintado petrel*	*LC*	*0*	
		Procellaria aequinoctialis	*White-chinned petrel*	*VU*	*0*	
	Oceanitidae	*Oceanites oceanicus*	*Wilson's storm-petrel*	*LC*	*0*	

Suliformes	*Phalacrocoracidae*	*Leucocarbo atriceps*	*Imperial cormorant*	*LC*	*1*	
		Leucocarbo magellanicus	*Magellanic cormorant*	*VU*	*1*	*Several small colonies over the coastal cliffs of Isla de los Estados and Observatorio Island*
		Nannopterum brasilianum	*Neotropic cormorant*	*LC*	*1*	
Pelecaniformes	*Ardeidae*	*Ardea ibis*	*Western cattle -egret*	*LC*	*0*	*New recorded species.*
		Nycticorax nycticorax	*Black-crowned night heron*	*LC*	*0*	
	Threskiornithidae	*Theristicus melanopis*	*Black-faced ibis*	*LC*	*1*	*Nesting on cliffs covered with coastal tussock stands on Observatorio Island.*
Anseriformes	*Anatidae*	*Tachyeres patachonicus*	*Flying steamer-duck*	*LC*	*0*	
		Tachyeres pteneres	*Flightless steamer-duck*	*LC*	*1*	
		Chloephaga poliocephala	*Ashy-headed goose*	*LC*	*1*	*Important breeding site on Observatorio Island, with a minimal number of 61 active nests (0.15 nests/ha). Also recorded breeding in Franklin Bay in lower densities.*
		Chloephaga hybrida	*Kelp goose*	*LC*	*1*	
		Chloephaga picta	*Upland goose*	*LC*	*1*	
		Lophonetta specularioides	*Crested duck*	*LC*	*1*	
		Anas flavirostris	*Yellow-billed teal*	*LC*	*1*	
Cathartiformes	*Cathartidae*	*Vultur gryphus*	*Andean condor*	*VU*	*0*	
		Cathartes aura	*Turkey vulture*	*LC*	*1*	*A nest found on the ground with two fledglings under an Evergreen beech (*Nothofagus betuloides*).*
Accipitriformes	*Accipitridae*	*Geranoaetus melanoleucus*	*Black-chested buzzard-eagle*	*LC*	*0*	
		Astur chilensis	*Chilean hawk*	*LC*	2	
		Circus cinereus	*Cinereous harrier*	*LC*	*0*	
		Geranoaetus polyosoma	*Variable hawk*	*LC*	*0*	

(continued)

Table 12.1 (continued)

Order	Family	Scientific name	Common name	IUCN status	Breeding	Observations
Falconiformes	*Falconidae*	*Falco sparverius*	*American kestrel*	*LC*	*0*	
		Caracara plancus	*Crested caracara*	*LC*	*1*	*Nests on Evergreen beech, at 4–6 m high.*
		Falco peregrinus	*Peregrine falcon*	*LC*	*0*	
		Phalcoboenus australis	*Striated caracara*	*NT*	*1*	
Gruiformes	*Rallidae*	*Pardirallus sanguinolentus*	*Plumbeus rail*	*LC*	2	*Presence of juveniles and territorial behavior on Observatorio Island.*
Charadriiformes	*Haematopodidae*	*Haematopus ater*	*Blackish oystercatcher*	*LC*	*1*	*Nest found on the rocks in the upper part of the intertidal zone on Observatorio Island.*
		Haematopus leucopodus	*Magellanic oystercatcher*	*LC*	*1*	*Nests found on sandy beaches in Franklin Bay.*
	Charadriidae	*Anarhyncus falklandicus*	*Two-banded plover*	*LC*	*0*	
	Scolopacidae	*Tringa flavipes*	*Lesser yellowlegs*	*LC*	*0*	
		Numenius phaeopus	*Whimbrel*	*LC*	*0*	
		Gallinago stricklandii	*Fuegian snipe*	*NT*	2	*Adult captured near San Juan cape.*
		Gallinago magellanica	*Magellanic snipe*	*LC*	*1*	*Fledgling found in October on Observatorio Island.*
		Calidris bairdii	*Baird's sandpiper*	*LC*	*0*	*New recorded species. Small groups observed on Observatorio Island.*
		Calidris fuscicollis	*White-rumped sandpiper*	*VU*	*0*	
	Thinocoridae	*Attagis malouinus*	*White-bellied Seedsnipe*	*LC*	*1*	*Adult with fledglings on high terrain in December.*
	Chionidae	*Chionis albus*	*Snowy sheathbill*	*LC*	*0*	
	Laridae	*Larus scoresbii*	*Dolphin Gull*	*LC*	*1*	*Small colonies of 10-40 breeding pairs with low interannual site fidelity.*
		Larus dominicanus	*Kelp Gull*	*LC*	*1*	*Small and scattered nesting sites across the coast of both Isla de los Estados and Isla Observatorio.*
	Stercorariidae	*Stercorarius chilensis*	*Chilean skua*	*LC*	*1*	*Breeding sites on Observatorio Island (minimum of 31 nests), San Juan cape (15–20 nests) and Franklin Bay (>30 nests).*

	Sternidae	*Sterna hirundinacea*	*South American tern*	*LC*	*2*	*Presence of juveniles and territorial behavior in Observatorio Island.*
Strigiformes	*Strigidae*	*Asio flammeus*	*Short-eared owl*	*LC*	*0*	
	Tytonidae	*Tyto furcata*	*American barn owl*	*LC*	*0*	*New recorded species.*
Passeriformes	*Furnariidae*	*Cinclodes oustaleti*	*Gray-flanked Cinclodes*	*LC*	*0*	
		Cinclodes patagonicus	*Dark-bellied Cinclodes*	*LC*	*1*	
		Cinclodes antarcticus	*Blackish Cinclodes*	*NT*	*2*	
		Cinclodes fuscus	*Buff-winged Cinclodes*	*LC*	*1*	
		Aphrastura spinicauda	*Thorn-tailed Rayadito*	*LC*	*1*	
	Tyrannidae	*Muscisaxicola capistrata*	*Spot-billed ground-tyrant*	*LC*	*0*	
		Lessonia rufa	*Austral Negrito*	*LC*	*1*	*Nest on the floor inside of a small* Gaultheria pumilia *shrub.*
		Muscisaxicola maclovianus	*Dark-faced ground-tyrant*	*LC*	*1*	
		Pyrope pyrope	*Fire-eyed Diucon*	*LC*	*0*	
		Elaenia albiceps	*White-crested Elaenia*	*LC*	*0*	
	Turdidae	*Turdus falklandii*	*Austral thrush*	*LC*	*1*	*Fledglings observed in December. Nests found on the ground in the forest on fallen Evergreen beeches.*
	Thraupidae	*Phrygilus patagonicus*	*Patagonian sierra finch*	*LC*	*1*	*Fledglings observed in December.*
	Passerellidae	*Zonotrichia capensis*	*Rufous-collared sparrow*	*LC*	*1*	
	Fringillidae	*Spinus barbatus*	*Black-chinned siskin*	*LC*	*0*	
	Troglodytidae	*Cistothorus platensis*	*Grass wren*	*LC*	*1*	*Nestlings in late November and fledglings in December; a nest found in a holly barberry* (Berberis ilicifolia).
		Troglodytes musculus	*Southern house wren*	*LC*	*1*	*Fledglings observed in December.*
	Hirundinidae	*Hirundo rustica*	*Barn swallow*	*LC*	*0*	
		Tachycineta leucopyga	*Chilean swallow*	*LC*	*1*	*Nesting inside the roof crevices of an abandoned building.*

Breeding status: 0 non-breeding; 1 confirmed breeding and 2 probably breeding

The archipelago supports globally important populations of the Magellanic Penguin, the Southern Rockhopper Penguin, and the Striated Caracara. It is also suspected to be an important area for several other globally threatened species, including the Fuegian Snipe (*Gallinago stricklandii*), the Blackish Cinclodes (*Cinclodes antarcticus*), the Andean Condor (*Vultur gryphus*), and various pelagic seabirds.

We recorded four new species for the archipelago and 17 new breeding species (Table 12.1). Most of these records likely represent increased and sustained field effort rather than recent changes in presence or breeding status.

The distribution of birds within Isla de los Estados remains poorly studied, except for seabird colonies. Two coastal species, the Kelp Goose (*Chloephaga hybrida*) and the Flightless Steamer-Duck (*Tachyeres pteneres*), are more abundant along the island's north-facing coasts (Parera et al. 1997), which coincide with areas of greater kelp forest cover (Kaminsky et al., Chap. 5, this volume) and higher diversity and abundance of some invertebrates (Fraysse and Perez, Chap. 8, this volume). This suggests that the northern coasts may offer greater food availability for some coastal bird species. In contrast, the more exposed and windier south-facing coasts (Ponce and Fernández, Chap. 2, this volume) are associated with higher diversity and utilization by pelagic seabirds, including albatrosses and large petrels (Fig. 12.1).

Forest bird communities on Isla de los Estados largely reflect a subset of the Fuegian temperate forest avifauna. This is consistent with patterns observed in other Subantarctic Forest ecosystems where temperate forest birds often do not exhibit strong island-specificity (Vuilleumier 1985). However, certain functional groups represented by only a single species on the main island of Tierra del Fuego are entirely absent from Isla de los Estados, including woodpeckers, parrots, and hummingbirds. These groups play key ecological roles in other Patagonian forests—such as seed dispersal and pollination (Aizen 2003), and their absence portrays a functional simplification of the island's forest bird community.

In contrast, the coastal areas of Isla de los Estados differ markedly from those of mainland Tierra del Fuego by hosting massive seabird colonies that resemble those found on other Subantarctic islands (Barroso et al. 2020; Humphrey et al. 1970; Sandvig et al. 2020). The number of breeding bird species and overall avian richness on the island is comparable to that of other islands of similar size and latitude, such as the Auckland Islands south of New Zealand (Miskelly et al. 2020). However, unlike more remote Subantarctic archipelagos, such as the Kerguelen Islands in the Indian Ocean, which lack resident raptors and passerines (Weimerskirch et al. 1989), Isla de los Estados supports a rich and well-structured terrestrial and marine food web, offering a more ecologically complex and interconnected system.

12.3 Species with Populations of Global Importance

Here, we summarize the status and update information of the species highlighted for defining the Importance Bird Area and other species of global importance and/or of relevance for the area (Fig. 12.2) (BirdLife International 2025).

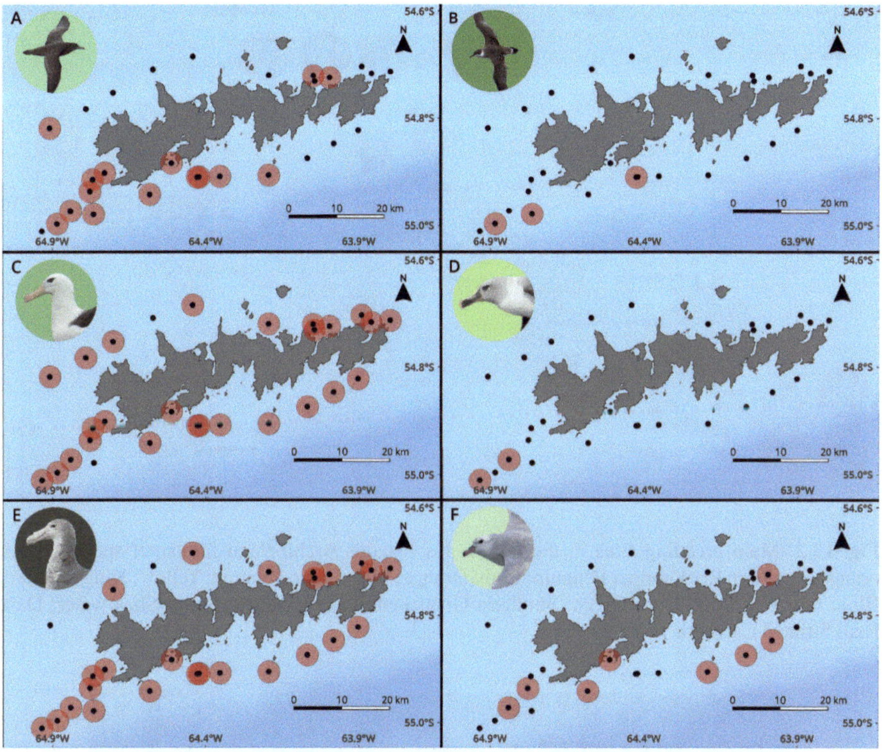

Fig. 12.1 Some of the most frequently recorded pelagic seabirds around Isla de los Estados during March and October 2024 surveys. (**a**) Sooty shearwater (*Ardenna grisea*); (**b**) Great shearwater (*A. gravis*); (**c**) Black-browed Albatross (*Thalassarche melanophris*); (**d**) Grey-headed Albatross (*T. chrysostoma*); (**e**) Southern Giant Petrel (*Macronectes giganteus*); (**f**) Southern Fulmar (*Fulmarus glacialoides*). Nearly all species were more frequent in the southern coast and/or the Le Maire Strait

12.3.1 Southern Rockhopper Penguin

Isla de los Estados hosts a population of Southern Rockhopper Penguins (*Eudyptes chrysocome*) estimated at approximately 140,000 breeding pairs, with over 90% concentrated in Franklin Bay and the rest at San Juan Cape (Raya Rey et al. 2014; Schiavini 2000; Figs. 12.2 and 12.3). This colony represents a significant proportion of the global population, estimated at roughly 750,000 breeding pairs distributed along the southern tip of South America (Pütz et al. 2013). Note that we are assuming former eastern and southern rockhopper penguins as separate entities (Frugone et al. 2018), although this is still not recognized by IUCN (BirdLife International 2020). While global population trends remain uncertain, regional data reveal contrasting dynamics, with some colonies stable, others declining (Pütz et al. 2013), and a few, such as the recently funded colony off the coast of Patagonia, increasing (Gandini et al. 2017). The causes behind current population declines remain poorly

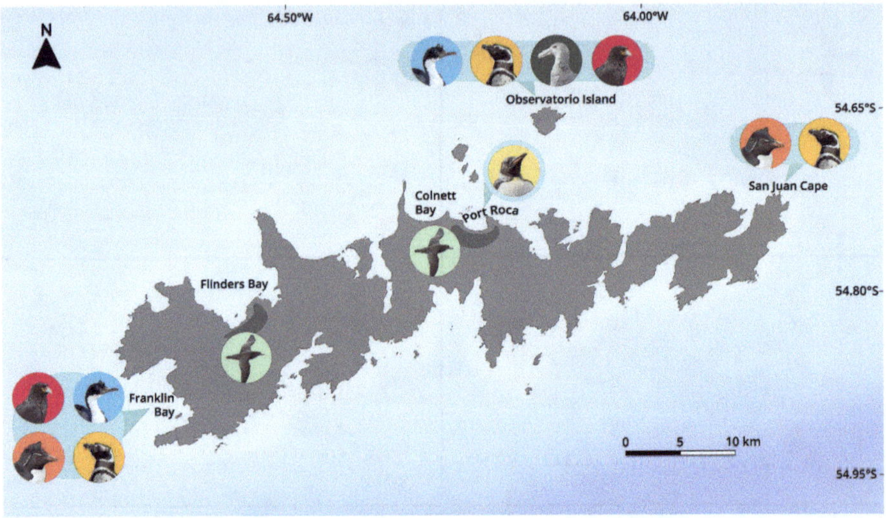

Fig. 12.2 Main breeding sites in the Isla de los Estados Archipelago for target species. Dark orange: Southern Rockhopper Penguin; Light Orange: Magellanic Penguin; Yellow: King Penguin; Blue: Imperial Shag; Dark Green: Southern Giant Petrel; Light Green: Sooty Shearwater; Dark Red: Striated Caracara

Fig. 12.3 Southern Rockhopper penguin at Bahía Franklin, Isla de los Estados. (Credit: A Raya Rey)

understood; however, oceanographic conditions have shown to strongly influence both breeding success and adult survival (Dehnhard et al. 2013; Morgenthaler 2018; Raya Rey et al. 2007a, b). Additionally, while the South American sea lion (*Otaria flavescens*) is known to prey on adult penguins at sea in the Isla de los Estados population (Raya Rey et al. 2012b), the overall impact of this predation on population dynamics has yet to be quantified. On land, the Striated Caracara also relies on penguin chicks as a key food source during its own breeding season (Balza et al. 2020; Liljesthröm et al. 2008), adding another potential source of pressure on reproductive success.

Over recent decades, studies have greatly advanced our understanding of the ecology of this population. Phenology, colony attendance patterns, and breeding success have been monitored using passive integrated transponders and fixed radiofrequency identification stations. Southern Rockhoppers on Isla de los Estados initiate reproduction roughly 2 weeks later and exhibit generally lower breeding success than other sites in the Southern Atlantic Ocean, but similarly to Pacific Ocean colonies (Pütz et al. 2006; Raya Rey et al. 2007a, b, 2009). This temporal mismatch likely contributes to the limited gene flow between Isla de los Estados and Pacific colonies with other Southern Atlantic colonies (Lois et al. 2020). Further, as crested penguins present a unique reversed hatching asynchrony, the lower breeding success on Isla de los Estados may be partially explained by significantly smaller A-egg volumes observed in this population (Demongin et al. 2010).

The species' diet has been studied using complementary approaches. Stomach content analyses conducted during chick-rearing revealed a diet dominated by gregarious crustaceans, small cephalopods, fish larvae, and juvenile fish, with crustaceans being the most abundant prey by number. However, high interannual variability in prey composition suggests an opportunistic feeding strategy during both early and late chick rearing (Raya Rey et al. 2007a, b; Raya Rey and Schiavini 2005). Stable isotope analyses were consistent with these patterns, showing a preference for oceanic and pelagic waters and lower trophic level prey, especially in females during chick-rearing and pre-molt periods (Rosciano et al. 2016, 2020, 2022). Mercury concentrations in feathers from individuals from Isla de los Estados were among the highest for all penguin species (Brasso et al. 2015; Lois et al. 2022), echoing differences in foraging niche and aligning with genetic structure of the species. The concordance among mercury levels, isotopic signatures, and genetic structure underscores the ecological and reproductive segregation of these populations, reinforcing the need to manage them as distinct conservation units.

Franklin Bay has also been the focus of studies on at-sea distribution across different life stages. During the breeding season, tracking during incubation and early chick-rearing revealed that foraging activity is largely confined to nearby marine areas, including the Le Maire Strait and waters southeast of Isla de los Estados (Pütz et al. 2006; Rosciano et al. 2016, 2022). Foraging and diving behavior studies showed that daily and overnight trips were 30% and 60% longer, respectively, and birds performed 53% more overnight trips (five to seven more) compared to Indian Ocean eastern populations (Schiavini and Rey 2004). In the non-breeding season, tracking data revealed wide dispersal southward and westward into the Pacific and Southern Oceans (Pütz et al. 2007; Raya Rey et al. 2007a, b). Recent modeling of winter at-sea distribution under climate change scenarios predicts significant shifts in suitable foraging habitat in the Southern Ocean (Green et al. 2023). Tracking during the pre-molt stage showed penguins moving predominantly southward, using

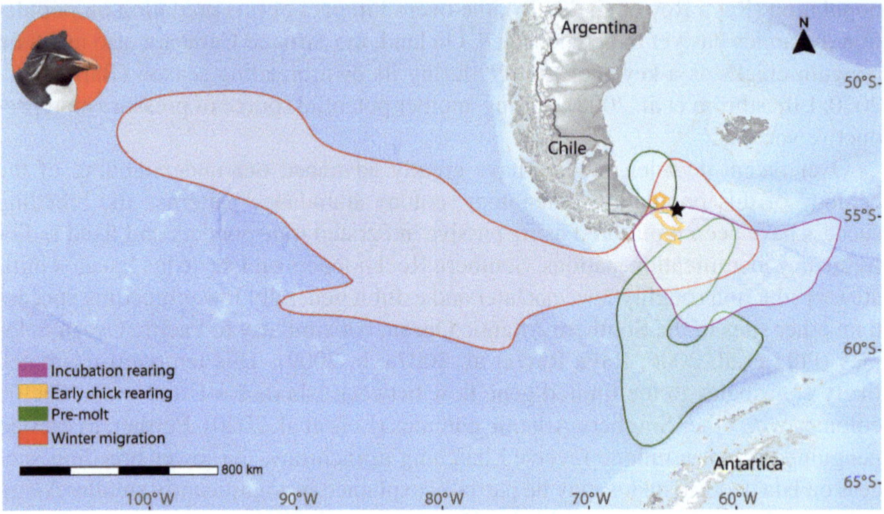

Fig. 12.4 At-sea distribution of Southern Rockhopper Penguins from Franklin Bay during different periods of its life cycle. The star indicates the location of the breeding colony

Marine Protected Areas as corridors to reach key foraging zones near the Antarctic Polar Front. However, the primary foraging grounds remain outside protected areas (Dodino et al. 2024). Marked seasonal changes in at-sea distribution (Fig. 12.4) underscore the challenges of conserving this wide-ranging species.

Home to ~20% of the global population and the largest known colony of the species, Isla de los Estados is a critical site for the conservation of Southern Rockhopper Penguins. Further, ongoing population declines, together with high mercury concentrations, raise serious concerns about the vulnerability of this population (Santaniello 2026). As the most intensively studied seabird in the archipelago, this species serves as a valuable early-warning indicator of anthropogenic pressures in the region. Ongoing research efforts are now leveraging camera traps and AI-assisted analysis to monitor reproductive success and phenology, aiming to detect early signs of global change impacts and guide timely conservation actions.

12.3.2 Magellanic Penguin

Isla de los Estados represents the most important breeding ground for Magellanic Penguins in the Fuegian archipelago (Schiavini et al. 2005, Fig. 12.5). Nest density ranges between 34 and 80 breeding pairs per hectare, which is lower than densities reported for other Patagonian regions (García-Borboroglu et al. 2022). In 2020, we estimated a total of 38,062 breeding pairs across the Isla de los Estados archipelago (95% CI: 22,337–53,788), with 87% of the population concentrated on Observatorio Island, that holds over 33,000 breeding pairs (Balza et al. 2022). Three smaller

Fig. 12.5 Magellanic penguin (*Spheniscus magellanicus*) from Franklin Bay. (Credit: P Petracci)

colonies include Goffré Island with approximately 3000 pairs, Franklin Bay with 1800 pairs, and the recently described Cape San Juan colony with fewer than 100 pairs (Raya Rey et al. 2014, 2022).

There is historical evidence of Magellanic Penguin colonies in the region dating back to at least the eighteenth century (e.g. Castellanos 1935; Payró 1898). The species appears to follow a dynamic pattern of colonization and extinction across its range. For example, a former colony at Flinders Bay no longer exists, while the colonies at Franklin Bay and Cape San Juan are likely of more recent origin (Castellanos 1935; Imshaug 1972; Raya Rey et al. 2022). These shifting patterns highlight the importance of long-term monitoring of colony dynamics across the entire archipelago, including migration rates, rather than focusing solely on individual sites.

Ecological and ecotoxicological studies conducted over the past decade have improved our understanding of the foraging ecology and environmental pressures affecting Magellanic Penguins in this region. At Franklin Bay, during the early chick-rearing period, diving data and stable isotope analyses ($\delta^{13}C$ and $\delta^{15}N$) showed that males and females foraged in similar areas and depths and occupied comparable trophic positions, suggesting an absence of sexual segregation in foraging behavior (Rosciano et al. 2016, 2018).

More recent research compared colonies along a longitudinal environmental gradient, from a coastal site in the Beagle Channel to the more offshore colonies on Isla de los Estados. During the chick-rearing stage, stable isotope analyses revealed colony-specific foraging strategies, with $\delta^{13}C$ and $\delta^{15}N$ values decreasing from inshore to offshore colonies. This pattern reflects a gradient from freshwater-influenced environments to South Atlantic oceanic waters (Dodino et al. 2022b). While all individuals fed on low-trophic-level prey such as the squat lobster

(*Grimothea gregaria*), adults and juveniles from Isla de los Estados also incorporated higher trophic-level prey (e.g., silversides and nototheniids). Chicks had more restricted isotopic signatures, suggesting trophic partitioning between adults and their offspring. Juveniles exhibited the broadest isotopic niches, indicating greater variability or exploratory foraging behavior. A complementary study conducted during the pre-molt period further explored spatial foraging differences and contaminant exposure through feather analyses of mercury concentrations and stable isotopes (Dodino et al. 2022a). Adults consistently exhibited higher mercury concentrations than juveniles, and individuals from Observatorio Island had significantly higher levels than those from Franklin Bay and the nearby Beagle Channel colony. Although isotopic niches partially overlapped among colonies, differences in contaminant profiles confirmed that penguins were exploiting distinct foraging areas.

Together, these findings emphasize the ecological significance of Franklin Bay and Observatorio Island and illustrate how contrasting environmental conditions shape foraging behavior, trophic interactions, and exposure to environmental stressors among Magellanic Penguin colonies in the Isla de los Estados archipelago.

12.3.3 Imperial Shag

Imperial Shag colonies are primarily located on Observatorio Island (~8800 breeding pairs) and in Franklin Bay (~4600 breeding pairs), ranking among the most important colonies globally (Raya Rey et al. 2014, 2022; Yorio et al. 2020). The Observatorio Island colony is at least ~90 years old (Castellanos 1937).

Despite the global significance of these colonies, detailed information on the species' ecology and life history remains limited. A study at Franklin Bay provided some of the first insights into spatial and temporal variation in diet and foraging behavior using a combination of stable isotope analysis and GPS tracking (Harris et al. 2016). Individuals foraged in shallow, coastal waters close to the colony and exhibited consistent foraging strategies. Feather stable isotope analyses revealed similar diets across sexes and years, while slight interannual differences in blood isotope values suggested some variability in prey availability. Overall, findings indicated a relatively stable trophic niche in this population.

12.3.4 Striated Caracara

The Striated Caracara population on Isla de los Estados represents the largest known population within the Fuegian archipelago and forms one of two genetically distinct lineages of the species (Fig. 12.6). Genetic evidence indicates that the Fuegian and Malvinas Islands lineages diverged approximately 50,000 years ago, during the last glacial period. The Fuegian lineage exhibits lower genetic diversity and shows no

Fig. 12.6 Striated caracara nestlings on Isla de los Estados. (Credit: Nicolás A. Lois)

evidence of recent connectivity with the Malvinas population, with gene flow likely having ceased around 23,000 years ago (Balza et al. 2025).

Each year, an estimated 40–50 nests are active, with the breeding population roughly equally distributed between Franklin Bay and Observatorio Island (Balza et al. 2017, 2022), although additional, scattered nesting sites have also been recorded. Individuals are highly philopatric and non-migratory: five birds fitted with GPS transmitters remained within the Isla de los Estados archipelago throughout the winter, dispersing to various coastal areas but never leaving the island group (Balza 2021). Likewise, none of the 19 resighting events of banded individuals (from a total of over 150) occurred outside the island where the bird was originally captured. The maximum confirmed longevity is 8 years, recorded in a breeding adult male, and the fledgling sex ratio in Franklin Bay is biased towards females (Balza et al. 2023).

The species' foraging ecology and breeding phenology are closely tied to seabird colonies, which provide pulses of food during the breeding season, especially for fledglings (Castellanos 1937; Strange 1996). However, because seabirds later disperse to their overwintering areas at sea, this resource is both spatially concentrated and seasonally variable, though predictable. This heavy reliance on seabird-derived resources may render the caracara population particularly vulnerable to fluctuations and long-term trends in seabird reproductive success and phenology. Moreover, marine-originated prey in the Isla de los Estados region exhibit elevated mercury concentrations compared to other Patagonian sites, and Striated Caracaras have some of the highest mercury levels recorded for raptors in the region, likely due to biomagnification through seabird-based diets (Balza et al. 2021).

Given its restricted range, genetic distinctiveness, and strong dependence on threatened marine prey, the conservation of the Striated Caracara is intimately linked to the effective management of Isla de los Estados. For example, the presence of introduced deer and goats (Salom and Schiavini, Chap. 14, this volume) is altering the structure of coastal tussac grasslands, critical nesting habitat for caracaras, potentially limiting nest site availability. This may partly explain the high number of non-breeding adults observed in areas like Franklin Bay (Balza et al. 2023). As a long-lived, resident top predator, the Striated Caracara also holds promise as a biomonitor for tracking increases in mercury and other environmental contaminants on the island. Additionally, its cognitive abilities make it a valuable model for future research in animal behavior and intelligence (Harrington et al. 2024), adding further scientific and conservation relevance to the world's southernmost bird of prey.

12.3.5 Other Globally and Regionally Important Species

Several lesser-studied species have promising populations in the area, making Isla de los Estados a priority for future research and conservation. Based on our field observations, we summarize their current status and highlight knowledge gaps to guide future efforts.

Isla de los Estados represents the easternmost known breeding area for the Fuegian Snipe (*Gallinago stricklandii*, Near Threatened) (Raimilla et al. 2023), which we found to be relatively common in grasslands at the island's eastern tip.

Upland Geese (*Chloephaga picta*) and Ashy-headed Geese (*C. poliocephala*) have been recorded nesting at Franklin Bay (Petracci et al. 2016), and large numbers of Ashy-headed Geese are observed annually on Observatorio Island. These findings reinforce the conservation significance of the archipelago for these globally declining species.

Andean Condors (*Vultur gryphus*, Vulnerable) appear to be abundant on Isla de los Estados, although no population estimate or assessment of connectivity with the rest of the Southern Patagonia conservation unit has been conducted (Wallace et al. 2022). In this region, condors primarily feed on marine-derived carrion (Balza et al. 2024). If the same applies to other scavengers in the area, such as the Turkey Vulture (*Cathartes aura*, now a confirmed breeder), then monitoring pinniped populations, which appear to be increasing (Schiavini and Valenzuela, Chap. 13, this volume), becomes essential for understanding the potential ecotoxicological risks facing these raptors.

The Chilean Hawk (*Astur chilensis*), listed as Data Deficient and a high-priority species for conservation in Argentina and Chile respectively (Argentinas et al. 2017; Pincheira-Ulbrich et al. 2008), was recorded at all field sites. Numerous juveniles observed between December and March suggest breeding occurs on the island, but no confirmed nests have been found. The extensive, relatively undisturbed evergreen forests (~140 km^2, Salom et al., Chap. 4, this volume) on the island could provide crucial habitat, but further research is needed to clarify the species' status on the island (Rivas-Fuenzalida et al. 2024).

The Southern Giant Petrel (*Macronectes giganteus*) breeds in small numbers on Observatorio Island and appears to be in decline since the colony's likely establishment after the 1930s (Castellanos 1935, 1937). Population estimates dropped from 500 breeding pairs in 2004 to 387 in 2016 (Balza et al. 2022; Quintana et al. 2005). A secondary colony near the southwestern tip of the island described in the 1970s (Imshaug 1972) persisted with at least eight pairs until 2012, after which it disappeared. Satellite-tracked adults foraged closer to their colony than those from northern Patagonia. While females utilized shelf and shelf-break waters, tagged males foraged in more coastal areas (Quintana et al. 2010). Stable isotope analyses of feathers, which integrate annual foraging behavior, indicated that individuals forage primarily over the continental shelf and subtropical waters (Raya Rey et al. 2012a).

The Sooty Shearwater (*Ardenna grisea*, globally Near Threatened) was historically reported in large numbers at night around Isla de los Estados (Castellanos 1935) and was recently confirmed as a breeding species (Raya Rey et al. 2024). A flock of 3000 individuals was observed in October 1975 on the southern coast, where the species was described as "locally abundant" (Rumboll and Jehl 1977). Other burrowing petrels, such as the Wilson's Storm-Petrel (*Oceanites oceanicus*), were also observed decades ago arriving at night (Castellanos 1935), although no breeding sites have been confirmed. Diving Petrels (*Pelecanoides* spp.) are frequently seen in nearby waters and may also nest on the island (Rumboll and Jehl 1977).

The King Penguin (*Aptenodytes patagonicus*) was extirpated from the region in the nineteenth century due to overexploitation (Zangrando et al., Chap. 3, this volume). While sporadic sightings continue along the island's coasts, in 2004 an adult was observed incubating an egg in Roca Port. Occasional visitors have since reported a few chicks at that location. During our surveys, the highest count was six adults in nearby Colnett Bay. Although it is unclear whether this reflects the foundation of a new colony, the species has expanded to new breeding sites elsewhere (Cristofari et al. 2018; Kusch and Marín 2012).

The surrounding waters of Isla de los Estados may serve as important feeding grounds for non-breeding threatened seabirds as well as native species. Records of globally threatened species include large albatrosses (*Diomedea* spp., Vulnerable), the Grey-headed Albatross (*Thalassarche chrysostoma*, Endangered), and the White-chinned Petrel (*Procellaria aequinoctialis*, Vulnerable) (Dellabianca et al. 2023). 81% of Black-browed Albatrosses (*Thalassarche melanophris*) tracked from their main South Atlantic breeding grounds visited the archipelago, and 35% established it as their area of residency (Riaz et al. 2024). Although the species was reported to breed on Isla de los Estados over a century ago (Dabbene 1923), and new colonies have recently been discovered in the broader Fuegian archipelago (Aguayo et al. 2003), no active breeding colonies are currently known on the island.

12.4 Conservation and Current Threats

12.4.1 *Invasive Species*

Invasive mammals pose growing but understudied threats to bird communities on Isla de los Estados archipelago (Salom and Schiavini, Chap. 14, this volume). Despite early warnings against further introductions (Imshaug 1972), species like red deer continued to be deliberately introduced. Field observations and camera traps document red deer and goats grazing within active seabird colonies, likely reducing vegetation cover and structural complexity—particularly of dominant species like *Poa flabellata*, with cascading effects on nesting sites and faunal refuge (Balza et al. 2017, 2023). Rats have been recorded at sooty shearwater burrows and striated caracara nests, recalling cases of rat predation on seabirds in other regions (Lyver et al. 1999). At Observatorio Island, rabbits may contribute to overgrazing and soil disturbance in maritime tussock grasslands (Salom and Schiavini, Chap. 14, this volume).

The evidence from other Subantarctic islands consistently shows that ungrazed and rodent-free islands support greater bird diversity and prey availability (Hall et al. 2001; Tabak et al. 2015). While eradication may carry some secondary effects, especially on predators (Carey 2018), the net benefits for native bird conservation are usually positive (de Brooke et al. 2018; Glen et al. 2013). One important gap of knowledge is that smaller islands (i.e., the rest of the Año Nuevo islands) may harbor breeding colonies of bird species not represented in the rest of the archipelago, especially if it is confirmed that they are rodent-free.

12.4.2 *Tourism*

Tourism on Isla de los Estados is currently a low-scale activity, with minimal direct impact on bird populations. Large commercial cruises carrying over 100 passengers visit the island only once or twice a year, mostly stopping at a few historical sites. Due to the inaccessibility of major seabird colonies, tourists typically observe seabirds from ships or boats without landing. In addition, a limited number of sailboats and small yachts carrying between five and ten passengers occasionally visit the island. Although the effects of tourism on birds have not been assessed, the current level of activity has likely low-impact. However, it would be advisable to establish monitoring protocols and indicators to detect potential impacts on wildlife, providing a baseline in case tourism increases in the future.

12.4.3 Commercial Fishing

Fishing pressure around Isla de los Estados is relatively low compared to other regions. The highest levels of fishing effort are concentrated in the deeper zones off the east and southeast coast of the island (Dellabianca et al. 2023). These areas are mainly targeted by bottom and midwater trawl vessels, which pose a significant threat to pelagic seabirds. Numerous albatrosses and petrels are incidentally caught each year in these fisheries, highlighting the need to implement and strengthen mitigation measures to reduce seabird bycatch. Beyond direct mortality, fisheries can also cause indirect impacts through the destruction of benthic habitats, leading to a decline in biodiversity and alterations in food web structure. These cascading effects, although less visible, may ultimately affect the availability of prey for seabirds and other marine predators, underscoring the importance of considering both direct and indirect consequences of fishing activity in conservation planning.

12.5 Future Directions

Future research on bird communities on Isla de los Estados faces the ongoing challenge of sustaining local scientific activity in the absence of clear political interest (Carignano and Jaworski 2019; Debat 2025). Limited funding and logistical difficulties can further hinder long-term efforts, yet such studies are crucial for understanding ecological dynamics in this remote and vulnerable ecosystem. Many key patterns, population trends, shifts in species composition, or the impacts of invasive species, only became evident after several years of monitoring; but at the same time, they are only available for a small subset of the island's avifauna. Strengthening long-term research programs is essential not only for detecting these trends but also for informing effective conservation strategies and guiding potential management actions in response to emerging threats. We hope the coming years will prioritize the active management of protected areas in Argentina, and on Isla de los Estados in particular, to ensure the future of one of the jewels of the South Atlantic.

Competing Interests We are deeply grateful to Dr. Adrián Schiavini, whose vision and ideas laid the foundation for the research conducted over the past 30 years. Throughout the years, many individuals have joined us in the field—as volunteers, students, nature enthusiasts, and friends. Most of them shared our love for the island, while others faced the harshness of its extreme weather; to all, and especially to those who endured difficult conditions, we extend our sincerest thanks.

Logistical support and safe transportation to Isla de los Estados were made possible thanks to the Argentine Navy and their dedicated crews, as well as a number of sailboats that assisted us over the years, including the ***Sarah Vorwerk***, ***Unicornio***, ***Fernande***, ***Spirit of Sydney***, and ***Ocean Tramp***. Special thanks to Henk Boersma "The Captain" of the ***Sarah*** who is also in love with the island, and it is happy to be part of most of our campaigns.

Funding over the years has been generously provided by CONICET, our home institution, along with the (sadly, now dissolved) Agencia Nacional de Promoción de la Investigación, el Desarrollo Tecnológico y la Innovación, the Wildlife Conservation Society, and the Antarctic Research Trust.

References

Aguayo A, Acevedo J, Acuña P (2003) Breeding new site of black-browed albatross, Diomedea Melanophris Temmick 1828, at Seno Almirantazgo, Tierra del Fuego. Anales Del Instituto de La Patagonia 31:91–96

Aizen MA (2003) Influences of animal pollination and seed dispersal on winter flowering in a temperate mistletoe. Ecology 84(10):2613–2627. https://doi.org/10.1890/02-0521

Argentinas A, de Ambiente M, de la Nación DS (2017) Categorización de las aves de Argentina según su estado de conservación año: 2015. https://avesargentinas.org.ar/sites/default/files/Categorizacion-de-aves-de-la-Argentina.pdf

Balza U (2021) Ecología y conservación del carancho austral (Phalcoboenus australis) en Isla de los Estados: Factores asociados a su vulnerabilidad a distintas escalas espacio-temporales. Doctoral, Universidad de Buenos Aires (UBA). http://rgdoi.net/10.13140/RG.2.2.30976.35841

Balza U, Lois NA, Raya Rey A (2017) Status and reproductive outcome of the breeding population of striated caracaras (*Phalcoboenus australis*) at Franklin Bay, Staten Island, Argentina. Wilson J Ornithol 129(4):890–898. https://doi.org/10.1676/16-189.1

Balza U, Lois NA, Polito MJ, Pütz K, Salom A, Raya Rey A (2020) The dynamic trophic niche of an Island bird of prey. Ecol Evol 10(21):12264–12276. https://doi.org/10.1002/ece3.6856

Balza U, Brasso R, Lois NA, Pütz K, Raya Rey A (2021) The highest mercury concentrations ever reported in a south American bird, the striated caracara (*Phalcoboenus australis*). Polar Biol 44(11):2189–2193. https://doi.org/10.1007/s00300-021-02938-w

Balza U, Liljesthröm M, Pimper L, Franco-Navarro I, Cañas-Barrovecchio S, Zunino F, Domato I, Acardi S, Iturraspe R, Cano J, Raya-Rey A (2022) Status of breeding birds at Observatorio and Goffré Islands, Argentina. *Polar Biol* 45(4):667–673. https://doi.org/10.1007/s00300-022-03019-2

Balza U, Lois NA, Pizzarello G, Lopez ME, Lois-Milevicich J, Raya-Rey A (2023) Variation in nest-site occupancy and breeding outcome in a threatened Subantarctic raptor. Ibis 166:ibi.13234. https://doi.org/10.1111/ibi.13234

Balza U, Dodino S, Lambertucci SA, Kusch A, Raya-Rey A (2024) Maintenance of a scavenger ancient food web in insular remote areas. Perspect Ecol Conserv 22:367–373. https://doi.org/10.1016/j.pecon.2024.10.001

Balza U, Lois NA, Harrington KJ, León F, Pütz K, Raya-Rey A, Ceballos SG (2025) Glacial history and ecological restrictions Shape Island-scale genetic structure and demography in the southernmost bird of prey. J Biogeogr 52:e15083. https://doi.org/10.1111/jbi.15083

Barroso O, Crego RD, Mella J, Rosenfeld S, Contador T, Mackenzie R, Vásquez RA, Rozzi R (2020) Colaboración científica con la Armada de Chile en estudios ornitológicos a largo plazo en el archipiélago Diego Ramírez: Primer monitoreo del ciclo anual del ensamble de aves en la isla Gonzalo. Anales del Instituto de la Patagonia 48(3):149–168. https://doi.org/10.4067/S0718-686X2020000300149

Benitez J, Lencinas MV, Huertas Herrera A, Martínez Pastur G (2019) Assessing the conservation value of nature reserves: terrestrial birds in Isla de los Estados (Staten Island) provincial reserve, Tierra del Fuego, Argentina. Commun Ecol 20(2):181–193. https://doi.org/10.1556/168.2019.20.2.8

BirdLife International (2020) Species factsheet: Southern Rockhopper Penguin Eudyptes chrysocome. Downloaded from https://datazone.birdlife.org/species/factsheet/southern-rockhopper-penguin-eudyptes-chrysocome on 30/06/2025

BirdLife International (2025) Site factsheet: Isla de los Estados—Banco Burdwood. https://datazone.birdlife.org/site/factsheet/isla-de-los-estados-banco-burdwood

Brasso RL, Chiaradia A, Polito MJ, Raya Rey A, Emslie SD (2015) A comprehensive assessment of mercury exposure in penguin populations throughout the southern hemisphere: using trophic calculations to identify sources of population-level variation. Mar Pollut Bull 97(1–2):408–418. https://doi.org/10.1016/j.marpolbul.2015.05.059

Carey PW (2018) Simultaneous rat, mouse and rabbit eradication on Bense and little Bense Islands, Falkland Islands. In: Veitch CR, Clout MN, Martin AR, Russell JC, West CJ (eds) Island invasives: scaling up to meet the challenge. IUCN, Gland, pp 108–113

Carignano HA, Jaworski JP (2019) Argentina's subpar investment in science. Science 363(6428):702–702. https://doi.org/10.1126/science.aaw3872

Castellanos A (1935) Observaciones de algunas aves de Tierra del Fuego e Isla de los Estados. El Hornero 6(1):22–35

Castellanos A (1937) Observaciones de algunas aves de Tierra del Fuego e Isla de los Estados. El Hornero 7(3):382–394

Chebez JC, Bertonatti C (1994) La avifauna de la Isla de los Estados, Islas de Año Nuevo y mar circundante (Tierra del Fuego, Argentina). L.O.L.A, Buenos Aires

Contador T, Rendoll J, Mackenzie R, Rosenfeld S, Barroso O, Rozzi R, Goffinet B, Kennedy J, Convey P (2020) Comunidades de invertebrados terrestres del archipiélago Diego Ramírez (56°31'S), el sitio de estudios ecológicos de largo plazo más austral de américa: Diversidad y afinidades con las islas subantárticas del océano austral. Anales del Instituto de la Patagonia 48(3):83–98. https://doi.org/10.4067/S0718-686X2020000300083

Cristofari R, Liu X, Bonadonna F, Cherel Y, Pistorius P, Le Maho Y, Raybaud V, Stenseth NC, Le Bohec C, Trucchi E (2018) Climate-driven range shifts of the king penguin in a fragmented ecosystem. Nat Clim Chang 8(3):245–251. https://doi.org/10.1038/s41558-018-0084-2

Dabbene R (1923) Los petreles y los albatros del Atlántico Austral. Hornero 3(4):311–329

de Brooke ML, Bonnaud E, Dilley BJ, Flint EN, Holmes ND, Jones HP, Provost P, Rocamora G, Ryan PG, Surman C, Buxton RT (2018) Seabird population changes following mammal eradications on islands. Anim Conserv 21(1):3–12. https://doi.org/10.1111/acv.12344

Debat H (2025) The collapse of Argentine science under Javier Milei: one year of unprecedented defunding. Info 3(1):e25007. https://doi.org/10.3145/infonomy.25.007

Dehnhard N, Ludynia K, Poisbleau M, Demongin L, Quillfeldt P (2013) Good days, bad days: wind as a driver of foraging success in a flightless seabird, the southern rockhopper penguin. PLoS One 8(11):e79487. https://doi.org/10.1371/journal.pone.0079487

Dellabianca NA, Torres MA, Ordoñez C, Fioramonti N, Raya Rey A (2023) Marine protected areas in the southern south-West Atlantic: insights from marine top predator communities. Aquat Conserv Mar Freshwat Ecosyst 33(5):472–487. https://doi.org/10.1002/aqc.3935

Demongin L, Poisbleau M, Rey AR, Schiavini A, Quillfeldt P, Eens M, Strange IJ (2010) Geographical variation in egg size dimorphism in rockhopper penguins. Polar Biol 33(4):469–476. https://doi.org/10.1007/s00300-009-0722-2

Dodino S, Riccialdelli L, Polito MJ, Pütz K, Brasso RL, Raya Rey A (2022a) Mercury exposure driven by geographic and trophic factors in Magellanic penguins from Tierra del Fuego. Mar Pollut Bull 174:113184. https://doi.org/10.1016/j.marpolbul.2021.113184

Dodino S, Riccialdelli L, Polito MJ, Pütz K, Rey AR (2022b) Intraspecific trophic variation during the early chick-rearing period in Magellanic penguins Spheniscus magellanicus: influence of age and colony location. Mar Biol 169(9):116. https://doi.org/10.1007/s00227-022-04100-4

Dodino S, Balza U, Riccialdelli L, Polito MJ, Pütz K, Raya Rey A (2024) Pre-molt dispersal and use of marine protected areas by Southern Rockhopper Penguins (Eudyptes chrysocome) at the southernmost oceanic regions of South America. Prog Oceanogr 229:103369. https://doi.org/10.1016/j.pocean.2024.103369

Flint OS (1971) R/V Hero Cruise 71-2 to Isla de los Estados. Antarct J US 6(6):263–265

Frugone MJ, Lowther A, Noll D, Ramos B, Pistorius P, Dantas GPM, Petry MV, Bonadonna F, Steinfurth A, Polanowski A, Raya Rey A, Lois NA, Pütz K, Trathan P, Wienecke B, Poulin E, Vianna JA (2018) Contrasting phylogeographic pattern among Eudyptes penguins around the Southern Ocean. Sci Rep 8(1):17481. https://doi.org/10.1038/s41598-018-35975-3

Gandini P, Millones A, Morgenthaler A, Frere E (2017) Population trends of the southern rockhopper penguin (Eudyptes chrysocome chrysocome) at the northern limit of its breeding range: Isla Pingüino, Santa Cruz, Argentina. Polar Biol 40(5):1023–1028. https://doi.org/10.1007/s00300-016-2026-7

García-Borboroglu P, Pozzi LM, Parma AM, Dell'Arciprete P, Yorio P (2022) Population distribution shifts of Magellanic penguins in northern Patagonia, Argentina: implications for conservation and management strategies. Ocean Coast Manag 226:106259. https://doi.org/10.1016/j.ocecoaman.2022.106259

Glen AS, Atkinson R, Campbell KJ, Hagen E, Holmes ND, Keitt BS, Parkes JP, Saunders A, Sawyer J, Torres H (2013) Eradicating multiple invasive species on inhabited islands: the next big step in Island restoration? Biol Invasions 15(12):2589–2603. https://doi.org/10.1007/s10530-013-0495-y

Green C, Green DB, Ratcliffe N, Thompson D, Lea M, Baylis AMM, Bond AL, Bost C, Crofts S, Cuthbert RJ, González-Solís J, Morrison KW, Poisbleau M, Pütz K, Rey AR, Ryan PG, Sagar PM, Steinfurth A, Thiebot J et al (2023) Potential for redistribution of post-moult habitat for *Eudyptes* penguins in the Southern Ocean under future climate conditions. Glob Chang Biol 29(3):648–667. https://doi.org/10.1111/gcb.16500

Hall JR, Woods RW, Brooke DLM, Hilton GM (2001) Factors affecting the distribution of landbirds on The Falkland Islands. Bird Conserv Int 12(2):151–167. https://doi.org/10.1017/S0959270902002095

Harrington KJ, Folkertsma R, Auersperg AMI, Biondi L, Lambert ML (2024) Innovative problem solving by wild falcons. Curr Biol 34(1):190–195.e3. https://doi.org/10.1016/j.cub.2023.10.061

Harris S, Samaniego RAS, Rey AR (2016) Insights into diet and foraging behavior of imperial shags (*Phalacrocorax atriceps*) breeding at Staten and Becasses Islands, Tierra Del Fuego, Argentina. Wilson J Ornithol 128(4):811–820. https://doi.org/10.1676/15-141.1

Humphrey PS, Bridge D, Reynolds PW, Peterson RT (1970) Birds of Isla Grande (Tierra del Fuego). Preliminary Smithsonian Manual, Museum of Natural History. University of Kansas, Lawrence

Imshaug HA (1972) R/V Hero Cruise 71-5 to Isla de los Estados. Antarct J US 7(2):42–44

Kühnemann O (1976) Observaciones ecológicas sobre la vegetación marina y terrestre de la Isla de los Estados (Tierra del Fuego, Argentina). Ecosur 3(6):121–248

Kusch A, Marín M (2012) Sobre la distribución del Pingüino Rey Aptenodytes Patagonicus (Aves: Spheniscidae) en Chile. Anales Del Instituto de La Patagonia 40(1):157–163. https://doi.org/10.4067/S0718-686X2012000100020

Liljesthröm M, Emslie SD, Frierson D, Schiavini A (2008) Avian predation at a southern rockhopper penguin colony on Staten Island, Argentina. Polar Biol 31(4):465–474. https://doi.org/10.1007/s00300-007-0372-1

Lois NA, Campagna L, Balza U, Polito MJ, Pütz K, Vianna JA, Morgenthaler A, Frere E, Sáenz-Samaniego R, Raya Rey A, Mahler B (2020) Metapopulation dynamics and foraging plasticity in a highly vagile seabird, the southern rockhopper penguin. Ecol Evol 10(7):3346–3355. https://doi.org/10.1002/ece3.6127

Lois NA, Balza U, Brasso R, Dodino S, Pütz K, Polito MJ, Riccialdelli L, Ciancio J, Quillfeldt P, Mahler B, Rey AR (2022) Mercury and stable isotopes portray colony-specific foraging grounds in southern rockhopper penguins over the Patagonian Shelf. Mar Pollut Bull 184:114137. https://doi.org/10.1016/j.marpolbul.2022.114137

Lysaght A (1959) Some eighteenth century bird paintings in the library of Sir Joseph Banks (1743–1820). Bull Nat His Museum (Nat His) His Ser 1:251–371. https://doi.org/10.5962/p.92313

Lyver PO, Moller H, Robertson CJ (1999) Predation at sooty shearwater Puffinus griseus colonies on the New Zealand mainland: is there safety in numbers? Pac Conserv Biol 5(4):347. https://doi.org/10.1071/PC000347

Miskelly CM, Elliott GP, Parker GC, Rexer-Huber K, Russ RB, Taylor RH, Tennyson AJD, Walker KJ (2020) Birds of the Auckland Islands, New Zealand subantarctic. *Notornis* 67(1):59–151

Morgenthaler A, Frere E, Rey AR, Torlaschi C, Cedrola P, Tiberi E, Millones A (2018) Unusual number ofSouthern Rockhopper Penguins, Eudyptes chrysocome, molting and dying along the Southern Patagonian coast of Argentina:pre-molting dispersion event related to adverse oceanographic conditions?. Polar Biol, 41(5):1041–1047

Parera A, Schiavini A, Frere E (1997) Relevamiento ecológico de la Isla de los Estados: Observaciones sobre su estado de conservación y sugerencias de manejo. Boletín Técnico 38. Fundación Vida Silvestre

Payró R (1898) La Australia Argentina: excursión periodistica a las costas patagónicas, Tierra del Fuego é Isla de los Estados; con una carta-prólogo del general Bartolmé Mitre. Centro Editor de America Latina, Buenos Aires

Petracci P, Sáenz Samaniego R, Raya Rey A (2016) Aspectos reproductivos y uso de hábitat del Cauquén Común (Chloephaga picta) y el cauquén real (Chloephaga poliocephala) en Isla de los Estados, Argentina. Hornero 31(2):73–81

Pincheira-Ulbrich J, Rodas-Trejo J, Almanza VP, Rau JR (2008) The conservation status of raptors in Chile. El Hornero 23(1):5–13. https://bibliotecadigital.exactas.uba.ar/collection/hornero/document/hornero_v023_n01_p005

Pütz K, Rey AR, Huin N, Schiavini A, Pütz A, Lüthi BH (2006) Diving characteristics of southern rockhopper penguins (Eudyptes c. Chrysocome) in the Southwest Atlantic. Mar Biol 149(2):125–137. https://doi.org/10.1007/s00227-005-0179-y

Pütz K, Schiavini A, Rey AR, Lüthi BH (2007) Winter migration of magellanic penguins (Spheniscus magellanicus) from the southernmost distributional range. Mar Biol 152(6):1227–1235. https://doi.org/10.1007/s00227-007-0770-5

Pütz K, Raya Rey A, Otley H (2013) Souhern rockhopper penguin. In: García-Borboroglu P, Boersma PD (eds) Penguins: natural history and conservation. University of Washington Press, Seattle

Quintana F, Schiavini A, Copello S (2005) Estado poblacional, ecología y conservación del Petrel Gigante del Sur (*Macronectes giganteus*) en Argentina. El Hornero 20(1):25–34. https://doi.org/10.56178/eh.v20i1.817

Quintana F, Dell'Arciprete OP, Copello S (2010) Foraging behavior and habitat use by the southern Giant Petrel on the Patagonian Shelf. Mar Biol 157(3):515–525. https://doi.org/10.1007/s00227-009-1337-4

Raimilla V, Díaz F, Barros R, Norambuena HV, Contreras G, Carrasco S, Torres D, Sandvig EM (2023) What is the northernmost breeding range of Fuegian Snipe Gallinago stricklandii? Bull Br Ornithol Club 143(3):350–361. https://doi.org/10.25226/bboc.v143i3.2023.a4

Raya Rey A, Schiavini A (2005) Inter-annual variation in the diet of female southern rockhopper penguin (Eudyptes chrysocome chrysocome) at Tierra del Fuego. Polar Biol 28(2):132–141

Raya Rey A, Trathan P, Pütz K, Schiavini A (2007a) Effect of oceanographic conditions on the winter movements of rockhopper penguins Eudyptes chrysocome chrysocome from Staten Island, Argentina. Mar Ecol Prog Ser 330:285–295. https://doi.org/10.3354/meps330285

Raya Rey A, Trathan P, Schiavini A (2007b) Inter-annual variation in provisioning behaviour of southern rockhopper penguins *Eudyptes chrysocome chrysocome* at Staten Island, Argentina. Ibis 149(4):826–835. https://doi.org/10.1111/j.1474-919X.2007.00718.x

Raya Rey A, Pütz K, Luna-Jorquera G, Lüthi B, Schiavini A (2009) Diving patterns of breeding female rockhopper penguins (Eudyptes chrysocome): Noir Island, Chile. Polar Biol 32(4):561–568. https://doi.org/10.1007/s00300-008-0550-9

Raya Rey A, Polito M, Archuby D, Coria N (2012a) Stable isotopes identify age- and sex-specific dietary partitioning and foraging habitat segregation in southern giant petrels breeding in Antarctica and southern Patagonia. Mar Biol 159(6):1317–1326. https://doi.org/10.1007/s00227-012-1912-y

Raya Rey A, Sáenz Samaniego R, Petracci PF (2012b) New records of south American sea lion Otaria flavescens predation on southern rockhopper penguins Eudyptes chrysocome at Staten Island, Argentina. Polar Biol 35(2):319–322. https://doi.org/10.1007/s00300-011-1072-4

Raya Rey A, Rosciano N, Liljesthröm M, Sáenz Samaniego R, Schiavini A (2014) Species-specific population trends detected for penguins, gulls and cormorants over 20 years in sub-Antarctic Fuegian Archipelago. Polar Biol 37(9):1343–1360. https://doi.org/10.1007/s00300-014-1526-6

Raya Rey A, Balza U, Domato I, Zunino F (2022) New Magellanic penguin Spheniscus magellanicus colony in a subantarctic Island. Polar Biol 45(10):1553–1558. https://doi.org/10.1007/s00300-022-03093-6

Raya Rey A, Balza U, Domato I, Salom A, Pütz K (2024) Breeding confirmation and habitat utilization of Sooty Shearwater (Ardenna grisea) on Isla de los Estados, Argentina. Polar Biol 47(10):1159–1164. https://doi.org/10.1007/s00300-024-03285-2

Riaz J, Orben RA, Gamble A, Catry P, Granadeiro JP, Campioni L, Tierney M, Baylis AMM (2024) Coastal connectivity of marine predators over the Patagonian Shelf during the highly pathogenic avian influenza outbreak. Ecography 2024(11):e07415. https://doi.org/10.1111/ecog.07415

Rivas-Fuenzalida T, Burgos-Andrade K, García Á, Castrilli S, Rosales V, Peña C, Ziehlmann E, Figueroa R (2024) Chilean Hawk (Astur chilensis). In: Billerman SM, Keeney BK, Rodewald PG, Schulenberg TS (eds) Birds of the World. Cornell Lab of Ornithology, Ithaca. https://doi.org/10.2173/bow.bichaw4.03.1

Rosciano NG, Polito MJ, Rey AR (2016) Do penguins share? Evidence of foraging niche segregation between but not within two sympatric, central-place foragers. Mar Ecol Prog Ser 548:249–262. https://doi.org/10.3354/meps11689

Rosciano NG, Pütz K, Polito MJ, Raya Rey A (2018) Foraging behaviour of Magellanic penguins during the early chick-rearing period at Isla de los Estados, Argentina. Ibis 160(2):327–341. https://doi.org/10.1111/ibi.12547

Rosciano NG, Polito MJ, Raya Rey A (2020) Seasonally persistent foraging niche segregation between sympatric southern rockhopper and Magellanic penguins breeding at Isla de los Estados, Argentina. J Ornithol 161(4):1093–1104. https://doi.org/10.1007/s10336-020-01800-w

Rosciano NG, Pütz K, Polito MJ, Raya Rey A (2022) Where's the best supermarket deal? Female Southern Rockhopper Penguins (*Eudyptes chrysocome*) show variable foraging areas during the guard stage at Isla de los Estados, Argentina. Can J Zool 100(1):46–55. https://doi.org/10.1139/cjz-2021-0020

Rumboll M, Jehl JR (1977) Observations on pelagic birds in the South Atlantic Ocean in the Austral Spring. Trans San Diego Soc Natural History 19(1):1–16

Salvadori T (1900) Contribuzione all' avifauna dell' America Australe (Patagonia, Terra del Fuoco, Isola Degli Stati, Isole Falkland). Annali Del Museo Civico Di Storia Naturale Di Genova 2(20):609–634

Sandvig EM, Quilodrán CS, Aguirre F, Rivero De Aguilar J, Barroso O, Vásquez RA, Rozzi R (2020) Patrones de distribución de la avifauna de los bosques de la Reserva de la Biosfera Cabo de Hornos: Un antecedente básico para la planificación del aviturismo sustentable. Anales del Instituto de la Patagonia 48(3):169–183. https://doi.org/10.4067/S0718-686X2020000300169

Santaniello C, Dodino S, Pütz K, Raya Rey A (2026) Comprehensive Stressor Analysis for Breeding Magellanic and SouthernRockhopper Penguins. J Nat Conserv. 89:127152. https://doi.org/10.1016/j.jnc.2025.127152

Schiavini A (2000) Staten Island, Tierra del Fuego: the largest breeding ground for southern rockhopper penguins? Waterbirds 23(2):286–291

Schiavini A, Rey A (2004) Long days, long trips: foraging ecology of female rockhopper penguins Eudyptes chrysocome chrysocome at Tierra del Fuego. Mar Ecol Prog Ser 275:251–262. https://doi.org/10.3354/meps275251

Schiavini A, Frere E, Yorio P, Parera A (1999) Las aves marinas de la Isla de los Estados, Tierra del Fuego, Argentina: Revision histórica, estado poblacional y problemas de conservación. Anales Instituto Patagonia 27:25–40

Schiavini A, Yorio P, Gandini P, Rey AR, Boersma YPD (2005) Los pingüinos de las costas argentinas: estado poblacional y conservación

Smith RIL, Prince PA (1985) The natural history of Beauchêne Island. Biol J Linn Soc 24(3):233–283. https://doi.org/10.1111/j.1095-8312.1985.tb00374.x

Strange I (1996) The Striated Caracara (Phalcoboenus australis) in the Falkland Islands. Philip Myers, Cheshire

Tabak MA, Poncet S, Passfield K, Goheen JR, Martinez del Rio C (2015) Rat eradication and the resistance and resilience of passerine bird assemblages in The Falkland Islands. J Anim Ecol 84(3):755–764. https://doi.org/10.1111/1365-2656.12312

Vuilleumier F (1985) Forest birds of Patagonia: ecological geography, speciation, endemism, and faunal history. Ornithol Monogr 36:255–304. https://doi.org/10.2307/40168287

Wallace RB, Reinaga A, Piland N, Piana R, Vargas FH, Zegarra RE, Alvarado S, Kohn S, Lambertucci SA, Alarcón P, Méndez D, Sáenz-Jiménez F, Ciri F, Álvarez J, Angulo F, Astore V, Cisneros J, Gálvez-Durand J, Vento R et al (2022) Defining spatial conservation priorities for the Andean Condor (Vultur gryphus). J Raptor Res 56(1):1–16. https://doi.org/10.3356/JRR-20-59

Weimerskirch H, Zotier R, Jouventin P (1989) The Avifauna of the Kerguelen Islands. Emu Austral Ornithol 89(1):15–29. https://doi.org/10.1071/MU9890015

Yorio P, Pozzi L, Herrera G, Punta G, Svagelj WS, Quintana F (2020) Population trends of Imperial cormorants (*Leucocarbo atriceps*) in northern coastal Argentine Patagonia over 26 years. Emu Austral Ornithol 120(2):114–122. https://doi.org/10.1080/01584197.2020.1730192

Chapter 13
The Day After: Native Mammals at Isla de los Estados

Adrián Schiavini and Alejandro Eduardo Jorge Valenzuela

Abstract Isla de los Estados hosts few native mammal species: the South American Fur Seal, the South American Sea Lion, the Southern Elephant Seal, the Southern River Otter, the Marine Otter (of uncertain presence), and the Olive Soft-haired Mouse.

Indigenous people reaching sporadically the island utilized the pinnipeds, but the sealing occurring between the eighteenth and the twentieth century depleted seal numbers severely. At the same time, otters might have been hunted by sealers and sailors for their fur.

Today, pinnipeds at Isla de los Estados are recovering. The last survey of 2012 reported ~9500 South American Fur Seals, increasing at a rate of 6.4%/year since 1995. The South American Sea Lion reached ~900 individuals, increasing at a rate of 8%/year since 1997. Southern Elephant Seals, on the other hand, could be developing a new breeding location. The Southern River Otter shows a stable presence and reproduction, but their main threat is habitat degradation from trampling by introduced Red Deer and Goats. The native mice may face threats from the introduced rat species.

A. Schiavini (✉)
Centro Austral de Investigaciones Científicas (CADIC-CONICET),
Ushuaia, Tierra del Fuego, Argentina

Wildlife Conservation Society, Representación Argentina, Buenos Aires, Argentina

Instituto de Ciencias Polares, Ambiente y Recursos Naturales, Universidad Nacional de Tierra del Fuego, Ushuaia, Tierra del Fuego, Argentina
e-mail: aschiavini@untdf.edu.ar

A. E. J. Valenzuela
Instituto de Ciencias Polares, Ambiente y Recursos Naturales, Universidad Nacional de Tierra del Fuego, Ushuaia, Tierra del Fuego, Argentina

Consejo Nacional de Investigaciones Científicas y Técnicas,
Ushuaia, Tierra del Fuego, Argentina

Otter Specialist Group, Species Survival Commission, International Union for Conservation of Nature, Gland, Switzerland

A. Raya Rey, U. Balza (eds.), *Biodiversity and Conservation of Isla de los Estados (Chuani-sin)*, Natural and Social Sciences of Patagonia,
https://doi.org/10.1007/978-3-032-13630-5_13

Although pinnipeds are currently protected at the island, the impact of fisheries around the archipelago remain poorly understood. The island's isolation helped in the recovery of the mammal species after their exploitation but contributed to the impact by uncontrolled introduced species. Its isolation also poses difficulties for scientific research and management.

Keywords Historical exploitation · Otaria byronia · Arctocephalus australis · Mirounga leonina · Lontra provocax

13.1 The Species

Isla de los Estados hosts few species of native mammals. On the one hand, the South American Sea Lion (*Otaria byronia*), the South American Fur Seal (*Arctocephalus australis*) and the Southern Elephant Seal (*Mirounga leonina*) are found on the coasts of the island. On the other hand, two species of otters were reported for the island: the Southern River Otter (*Lontra provocax*) and the Marine Otter (*L. felina*). Finally, the Olive Soft-haired Mouse (*Abrothrix olivacea llanoi*) is the only terrestrial mammal native to the Island.

13.1.1 South American Fur Seal

The South American Fur Seal is distributed along the coasts and waters of South America, from center Brazil to northern Peru, including the Malvinas Islands (Cárdenas-Alayza et al. 2016a, Fig. 13.1). Adult males reach 90–160 kg and females on average 60 kg. Their colonies and haul out sites occupy rocky coasts, ledges above the shoreline or boulder-strewn areas (Cárdenas-Alayza 2018a). Their breeding season occurs from October through January.

13.1.2 South American Sea Lion

The South American Sea Lion is distributed also along the coasts and waters of South America, from southern Brazil to northern Peru, including Malvinas Islands (Cárdenas-Alayza et al. 2016b, Fig. 13.2). They are strongly sexually dimorphic, with males reaching 300–350 kg and females on average 170 kg. Their colonies and haul out sites can be located on rocky coasts including pebble beaches. Breeding takes place during the austral summer, starting around mid-December (Cárdenas-Alayza 2018b).

Fig. 13.1 South American Fur Seal (*Arctocephalus australis*) on Observatorio Island. (Credit Emiliano Arona)

13.1.3 Southern Elephant Seal

The Southern Elephant Seal is distributed with breeding sites on islands around the subantarctic and the Antarctic Peninsula, including Peninsula Valdés and the Malvinas Islands, although they use the Southern Ocean waters from the north of the Antarctic Polar Front to the Antarctic pack ice (Hindell 2018). They are strongly sexual dimorphic, with males reaching 3700 kg and females between 400 and 800 kg. Their colonies and haul out sites can be located on pebble beaches. Breeding

Fig. 13.2 South American Sea Lion (*Otaria byronia*) on Observatorio Island. (Credit Samanta Subires)

occurs during the austral spring, starting by August, including a short lactation period of about 3 weeks.

13.1.4 Southern River Otter

The Southern River Otter is distributed from northern Patagonia to Isla de los Estados, in Argentina and Chile. It inhabits rivers and lakes banks on the northern part of its distribution and marine coasts of fjords and channels at their southern part of their distribution (Valenzuela et al. 2019, Fig. 13.3). With no strong differences between sexes, adults could reach 10 kg and 1.4 m long. As territorial solitary individuals, they establish their territories along rocky vegetated coasts protected from winds and associated to macroalgal forests (Valenzuela et al. 2013; Claverie 2025). Breeding season starts at the end of winter with newborns between October and December.

13.1.5 Marine Otter

The Marine Otter inhabits the rocky Pacific coasts from Perú to Cape Horn, Chile (Mangel et al. 2022). Historical records suggest their presence in Isla de los Estados and the habitat in the southern coasts of the island is like the one used by the species

Fig. 13.3 Southern River Otter (*Lontra provocax*). (Credit Joel Reyero)

elsewhere. However, there are doubts about whether they currently or ever existed on the island, and thus in Argentina (Cassini 2008; Valenzuela 2019). This smaller otter has adults weighing between 5 and 6 kg (Valenzuela 2019).

13.1.6 Olive Soft-Haired Mouse

The Olive Soft-haired Mouse is an endemic rodent of Argentina and Chile, found from central Chile and Mendoza Province in Argentina to Isla de los Estados. It averages 28 g in weight with highly variable population fluctuations (Teta and D'Elía 2019).

13.2 The Past

13.2.1 Hunter-Gatherer Times

Before the twentieth century, the only available information on the pinnipeds of Isla de los Estados comes from archaeological findings and, later, from reports and records associated with sealing and exploration activities. All three pinniped species were utilized by the indigenous people of the southernmost tip of South America, including both terrestrial and maritime hunter-gatherer groups. The indigenous communities of the Fuegian archipelago, known as the Yamana, Alakaluf, Selk'nam, and Haush, made use of pinnipeds. The Yamana and Alakaluf developed maritime technologies that enable the exploitation of marine resources, including pinnipeds

(Orquera and Piana 2009). Archaeological evidence indicates that Isla de los Estados was intermittently occupied by hunter-gatherer-fisher groups between approximately 2800 and 1375 cal. yr BP, likely as part of seasonal movements targeting seabirds and marine mammals (Zangrando et al. 2025, this volume). Thus, pinniped populations on Isla de los Estados were accessible to humans during the Late Holocene. However, current archaeological evidence does not support a significant ecological human impact on pinnipeds. Similarly, even when otters are present in the archeological sites in the rest of the Tierra del Fuego archipelago, no otters were recorded in archeological contexts from Isla de los Estados (Zangrando, personal communication). Further details are discussed in Zangrando et al. (2025, this volume).

13.2.2 The Sealing Industry

The European discovery of Cape Horn by Schouten and Le Maire in 1616 integrated the nearby Isla de los Estados into what was then considered one of the "corners" of the world. This event marked a turning point not only for the pinnipeds of Isla de los Estados but also for the broader region of South America and Antarctica. Once Cape Horn was established as a navigable passage, sailors began to frequent Isla de los Estados to seek refuge, fresh water and wood. Later, the island became a hunting ground for pinnipeds.

It has been noted that several biological traits of pinnipeds made them particularly vulnerable to exploitation by humans (Bonner 1982). Pinniped fur typically consists of a guard hair and numerous short, soft fibers, often more than 50 per pilosebaceous unit in fur seals (*Arctocephalus* sp.). Alongside their thick blubber layer, typical of marine mammals, these features made pinnipeds valuable for their pelt, fur and oil. Their large size, gregarious behaviour, and predictable presence on shorelines due to their breeding habits, rendered them ideal targets for human exploitation, especially under expanding capitalist economies.

In the eighteenth century, the growing economies of Europe and North America demanded oil, leather, and fur. Additionally, the Pacific and Indian Oceans were opened to navigation for British colonial interests, following previous restrictions that lasted until the American Revolutionary War (Kirker 1970). The depletion of the whale stocks in the northern hemisphere due to overexploitation drove the expansion of whaling activities into the southern hemisphere. At that time, the southwestern Atlantic was under Spanish control through the *Virreinato del Río de la Plata*. However, enforcement of regulations over these lands and waters was weak, particularly at southern latitudes.

As a result, the southwestern Atlantic became a contested region, where Spaniards, English, French, and North Americans disputed access to "fishing" resources (as seal and whale hunting was called). In response, the Spanish crown established control measures, including a ban on whale and seal hunting in Spanish waters in 1776 (Destéfani 1981), and the creation of the Royal Maritime Company in 1789 (Destéfani 1987). However, the 1790 Nootka Sound Treaty between Spain and England allowed English ships to navigate and hunt in southern seas and coasts, provided they remained more than ten miles from Spanish-claimed shores. This agreement also enabled temporary hunting settlements on the mainland, with long-term consequences for seal populations.

One of the first references to pinnipeds on Isla de los Estados comes from James Cook's second voyage to the southern hemisphere. In 1775, he recorded the presence of South American Fur Seals and of South American Sea Lions and hunted some specimens on Observatorio Island (called New Year's Island by Cook). Cook wrote: "*it mattered not where we landed; for the whole shore was covered with them; and, by the noise they made, one would have thought the island was stocked with cows and calves.... They were, in general, so tame, or rather stupid, as to suffer us to come near enough to knock them down with sticks; but the large ones we shot, not thinking it safe to approach them*".

By 1774, a branch of the whaling industry had emerged: seal hunting. That year, Collins and Smith traveled to the Malvinas Islands in search of sperm whales. Instead, they encountered large colonies of sea lions, which were easy to hunt and yielded high-quality oil (Caviglia 2015). By 1790, whaling expeditions had begun to move towards the Pacific, using Cape Horn as a critical route. During his second voyage, Cook not only visited Isla de los Estados but also the South Georgia Islands, where he observed numerous concentrations of the Antarctic Fur Seal (*Arctocephalus gazella*). News of these rich harvesting grounds quickly spread, boosting seal-hunting operations. His observations on whales' distribution also paved the way for the arrival of more whaling vessels in these latitudes. These developments marked the beginning of intensive whale and seal exploitation across the southern Atlantic and Pacific oceans and the Antarctic waters.

How many seals once inhabited Isla de los Estados before the European sealing activities? This can only be estimated retrospectively through sealing records. However, captains of sealing ships were often cautious about revealing their harvesting locations, fearing competitors might gain valuable information. As a result, records were frequently and deliberately vague, obscuring specific locations and harvest yields (Caviglia 2015).

With the arrival of sealers and whalers, the Tierra del Fuego archipelago became both a logistical hub and a primary site for seal harvesting. Conditions were ideal for overexploitation: abundant yields of oil and hides, a declining colonial administration with limited enforcement capacity, a multi-species industry able to shift targets, and idle capital flowing from the Northern to the Southern Hemisphere. This scenario led to what Hardin (1968) termed a "tragedy of the commons", a situation where open-access resources are depleted due to overuse, a well-documented phenomenon in natural resource exploitation (Hilborn et al. 1995). The pressure was especially intense between the 18th and 19th centuries, when three closely related *Arctocephalus* species were hunted: *A. australis* and *A. phillippii* in southern South America and *A. gazella* in Antarctica. This extractive process resembled a multispecies fishery (May et al. 1979), where the depletion of one species did not prevent continued exploitation of others.

In the Southwestern Atlantic, fur seals were particularly sought after because they provided both pelt and oil. Fur was easier to obtain than oil, required less storage space on board, and was more valuable. Generally, when fur seal stocks were depleted, hunters switched to the South American Sea Lion and to the Southern Elephant Seal. Elephant sealing often preceded the fur sealing season, and when fur seals became scarce, oiling for elephant seals offered a profitable activity while searching for remaining fur seal colonies (Bonner 1982). Many sealing vessels were originally outfitted for whaling and thus already equipped to process blubber. Some whaling ships carried partial loads of elephant seal oil and fur seal skins. Hunting took place along the coasts of the Malvinas Islands, continental Patagonia, the Tierra del Fuego archipelago, the Pacific coast of South America, and later the Antarctic shores. The location of hunting grounds shifted dynamically depending on objectives, weather conditions, and yields (Haller 2023).

Sealing activities were primarily focused on breeding colonies. The predictable presence and concentration of animals enabled sealers to focus their efforts and increase efficiency. A sealing captain, quoted in Bonner (1968), describes this pattern, one that contributed significantly to the rapid depletion of populations:

> I don't think it is a good plan to commence killing until they get well into pupping. Don't kill any of the old wigs until you have worked off most of the clapmatches, or near the end of the pupping season, as they hold the other seals and will not let them go off the rocks if they can prevent it. If you want all the seals that are on the rocks to make up your cargo, you can all through the pupping season be working off the young wigs, which are always hauled in small rookeries near the pupping seal, driven there by the old wigs.
>
> You will bear in mind that the full grown clapmatches or young wigs that are larger, are the most valuable. The next in value are the two and three year old seals, that haul at different times on the rocks. The yearlings that are driven off in November are prime skins, and taking into account how much less work it is to skin them, and how much less salt, and room they take, they are worth striving for. The large old wigs are of the least value, taking more work, salt and room. Still they are better than nothing. You must try to get to the island in time to get all of last year's pups with the clapmatches. If you can get there thus early it will give you a great start and make everything smooth for you; but if you cannot get there in time to take the young seal season, get there as soon as possible, and be sure that you examine the islands and rocks thoroughly.
>
> You will remember that you cannot get all the seal from a rookery in one season. If you get 3000 the first season you may expect to get 1000 or more the next, and in the same proportion for a larger or smaller number.

During the nineteenth century, as Argentina was establishing itself as a nation, it started to look after the activities that were being carried out in the southwestern Atlantic. In 1828, the Government of Buenos Aires granted Luis Vernet control over Isla de los Estados, where he established a sea lion exploitation operation in Port Hoppner, along with timber extraction activities. Later, in 1829, Vernet was appointed Political and Military Commander of the Malvinas Islands, assuming responsibilities including the control and exploitation of the sea lion populations on Isla de los Estados. Based on available historical records and evidence found on the island itself, Vernet is believed to have harvested fur seals in, at least, Port Hoppner and Port Basil Hall (Vairo 1997; Niekisch and Schiavini 1998). When the Malvinas Islands were seized by England in 1833, Luis Vernet abandoned the area, leading to a decline in the regulation of sea lion exploitation by foreign fleets. By the mid-nineteenth century, Luis Piedrabuena had emerged as a prominent figure, engaging in sealing and other activities on Isla de los Estados. In 1869, the National Congress granted Piedrabuena ownership of the island. A decade later, in 1879, he established facilities specifically for sea lion hunting (Vairo 1997, Fig. 13.4).

By the early twentieth century, populations of southern fur seals had been drastically reduced in the southwestern Atlantic Ocean. The seal hunter Delano (1817)

Fig. 13.4 Hunting of southern sea lions at Cape Colnett, Isla de los Estados, in 1937. (Credit Vairo (1997)

visited Isla de los Estados in 1800 and wrote "*All that can be obtained from this island; is wood, water, and some few seals. The islands around Tierra del Fuego afford a few seals of both the fur and hair kinds*". According to Carrara (1964), between 1825 and 1830, seal hunting activities began to become uneconomical due to the overexploitation of all stocks. There were practically no fur seal colonies left to be discovered; and those already discovered had been plundered or would be in the years that followed. By 1840, the South American Fur Seal was almost extinct along the Argentine coast.

The final phase of fur seal exploitation reveals clear signs of scarcity. Bove (1883) observed that the Southern Fur Seal, once extremely abundant, had nearly disappeared, making it difficult for hunters to recover the costs of outfitting expeditions. Similarly, Hyades (2020, originally published in 1885), recounts that in October 1882, the American schooner Thomas Hunt captured 300 seals in 5 days on the Diego Ramírez Islands, a steep decline from the 2 years prior, when daily harvests reached up to 500. The schooner's captain, then in his seventh year of hunting, noted that each year brought diminishing returns due to the unsustainable methods employed.

By this time, demand for fur had begun to decline, although the market for seal oil persisted. As a result, the sealing industry turned to previously "overlooked" species, such as the South American Sea Lion (which provided both leather—albeit of lower quality than fur seal leather—and oil) and the Southern elephant seal (that provided only oil). By the end of the nineteenth century, the South American Sea Lion had replaced the South American Fur Seal exploitation in Isla de los Estados (Payró 1898). Even as late as 1932, commercial hunting of sea lions continued through settlements such as those in Colnett Bay. In 1937, the growing concern over declining populations led to a complete ban on fur seal hunting. However, this prohibition did not extend to the sea lion, which continued to be harvested until the 1950s.

Even when there are no actual records of otters in the literature related to the sealing industry, it is likely that otters where overhunted for its pelts trade as everywhere else (Chehebar 1985; Sepúlveda et al. 2021). The otters could be hunted opportunistically by sealing groups when they set up their camps along the island's coasts.

13.3 The Late Sealing Numbers

The information about pinnipeds from Isla de los Estados during the first half of the twentieth century is limited. Dabbene (1904) recorded in 1902 the presence of around 500 individuals of the South American Sea Lion on Isla Observatorio. The first population survey of sea lions and elephant seals along the entire Argentine coast was conducted between 1946 and 1947 (Carrara 1952). By then, exploitation in Argentina was focused on the South American Sea Lion. According to Carrara's survey, Península Mitre, located west of Isla de los Estados, housed the largest concentrations of South American Sea Lions, along with the Península Valdés area at northern Patagonia. Carrara pointed out that Isla de los Estados could not be fully surveyed due to weather conditions; the survey was conducted from an airplane by taking photographs.

Carrara's survey reported counts for three sites on Isla de los Estados (Table 13.1), although the published map showed five locations with South American Sea Lions and one with South American Fur Seals (Fig. 13.5). The exact route followed by Carrara during his survey remains unknown. As a result, it is unclear whether the absence of data from other sites is due to a lack of survey effort, as may be the case for the sea lion rookery under exploitation in 1937 shown in Fig. 13.4. Following Carrara's survey, the only additional record is from Pine et al. (1978), who reported 10 Southern fur seals in Capitán Cánepa Bay and 76 South American Sea Lions across various locations, including 34 individuals in a cave in Crossley Bay.

The presence of both otter species was described for Isla de los Estados (Massoia and Chebez 1993; Parera 1997), however, these historical records usually referred to "otters" in general, assigning the spraints found on the marine coast to Marine Otter and the ones found on the freshwater to Southern River Otter (Valenzuela 2019). Cassini (2008) sustained that the presence and distribution of the marine otter in Argentina is not clear.

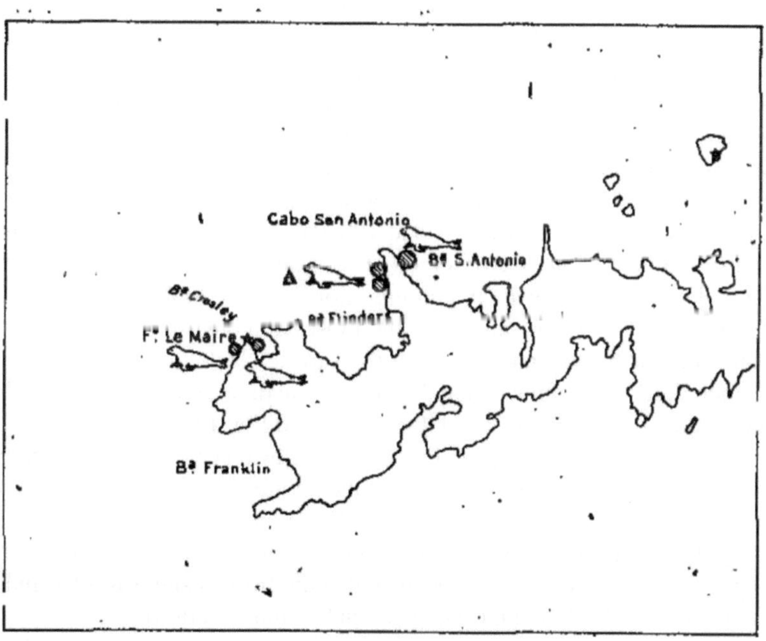

Fig. 13.5 Sea lion locations at Isla de los Estados reported by Carrara (1952). Original draw. Triangle: South American Fur Seal. Circle: South American Sea Lion

13.4 The Present

Beginning in the late 1980s, several expeditions began collecting data on the distribution and abundance of pinnipeds on Isla de los Estados (Schiavini 1987; Parera et al. 1997; Schiavini et al. 1998; Schiavini et al. 2004; Crespo et al. 2015; Milano et al. 2020a, b). These expeditions, conducted for varying purposes, sometimes opportunistically, used different platforms and occurred at different times of the year, thus, limiting the comparison of results. Nevertheless, based on the records, Milano et al. (2020a, b) were able to reconstruct a comprehensive overview of the current pinniped colonies on the island. The most recent data on distribution and abundance allows a direct comparison between Carrara's historical survey with and those conducted in 1995, 1997, and 2012 (Tables 13.1 and 13.2).

13.4.1 South American Fur Seal

The South American Fur Seal has increased in numbers since the Carrarra's 1946 survey (Milano et al. 2020a). However, as the 1946 survey did not encompass the entire island, therefore comparison with more recent data should be interpreted with caution. More precise counts revealed that the fur seal population on Isla de los Estados rose from an estimated 3334 individuals in the 1990s (Crespo et al. 2015; Parera et al. 1997) to 9519 individuals in 2012 (Milano et al. 2020a). Based on Table 13.1, the finite population growth rate on Isla de los Estados between 1995 and 2012 was 1.064 for all individuals and 1.041 for pups, giving to an annual change of 6.4% and 4.1%, respectively. Shifts in the spatial distribution of colonies are also evident: seven new sites were recorded in 2012, while four previously occupied sites were found to be abandoned. These dynamics are consistent with patterns observed across the southwest Atlantic, particularly the population increase in the Malvinas Islands as noted by Milano et al. (2020a). With fur seals appearing to have declined in southern Chile, the observed growth on Isla de los Estados may reflect both natural population growth and immigration from the Malvinas Islands.

Point Jira and Cape Furneaux were the most populous sites in 1995, 1997 and 2012, and held the highest numbers of pups. Across Isla de los Estados, the eastern region, including the Dampier Islands and surrounding areas, harbors the largest number of individuals (sites 6–18 of Table 13.1 and Fig. 13.6).

In recent years, South American Fur seals have also been observed at the northeastern tip of Observatorio Island. None of the previous studies, including our own earlier work reported their presence there. Therefore, this new record should be interpreted in the light of the overall population expansion and spatial redistribution documented for the species. The most recent count for this site was ~200 individuals in December 2024 (Joaquín Cano, personal communication).

Of the 13 locations where South American Fur Seals were recorded in 2012 on Isla de los Estados, 10 were shared with the South American Sea Lions. Despite the available data on numbers and colony distribution, virtually nothing is known about the biology of the species at Isla de los Estados.

Table 13.1 South American Fur Seal colonies of Isla de los Estados, comparing sites recorded in different surveys

					Summer 1997[c]		Summer 2012[d]								
	Location		Summer 1948[a]	Summer 1995[b]	Non pups	Pups	AM	SAM	F	J	F + J	Pups	Total 2012	Type of colony
1	Flinders Bay	54.73°S 64.55°W	1500											
2*	Cape San Antonio	54.73°S 64.54°W	100											
3	E of Point Shank	54.73°S 63.96°W		100	31								–	
4	Point Dorgambide and Point Shank	54.73°S 63.93°W		86									–	
5*,#	Point Brizuela	54.72°S 63.90°W					0	2	0	0	14	1	17	Mixed
6*	Cape Furneaux	54.72°S 63.89°W		318	734	0	15	103	32	55	1130	183	1518	Breeding
7*,#	Point Zapata	54.72°S 63.83°W					1	14	1	1	185	4	206	Mixed
8*	Cape San Juan	54.72°S 63.81°W		20			3	36	1	4	133	0	177	Haul-out
9*	Islet San Juan	54.72°S 63.81°W					11	15	5	7	191	22	251	Mixed
10	Ojeda Cove	54.73°S 63.81°W		205	333	0							–	
11*	Point Leguizamo	54.73°S 63.81°W		315	155	45	12	64	24	47	624	56	827	Mixed
12*	Point Jira	54.76°S 63.81°W		1758	1385	358	75	339	178	183	3863	701	5339	Breeding

13*,#	South of Point Pañuelo	54.77°S 63.83°W					8	21	0	0	13	0	42	Haul-out
14#	Tevez Rock	54.79°S 63.84°W					7	11	1	30	215	10	274	Mixed
15*	Point Fallows	54.79°S 63.85°W					4	34	1	0	73	2	114	Mixed
16#	Islet Siri	54.84°S 64.17°W					7	23	4	3	236	5	278	Mixed
17	Barrionuevo Island	54.85°S 64.16°W		130	168	54	0	21	3	4	216	9	253	Mixed
18	Point Achaval	54.84°S 64.18°W		400	311	89	1	21	2	1	193	5	223	Mixed
19	Menzies Islands	54.84°S 64.33°W		2									–	
	Grand total		1600	3334	3117	546	144	704	252	335	7086	998	9519	

Data taken from Milano et al. (2020a)

Note: Survey data by age classes in 2012: *AM* adult males (territorial and peripheral), *SAM* subadult males, *F* females, *J* juveniles, *F + J* indeterminate individuals. Numbers on the left refer to the position in Fig. 13.6

[*] Location with presence of *Otaria flavescens* in 2012; [#] New location recorded for SAFS in Milano et al. (2020a); [–] Site not visited during the survey. A value of 0 represents the colony was visited but no animals or age class were not recorded

[a]Data from Carrara (1952)

[b]Data from Parera et al. (1997) and Crespo et al. (2015)

[c]Data from Crespo et al. (2015)

[d]Data from Milano et al. (2020a)

Table 13.2 South American Sea Lion colonies of Isla de los Estados, comparing sites recorded in different surveys

			Summer 1948[a]		Summer 1997[b]		Summer 2012[c]		
	Location		Non pups	Pups	Non pups	Pups	Non pups	Pups	Type of colony
1	Flinders Bay	54.79°S 64.55°W	1500	200	51	20			
2	Cape San Antonio	54.73°S 64.54°W	300		+		129	0	Haul-out
5[*,#]	Point Brizuela	54.72°S 63.90°W					16	0	Haul-out
6[*,#]	Cape Furneaux	54.72°S 63.89°W					23	0	Haul-out
7[*,#]	Point Zapata	54.72°S 63.83°W					3	0	Haul-out
8[*]	Cape San Juan	54.72°S 63.81°W			20		9	0	Haul-out
9[*,#]	Islet San Juan	54.72°S 63.81°W					11	0	Haul-out
10[*]	Ojeda Cove	54.73°S 63.81°W			17				
11[*,#]	Point Leguizamo	54.73°S 63.81°W					4	0	Haul-out
12[*,#]	Point Jira	54.76°S 63.81°W					11	0	Haul-out
14[*,#]	Tevez Rock	54.79°S 63.84°W					7	0	Haul-out
15[*]	Point Fallows	54.79°S 63.85°W			10		5	0	Haul-out
20	Crossley Bay	54.79°S 64.69°W	500	50	+				
21	Goffré Island	54.70°S 64.24°W			50	D	155	10	Mixed
22	Observatorio Island	54.65°S 64.12°W			100+		509	0	Haul-out
	Grand total		2300	250	248	20	882	10	

Data taken from Milano et al. (2020b)

Numbers on the left refer to position in Fig. 13.7. Survey data grouped as pups and non-pups and Total (mean ± SD)

[*] Sites with the presence of *Arctocephalus australis* in 2012; [#] New location recorded for SASL in 2012; D: sea lions detected but not counted; [+] Colony not found during the survey. A value of 0 represents the site was visited but no animals or age class were not recorded

[a]Data from Carrara (1952)

[b]Data from Schiavini et al. (2004)

[c]Data from Milano et al. (2020b)

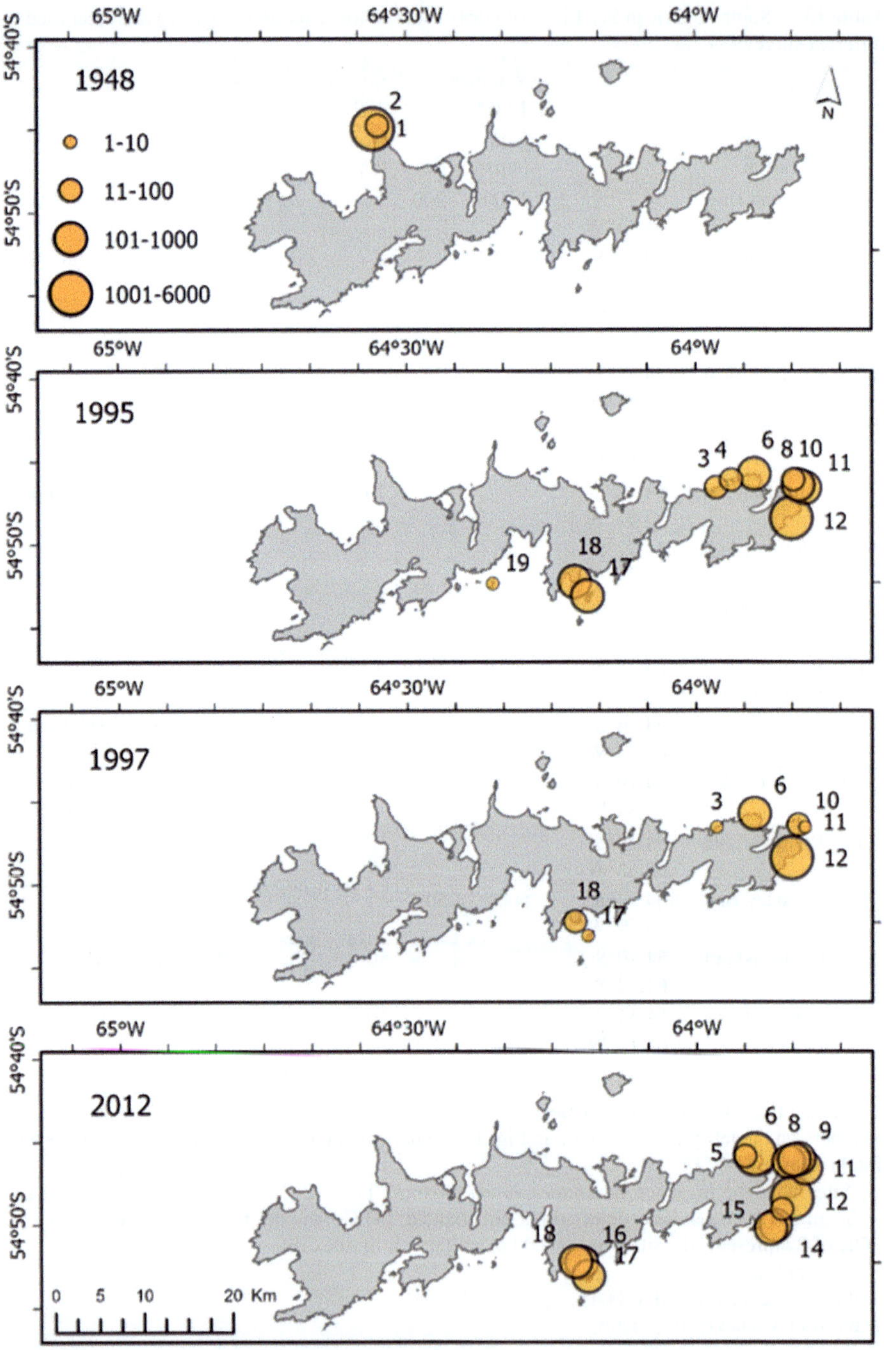

Fig. 13.6 South American fur seal colonies reported in Milano et al. (2020a) for different years. Numbers correspond to sites mentioned in Table 13.1

13.4.2 South American Sea Lion

The South American Sea Lion have declined in numbers on Isla de los Estados since the Carrarra's 1946 survey (Crespo et al. 2015; Parera et al. 1997; Milano et al. 2020b). As with the fur seal, caution is needed when comparing these data, since the 1946 survey did not cover the entire island. Numbers declined from 2550 in 1946 to 268 in 1997, followed by a partial recovery to 892 by 2012. However, data from nearby Isla Grande de Tierra del Fuego indicate a clearer trend of population recovery (Milano et al. 2020b).

Observatorio Island hosted the largest numbers of sea lions in 2012 (Table 13.2). Since 1997, sea lions have been regularly recorded in the northeastern point of Observatorio Island, which is now the largest rookery on Isla de los Estados. On a broader scale, the central portion of the island's northern coast holds the largest concentrations of sea lions (sites 2, 21 and 22 of Table 13.2 and Fig. 13.7), and this area has also seen the greatest increase in numbers since 1997.

According to Table 13.2, the finite population growth rate of the South American Sea Lion at Isla de los Estados between 1997 and 2012 was of 1.08 for all animals and 0.97 for pups. This represents an annual increase of 8% for the whole island but a decrease of 3.3% for pups. The increase between 1997 and 2012 was also accompanied by changes in the number and distribution of haul-out and colony sites. Of the two sites reported in 1948, only one remained occupied in 1997, but five new sites were identified that year. In 2012, sea lions were recorded at 12 sites, three of which had been reported in 1997, along with nine newly occupied locations. These patterns reflect a colonization process and expansion into new areas compared to 1997. Such processes are consistent with what has been observed on the Isla Grande de Tierra del Fuego and across the southwestern Atlantic Ocean, where South American Sea Lion have generally increased. These increases have been accompanied by the emergence of new breeding colonies and haul-out sites, as well as changes in the social structure (Milano et al. 2020b). The largest colonies on Isla Grande de Tierra del Fuego are located at the eastern tip of the island, approximately 70 km from the sea lion locations on Isla de los Estados. Most colonies on Isla de los Estados are haul-out sites, while Isla Grande de Tierra del Fuego is considered a reservoir area for non-reproductive age classes, with only small breeding groups (Milano et al. 2020b). In 2012, ten sites on Isla de los Estados were shared by both otariids species.

Despite the available data on colony size and distribution, virtually nothing is known about the biology of the species on Isla de los Estados. It has been reported that sub-adult males predate on southern rockhopper penguins (*Eudyptes chrysocome*) at the colony located at Franklin Bay, located on the western tip of Isla de los Estados (Raya Rey et al. 2012). Sea lions were also observed predating on Magellanic penguins (*Spheniscus megallanicus*) in Observatorio Island (Cano, personal communication). Although only a few individuals appear to engage in this behavior, the ecological significance of this predation remains unassessed.

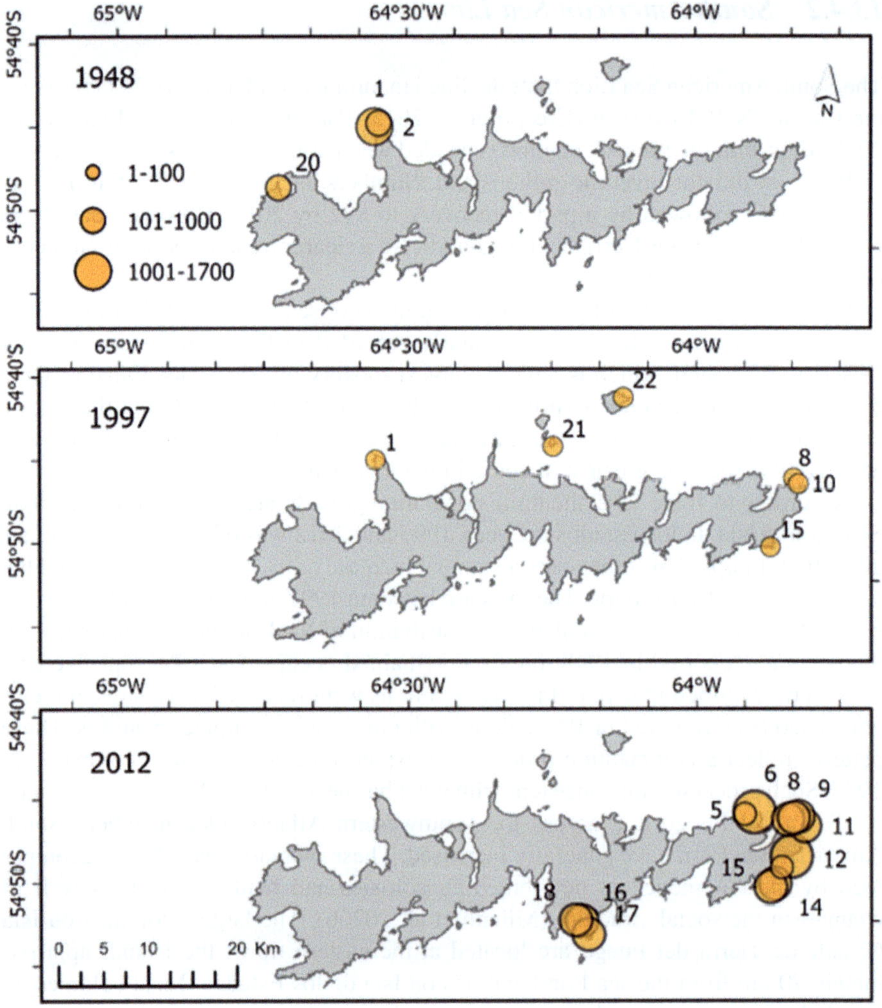

Fig. 13.7 South American sea lion colonies reported in Milano et al. (2020b) for different years. Numbers correspond to sites mentioned in Table 13.2

13.4.3 Southern Elephant Seal

The Southern Elephant Seal is present in the Año Nuevo archipelago. Parera et al. (1997) reported 44 individuals at Goffré Island and five individuals at Elizalde Island in 1995. The presence of pups was incidentally recorded (Raya Rey, personal communication) in the year 2020. An analysis from scats of individuals from Elizalde Island in 2013 revealed the presence of four helminth parasites (Hernandez-Orts et al. 2015). Beyond these observations, nothing else is known about the species on Isla de los Estados.

13.4.4 *Southern River Otter*

The Southern River Otter is present in all bays and fjords of Isla de los Estados, mostly occupying the marine coasts but also freshwater ecosystems (Claverie 2025). In Argentina, the coastal-marine population of the Tierra del Fuego Archipelago is categorized as critically endangered, mostly because their overall numeric reduction and their low numbers estimated for the region (Valenzuela et al. 2019). Since 2016, camera trap monitoring on Isla de los Estados has documented a relatively stable presence of the species, including evidence of reproduction. In the absence of significant human activity, the main threats identified are trampling of otters by invasive Red Deer (*Cervus elaphus*) and Domestic Goat (*Capra aegagrus hircus*), as well as potential impact of invasive rat species (*Rattus sp.*). Recently, Claverie and Valenzuela (2023) highlighted Península Mitre, at the eastern tip of Isla Grande de Tierra del Fuego, as a key area serving as corridor that could connect otter populations on Isla de los Estados with the rest of Tierra del Fuego archipelago.

13.4.5 *Marine Otter*

As previously mentioned, the presence of the Marine Otter on Isla de los Estados was suggested by historical accounts (Massoia and Chebez 1993; Parera et al. 1997) but remain controversial. These records assumed that spraints found along the coast belonged to this species, despite the fact that Southern River Otters are also known to use marine environments (Valenzuela 2019). Furthermore, genetic analysis of otter spraints collected in the area identified only *L. provocax* (Centrón et al. 2008). No sightings or detections through camera traps have been recorded in at least the past 20 years, making the past and current presence of the Marine Otter doubtful (Valenzuela 2019). As a result, the species is classified nationally as Data Deficient (Valenzuela 2019).

13.4.6 *Olive Soft-Haired Mouse*

Olive soft-haired mice from Isla de los Estados are not genetically distinct from *Abrothrix olivacea* populations on Isla Grande de Tierra del Fuego. However, individuals from Isla de los Estados are significantly larger in size, likely due to the effects of isolation (Sanchez et al. 2022). Camera trap surveys across the island have confirmed the widespread presence of this rodent. Nevertheless, the high abundance of invasive rats (*Rattus* sp.) detected by the same camera traps may pose a threat to this insular population.

13.5 Conclusions

The mammalian marine fauna of Isla de los Estados presents two contrasting aspects. On one hand, there is evidence of the recovery of pinniped species that were heavily exploited in the past. On the other hand, native species could be affected by the presence of introduced species with varying impacts.

The Tierra del Fuego archipelago was one of the three main regions of the Southwest Atlantic Ocean where fur seals and sea lions were severely reduced by sealing, along with the Valdés Peninsula and the Malvinas Islands. Following the cessation of sealing, populations of both the South American Fur Seal and the South American Sea Lion began to increase. This recovery trend is also evident in the Tierra del Fuego archipelago. As highlighted by Milano et al. (2020a, b), changes in the social structure of South American Fur Seals and South American Sea Lions colonies typically take the form of the establishment of haul-out sites and their gradual transition into breeding rookeries, patterns characteristic of a "recolonization" phase. These have also been observed on Isla de los Estados. Notably, in 2012, the South American Fur Seal was present in more breeding colonies than the South American Sea Lion.

For the South American Fur Seal, Milano et al. (2020a) proposed that immigration from colonies outside Isla de los Estados may be a stronger drive of population increase than local reproduction. In terms of pup production, the Malvinas Islands host the largest population of the species (Baylis et al. 2019). Located just 400 km from Isla de los Estados, these islands are likely the primary source of immigrants contributing to population growth on Isla de los Estados.

While the increase in South American Sea Lion numbers is among the largest recorded for the species in the southwest Atlantic Ocean, it is still lower than that observed for the South American Fur Seal (Milano et al. 2020b). Given the population trajectories seen along the Atlantic coast and in southern Chile, these authors suggest that immigration from these areas may be contributing to the increase in non-pup's individuals in the Fuegian Archipelago. However, the recovery of the South American Sea Lion has been delayed since the end of sealing in the mid-twentieth century. The causes of this delay remain unclear, though Milano et al. (2020b) emphasize that the decline of sea lion populations triggered a profound reorganization of the marine ecosystem (Koen-Alonso and Yodzis 2005), which may have impacted the timing and process of recovery.

The Southern elephant seal deserves a special mention. The individuals recorded at the Año Nuevo Islands suggest the potential emergence of a new breeding site. If confirmed, this would represent an additional extra Antarctic breeding center for the species.

All pinniped species are currently protected by national and provincial legislation, both on land and at sea. National Decree 1216 of 1974 and Provincial Law 101 of 1993 prohibit the hunting and harvesting of pinnipeds, among other species. Isla de los Estados and its surrounding islands and islets have been a protected area since 1991 under the Constitution of the Provincia de Tierra del Fuego, Antártida e Islas del Atlántico Sur. Additionally, the surrounding waters of the island are protected up to four nautical miles under Provincial Law 1461, which established the Península Mitre protected area in 2022. To the south of Isla de los Estados, the marine protected area "Yaganes" was created by National Law 27,490 in 2018. Despite this protection, the impact of high seas fisheries operating around the Tierra del Fuego archipelago remain unknown.

Milano et al. (2020b) underscore the importance of considering the potential "rescue effects", a concept from metapopulation theory, in understanding the spatial dynamics of recolonization and recovery of pinnipeds on Isla de los Estados. This highlights the need for long-term and detailed monitoring studies to evaluate the role of local population dynamics in regional recovery. Furthermore, considering the potential for broader ecosystem reorganization (Koen-Alonso and Yodzis 2005), studies on the trophic ecology of the species at Isla de los Estados are also essential.

Another consequence of human presence in the island was the introduction of Red Deer, Domestic Goats and rat species (Valenzuela et al. 2023). Salom and Schiavini (Chap. 14, this volume) found that both artiodactyls have impacts on native vegetation, but also, together with the introduced rats, might be affecting both the Southern River Otter and the Olive Soft-haired Mouse.

The island's isolation and inaccessibility helped facilitate the recovery of pinniped populations after the decline of the sealing in the nineteenth and twentieth century. The same holds for the recovery of the Southern River Otter. This isolation also challenges efforts to manage introduced species and for conducting scientific research in this remote region.

References

Baylis AMM, Orben RA, Arkhipkin AI, Barton J, Brownell RL Jr, Staniland IJ, Brickle P (2019) Re-evaluating the population size of South American fur seals and conservation implications. Aquat Conserv Mar Freshw Ecosyst 11: 1988–1995

Bonner WN (1968) The fur seal of South Georgia. Brit Antarct Surv Sci Rep 56(1):95

Bonner WN (1982) Seals and man. A study of interactions. Washington Sea Grant Publication, Seattle

Bove G (1883) Expedición Austral Argentina: Informes preliminares presentados a SSEE los Ministros del Interior y de Guerra y Marina de la República Argentina. Imprenta del Departamento Nacional de Agricultura, Buenos Aires

Cárdenas-Alayza S (2018a) South American sea lion: *Otaria byronia*. In: Encyclopedia of marine mammals. Academic, London, pp 907–910

Cárdenas-Alayza S (2018b) South American fur seal: *Arctocephalus australis*. In: Encyclopedia of marine mammals. Academic, London, pp 905–907

Cárdenas-Alayza S, Crespo E, Oliveira L (2016a) Otaria byronia. The IUCN red list of threatened specie 2016: e.T41665A61948292. https://doi.org/10.2305/IUCN.UK.2016-1.RLTS.T41665A61948292.en. Accessed 18 Apr 2025

Cárdenas-Alayza S, Oliveira L, Crespo E (2016b) Arctocephalus australis. The IUCN red list of threatened species 2016: e.T2055A45223529. https://doi.org/10.2305/IUCN.UK.2016-1.RLTS.T2055A45223529.en. Accessed 18 Apr 2025

Carrara IS (1952) Lobos marinos, pingüinos y guaneras del litoral marítimo e islas adyacentes de la República Argentina. Publicación especial Cátedra de Higiene e Industrias, Facultad de Ciencias Veterinarias, Universidad Nacional de La Plata, La Plata, p 171

Carrara IS (1964) Zoorecursos naturales de la Antártida. Mamíferos marinos (Pinnipedios). Universidad Nacional de La Plata. Facultad de Ciencias Veterinarias, La Plata, p 100

Cassini MH (2008) Present status of *Lontra felina* in Argentina. Endangered Species Update 25:57–60

Caviglia SE (2015) Malvinas: Soberanía, Memoria y Justicia, vol II: Balleneros – Loberos – Misioneros. S. 18th–19th, 1st edn. Ministerio de Educación de la Provincia de Chubut, Rawson, p 300

Centrón D, Ramírez B, Fasola L, Macdonald D, Chehebar C, Schiavini A, Cassini M (2008) Diversity of mtDNA in Southern River otter (*Lontra provocax*) from Argentinean Patagonia. J Hered 99:198–201

Chehebar C (1985) A survey of the southern river otter *Lutra provocax* Thomas in Nahuel Huapi National Park, Argentina. Biol Conserv 32:299–307

Claverie AÑ (2025) Abordaje socio-ecológico para el estudio y conservación del huillín (*Lontra provocax*) en el Archipiélago Fueguino. PhD thesis, National University of Comahue

Claverie AÑ, Valenzuela AEJ (2023) New records for the marine population of southern river otter *Lontra provocax*: confirming huillín presence on mitre peninsula, Tierra del Fuego, Argentina. IUCN Otter Spec Group Bull 40(3):135–142

Crespo EA, Schiavini A, García NA, Trecu VF, Goodall RNP, Rodríguez D, Morgante JS, De Oliveira LR (2015) Status, population trend and genetic structure of south American fur seals *Arctocephalus australis* in southwestern Atlantic waters. Mar Mamm Sci 31(3):866–890

Dabbene R (1904) Viaje a Tierra del Fuego y a la Isla de los Estados. Boletín del Instituto Geográfico Argentino 21:3–78

Delano A (1817) Narrative of voyages and travels in the northern and southern hemispheres: comprising three voyages round the world; together with a voyage of survey and discovery, in the Pacific Ocean and oriental islands. Printed by E. G. House, for the author, Boston

Destéfani LH (1981) Las Malvinas en la época hispana (1600–1811). Corregidor, Buenos Aires, p 424

Destéfani LH (1987) Historia Marítima Argentina. Tomo V. Departamento de Estudios Históricos Navales, Buenos Aires

Haller SC (2023) Balleneros, loberos y guaneros en Patagonia y Malvinas: Una historia socioambiental del mar 7:1800–1914. Sb editorial. Buenos Aires

Hardin G (1968) The tragedy of the commons. Science 162:1243–1248

Hernández Orts JS, Paso Viola MN, Garcia NA, Crespo EA, González RAC, Garcia Varela M, Kuchta R (2015) A checklist of the helminth parasites of marine mammals from Argentina. Zootaxa 3936(3):301–334

Hilborn R, Walters CJ, Ludwig D (1995) Sustainable exploitation of renewable resources. Annu Rev Ecol Syst 26:45–67

Hindell MA (2018) Elephant seals. *Mirounga angustirostris* and *M. leonina*. In: Encyclopedia of marine mammals. Academic, London, pp 303–307

Hyades P (2020) Een Jaar aan Kaap Hoorn. De Aarde en haar Volken, 1886. Available at https://www.gutenberg.org/ebooks/15037

Kirker J (1970) Adventures to China; Americans in the southern oceans, 1792–1812. Oxford University Press, New York

Koen-Alonso M, Yodzis P (2005) Multispecies modelling of some components of the marine community of northern and central Patagonia, Argentina. Can J Fish Aquat Sci 62(7): 1490–1512

Mangel J, Alfaro-Shigueto J, Verdi R, Rheingantz M (2022) *Lontra felina*. The IUCN red list of threatened species 2022: e.T12303A215395045. https://doi.org/10.2305/IUCN.UK.2022-2.RLTS.T12303A215395045.en. Accessed on 13 June 2025

Massoia E, Chébez JC (1993) Mamíferos silvestres del archipiélago fueguino. LOLA. Buenos Aires

May RM, Beddington JR, Clark CW, Holt SJ, R.M. & Laws. (1979) Management of multispecies fisheries. Science 205(4403):267–277

Milano VN, Grandi MF, Schiavini ACM, Crespo EA (2020a) Recovery of south American fur seals from Fuegian Archipelago (Argentina). Mar Mamm Sci 36:1022–1032

Milano VN, Grandi MF, Schiavini A, Crespo E (2020b) Sea lions (*Otaria flavescens*) from the end of the world: insights of a recovery. Polar Biol 43:695–706. https://doi.org/10.1007/s00300-020-02672-9

Niekisch M, Schiavini A (1998) Desarrollo y Conservación de la Isla de los Estados (Tierra del Fuego, Argentina). Unpublished report produced for the Government of Tierra del Fuego, Antártida e Islas del Atlántico Sur. Deutsche Gesellschaft für Technische Zusammenarbeit (GTZ) GmbH

Orquera LA, Piana EL (2009) Sea nomads of the Beagle Channel in southernmost South America: over six thousand years of coastal adaptation and stability. J Island Coast Archaeol 4(1):61–81. https://doi.org/10.1080/15564890902789882

Parera A, Schiavini A, Frere E (1997) Relevamiento ecológico de la Isla de los Estados. Observaciones sobre su estado de conservación y sugerencias de manejo. In: Boletín Técnico de la Fundación Vida Silvestre Argentina, vol 38. Buenos Aires, p 37

Payró RJ (1898) La Australia Argentina: excursión periodística a las costas patagónicas, Tierra del Fuego e Isla de los Estados; con una carta-prólogo del general Bartolomé Mitre. La Nación. Buenos Aires

Pine RH, Angle JP, Bridge D (1978) Mammals from the sea, mainland and islands at the southern tip of South America. Mammalia 42 (1):105–114

Raya Rey A, Sáenz Samaniego R, Petracci PF (2012) New records of south American sea lion *Otaria flavescens* predation on southern rockhopper penguins *Eudyptes chrysocome* at Staten Island, Argentina. Polar Biol 35:319–322

Sanchez J, Poljak S, Teta P, Lanusse L, Lizarralde MS (2022) A contribution to the knowledge of the subgenus *Abrothrix* (*Angelomys*) (Rodentia, Cricetidae) in southernmost South America. Polar Biol 45:601–614

Schiavini A (1987) Avances en el conocimiento del status del lobo de dos pelos sudamericano *Arctocephalus australis* en Tierra del Fuego, Argentina. In: En Siciliano S, Lodi L (eds) Anais da 2a. Reuniao de Trabalho de Especialistas em Mamíferos Aquáticos da América do Sul. Fundaçao Brasileira para a Conservaçao da Natureza, Río de Janeiro

Schiavini ACM, Crespo EA, Szapkievich V (1998) Status de los apostaderos del lobo marino de un pelo (*Otaria flavescens*) en las provincias de Santa Cruz y Tierra del Fuego. Informes Técnicos del Plan de Manejo Integrado de la Zona la Zona Costera Patagónica (Puerto Madryn, Argentina) 40. Fundación Patagonia Natural, 27 pp

Schiavini ACM, Crespo EA, Szapkievich V (2004) Status of the population of south american sea lion (*Otaria flavescens* Shaw, 1800) in southern Argentina. Mamm Biol 69:1–11

Sepúlveda MA, Valenzuela AEJ, Pozzi C, Medina-Vogel G, Chehébar C (2021) *Lontra provocax*. The IUCN red list of threatened species 2021: e.T12305A95970485. https://doi.org/10.2305/IUCN.UK.2021-3.RLTS.T12305A95970485.en. Accessed on 13 June 2025

Teta P, D'Elía G (2019) Abrothrix olivacea. En: SAyDS–SAREM (eds.) Categorización 2019 de los mamíferos de Argentina según su riesgo de extinción. Lista Roja de los mamíferos de Argentina. Versión digital: http://cma.sarem.org.ar

Vairo CP (1997) La Isla de los Estados y el Faro del Fin del Mundo. Zagier y Urruty. Buenos Aires

Valenzuela AEJ (2019) Lontra felina. En: SAyDS–SAREM (eds.) Categorización 2019 de los mamíferos de Argentina según su riesgo de extinción. Lista Roja de los mamíferos de Argentina. Versión digital: http://cma.sarem.org.ar

Valenzuela AEJ, Raya Rey A, Fasola L, Schiavini A (2013) Understanding the inter-specific dynamics of two co-existing predators in the Tierra del Fuego archipelago: the native southern river otter and the exotic American mink. Biol Invasions 15:645–656

Valenzuela AEJ, Fasola L, Pozzi C, Chehébar C, Ferreyra N, Gallo E, Massaccessi G (2019) Lontra provocax. En: SAyDS–SAREM (eds.) Categorización 2019 de los mamíferos de Argentina según su riesgo de extinción. Lista Roja de los mamíferos de Argentina. Versión digital: http://cma.sarem.org.ar

Valenzuela AEJ, Anderson CB, Ballari SA, Ojeda RA (2023) Introduced invasive mammals of Argentina. SAREM Series A Mammalogical Research, p. 340. https://www.sarem.org.ar/libros/#introducedinvasive

Zangrando AFJ, Ávila S, Gutiérrez S, Landa C, Raies A, Tivoli AM (2025) Chapter 3: Human impact on the Isla de los Estados environment. In: Biodiversity and conservation of Isla de los Estados (Chuani-sin). Springer, Cham

Chapter 14
Invasive Alien Mammals in Isla de los Estados

Amira Salom and Adrián Schiavini

Abstract Invasive species pose major threats to island ecosystems, often disrupting ecological dynamics and driving the decline or extirpation of native species. Their introduction typically brings novel functional groups to which island biotas are particularly vulnerable, due to their low diversity and unique ecological traits. Isla de los Estados, a protected area, was originally inhabited by only two native terrestrial mammal species prior to European arrival. Although it currently lacks permanent human settlements, the island continues to face pressure from several exotic mammal species historically introduced by sealers, whalers, sailors, and early ranching efforts. This chapter reviews the history of non-native mammal introductions and establishment on the island, alongside documented impacts from similar invasions globally. Drawing on nearly a decade of fieldwork, including vegetation surveys, diet studies, camera trapping, and exclusion experiments, we summarize the current understanding of introduced mammal population and their ecological impacts on native species and habitats. We also explore the potential for interactions among invasive species to amplify their effects across trophic levels and ecosystems. Finally, we identify key research needs and propose future steps for managing invasive species and restoring native ecosystems on this unique sub-Antarctic Island.

Keywords Invasive species · Red deer · Goats · History of exotic species introduction · Restoration

---

A. Salom (✉)
Laboratorio de Ecología y Conservación de Vida Silvestre, Centro Austral de Investigaciones Científicas (CADIC-CONICET), Ushuaia, Tierra del Fuego, Argentina
e-mail: amirasalom@cadic-conicet.gob.ar

A. Schiavini
Laboratorio de Ecología y Conservación de Vida Silvestre, Centro Austral de Investigaciones Científicas (CADIC-CONICET), Ushuaia, Tierra del Fuego, Argentina

Universidad Nacional De Tierra Del Fuego, AeIAS (UNTDF), Ushuaia, Tierra del Fuego, Argentina

Wildlife Conservation Society Representación Argentina, Buenos Aires, Argentina

A. Raya Rey, U. Balza (eds.), *Biodiversity and Conservation of Isla de los Estados (Chuani-sin)*, Natural and Social Sciences of Patagonia,
https://doi.org/10.1007/978-3-032-13630-5_14

14.1 Introduction

Biological invasions are among the primary drivers of contemporary ecological change and biodiversity loss worldwide (Vitousek et al. 1997; Mack et al. 2000; Bellard et al. 2016). These changes often result in socio-economic and cultural impacts that compromise human well-being (Nuñez and Simberloff 2005; Reaser et al. 2007). Invasive species are defined as organisms that, upon dispersing beyond their natural distribution, succeed in establishing and spreading in a new environment, effectively overcoming geographic, environmental and dispersal barriers (Mack et al. 2000). While species dispersal beyond native range is neither new nor exclusively human-driven, humans have historically played a central role in facilitating such movements, whether intentionally or unintentionally, through translocations, habitat alteration, or disruption of ecological interactions (Boivin et al. 2016). Over the past two centuries, the scale and frequency of biological invasions have increased dramatically, largely driven by globalization and the expansion of transport and trade (Boivin et al. 2016).

Once established, invasive species can act as predators, competitors, prey, seed dispersers, parasites, disease vectors or ecosystem engineers (Ricciardi et al. 2013; Simberloff et al. 2013). Their presence can disrupt food webs, alter nutrient cycling, modify disturbance regimes, facilitate additional invasions, and cause native species declines or extinctions (Ebenhard 1988; Mack et al. 2000; Towns et al. 2006; Simberloff et al. 2013; Bellard et al. 2016). These impacts, ultimately lead to shifts in community structure and composition, and the loss of critical ecosystem functions (Mack et al. 2000).

Island ecosystems, due to their isolation and limited size, are particularly vulnerable to biological invasions. Historically, islands played a key role in maritime exploration and colonization, and many have been deliberately or accidentally populated with non-native species for food (e.g., sheep, chicken, goats and cattle), companionship (e.g., dogs, cats), or by unintentional introductions (e.g., rodents) (Clout and Russell 2008; Boivin et al. 2016). As a result, islands have received 80% of all documented bird and mammal introductions, and only a few remain free from invasive species today (Ebenhard 1988). The introduction of invasive species to islands often exposes native biotas to novel functional groups absent from their evolutionary history (Courchamp et al. 2003; Sax and Gaines 2008; Simberloff et al. 2013; McCreless et al. 2016).

Island ecosystems are recognized as high-priority areas for conservation due to their high level of endemism (Kier et al. 2009) and their importance as breeding sites for seabirds and marine mammals (Chown et al. 2001; Reaser et al. 2007; Tershy et al. 2015). However, their restricted size and isolation often means limited taxonomic and functional diversity, as well as simplified food webs compared to continental systems (Reaser et al. 2007). These factors make islands especially vulnerable to biological invasions. Indeed, island species account for 61% of all known extinctions and 37% of all species currently listed as critically endangered species (Tershy et al. 2015).

Isla de los Estados is a protected area with no permanent human settlement apart from a small Argentine Navy post inhabited by four personnel. Despite this, the island faces ongoing pressure from various exotic mammals historically introduced by sealers, whalers, sailors and ranchers (Massoia and Chebez 1993). Prior to European arrival, only two native terrestrial mammal species were present: the Southern river otter (*Lontra provocax*) and the mouse of Chuani-sin (*Abrothrix xanthorhinus llanoi*), both of which are currently known from few records (Claverie et al. 2021; Sánchez et al. 2020). Today, the island's mammal community includes several established non-native species, such as the European rabbit (*Oryctolagus cuniculus*), red deer (*Cervus elaphus*), feral goat (*Capra hircus*), black rat (*Rattus rattus*) and brown rat (*Rattus norvegicus*) (Valenzuela et al. 2023).

In this chapter, we examine the global documented impacts of invasive mammal species on islands. Our guiding rationale is that these species tend to generate comparable ecological effects across similar insular ecosystems. Afterwards, we review current knowledge on the introduction and establishments of invasive mammal species on Isla de los Estados, and summarize updated information on the population status of these species and highlight both their known and potential impacts on native flora, fauna, and habitats. Finally, we propose future research directions and outline necessary steps for the development of effective management strategies and the restoration of native ecosystems in this unique protected area.

14.2 Ecological Disruption: Global Lessons from Invasive Mammals

Introduced mammals have caused more ecological disruption than any other group of vertebrates (Ebenhard 1988; Long 2003; McCreless et al. 2016). They are also responsible for some of the most thoroughly documented and dramatic ecological disturbances resulting from biological invasions (Long 2003; Courchamp et al. 2003; Clout and Russell 2008). Among the most problematic are feral goats, red deer, black rat and the European rabbits (Lowe et al. 2000). The European rabbit has been introduced to more than 800 islands worldwide, and now inhabits every continent except Antarctica (Flux and Fullagar 1992). Similarly, rodents, particularly black rat, brown rat and house mouse (*Mus musculus*), have achieved near-global distributions due to their ability to travel with humans by ships and canoes (Atkinson 1985). According to Ebenhard (1988), introduced species may impact native species in six different ways. In this section we highlight four key mechanisms by which ungulates, rabbits, and rats disrupt island ecosystems: (i) impacts on plant communities and the consequence for the species living in the habitats they structure, (ii) interference or resource exploitation competition, (iii) introduced species acting as prey, and (iv) introduced species acting as predators. The examples provided here focus on patterns relevant for Isla de los Estados.

Introduced herbivores and omnivores disrupt plant communities through seed consumption, browsing, grazing, trampling, and soil disturbance. In extreme cases, herbivores can cause local extinctions, as on Great Island, New Zealand, where introduced goats reduced native plant species from 143 to just 70 within 35 years (Turbott 1948). These impacts cascade through trophic webs, with effects in ecosystem function and integrity (Coblentz 1978; Mark et al. 1991; Courchamp et al. 2003; Spear and Chown 2009; Chynoweth et al. 2013; Harper and Bunbury 2015; Barbar and Lambertucci 2018). Likewise, selective browsing can favor unpalatable or browse-tolerant species, altering successional trajectories (Veblen et al. 1989; Augustine and McNaughton 1998; Barrios-Garcia et al. 2012; Barrette et al. 2017). In the temperate forests of northern Patagonia, red deer reduced growth in dominant tree seedlings and shifted understory composition towards thorny shrubs like *Berberis* spp. (Veblen et al. 1989; Barrios-Garcia et al. 2012). These changes in vegetation, often reduce structural complexity and resource availability for other taxa, impairing food resources, nesting and shelter, and in turn altering species interactions (Weimerskirch et al. 1989; Martin et al. 2010; Marchant et al. 2011; Rodriguez-Cabal et al. 2013). For instance, on the Kerguelen Islands, rabbit herbivory reduced nesting habitat availability for ground-nesting birds, likely lowering their reproductive success (Weimerskirch et al. 1989).

Herbivory can also facilitate plant invasions. Rats, as fruit and seed consumers, contribute to invasive plants dispersal (Shiels 2011). On sub-Antarctic islands, overgrazing by introduced herbivores reduces the dominance of tussock grasses, often leading to poor regeneration or local eradication (Campbell and Rudge 1984; Leader-Williams et al. 1987; Scott and Kirkpatrick 2008; Shaw et al. 2011). In many cases, disturbed soils and grazing pressure promoted the establishment of invasive grasses such as *Poa annua*, which replaced native megaherbs and tussocks (Leader-Williams et al. 1987; Scott and Kirkpatrick 2008; Shaw et al. 2011).

Introduced herbivores also compete with native fauna through exploitation or interference competition, especially in island ecosystems with limited resources. European rabbits reduce plant biomass, limiting food availability for granivorous and insectivorous birds (Gillham 1963) and competing for burrowing space with seabirds (Young 1981). Their burrowing activity can also damage nesting sites and increase erosion (Kossoff et al. 2024). On other islands, invasive rodents have displaced native species through aggressive behavior and competitive exclusion, as seen with *Nesoryzomys swarthi* in the Galapagos (Harris and Macdonald 2007).

Introduced species can act as prey and destabilize food webs. Rabbits, for example, have homogenized predator diets and increased niche overlap in regions like Australia and Patagonia (Barbar and Lambertucci 2023). On Auckland Island, rabbits supported large population of southern skuas (*Stercorarius antarcticus*), leading to hyperpredation on native birds. This effect was compounded by other introduced herbivores that reduced plant cover and therefore refuges (Russell et al. 2020). Such trophic shifts complicate conservation: the removal of one species may unintentionally harm others (Zavaleta et al. 2001). For example, the eradication of goats in the Galapagos led to declines in the endemic Galapagos hawk (*Buteo galapagoensis*), which had fed on goat carrion and used goat-altered habitats for hunting (Rivera-Parra et al. 2012; Jaramillo et al. 2016). With goats gone, hawks began preying on introduced rats, posing new management challenges (Jaramillo et al. 2016).

Introduced predators, especially cats and rodents, are a primary cause of seabird population decline on islands (Towns et al. 2006; Jones et al. 2008; Spatz et al. 2023). Rats prey on seabird eggs and chicks, particularly affecting small, burrow-nesting species (Lyver et al. 2000; Jones et al. 2008; Latorre et al. 2013). While larger seabirds such as albatrosses or giant petrels have been considered to be more affected by mousses than rats, the latter have been increasingly recognized as a major breeding-season threat (Jones et al. 2008; Davies et al. 2015; Phillips et al. 2016; Connan et al. 2024). Additionally, rats have contributed to the decline or extinction of flightless invertebrates, ground-dwelling reptiles, and land birds (Towns et al. 2006; Harper and Bunbury 2015). Even ungulates may act as predators in rare occasions: red deer has been observed preying on artificial nests in Patagonian wetlands (Vazquez et al. 2023) and on Manx Shearwater (*Puffinus puffinus*) in the Isle of Rhum (Furness 1988).

Complex communities of introduced species often produced synergistic effects that exceed the impact of any single species. Research in the Aegean Islands illustrates how multi-species invasions can cascade effects across ecological levels. At the population level, combined rat predation and habitat degradation by rabbits and goats reduced nest densities of Yellow-legged Gull (*Larus michahellis*) (Carlberg et al. 2022). At the community level, rabbit herbivory reduces herbaceous cover and arthropod abundance and diversity (Kossoff et al. 2024). At the ecosystem level, vegetation loss, increased bare ground, and erosion lead to reduced seabird nesting, decreased nutrient inputs, and further soil instability (Carlberg et al. 2022; Kossoff et al. 2024). These examples highlight the importance of studying and managing invasive species not in isolation, but as interacting assemblages.

14.3 Arrival of the Outsiders: A Historical Account of Mammalian Introductions

Reconstructing the history of mammalian introductions to Isla de los Estados is challenging. While some events are documented, others, particularly early introductions, were likely accidental or went unrecorded. Among the earliest known invaders, the brown rat (*Rattus norvegicus*) and the black rat (*Rattus rattus*) probably arrived with European and North American sealers, whalers, and explorers navigating the southwest Atlantic. The Fuegian archipelago, including Isla de los Estados, has a long maritime history and is well known for the numerous shipwrecks that have occurred along its coasts (Vairo 1997). Historical accounts, such as that of Payró (1898), describe the presence of rats at the San Juan de Salvamento prison: "*In that weather, not even the mice, those friendly little creatures that grow wild on the island until they reach enormous dimensions, infesting it from one end to the other, showed their snouts. They are so abundant and harmful that they have made rabbit breeding impossible*." Brown rats have since been recorded at multiple locations (Kühnemann 1976; Pine et al. 1978; Massoia and Chebez 1993), while black rats have been reported on Observatorio island (Pine et al. 1978) and observed in Franklin Bay and Port Roca (Fig. 14.1). Today both species appear to be widespread across the island.

Goats were introduced to Crossley Bay in 1856 by Argentine patriot and sailor Luis Piedrabuena, who aimed to provide a future food source for shipwreck survivors. Recognized for his rescue efforts and his role in promoting Argentine sovereignty in the southern region, Piedrabuena was granted official custody of the island in 1868, a responsibility he maintained until 1913. During the late nineteenth century, the Argentine government established a prison and a lighthouse on the island, facilities that later inspired Jules Verne's *The Lighthouse at the End of the World.*

Rabbits were likely introduced by the inhabitants of the prison at San Juan de Salvamento sometime during its operation between 1884 and 1898, when Roberto Payró visited the site (Payró 1898). He noted that rat predation on young rabbits rendered breeding efforts unsuccessful. In 1899, the prison was relocated to Port Cook, and the lighthouse at San Juan de Salvamento was decommissioned. A new lighthouse was established at Observatorio Island. Notably, Dabbene (1902, 1904), who visited the island in 1902, did not report the presence of rabbits, suggesting they had not yet been introduced. However, a lighthouse keeper reported

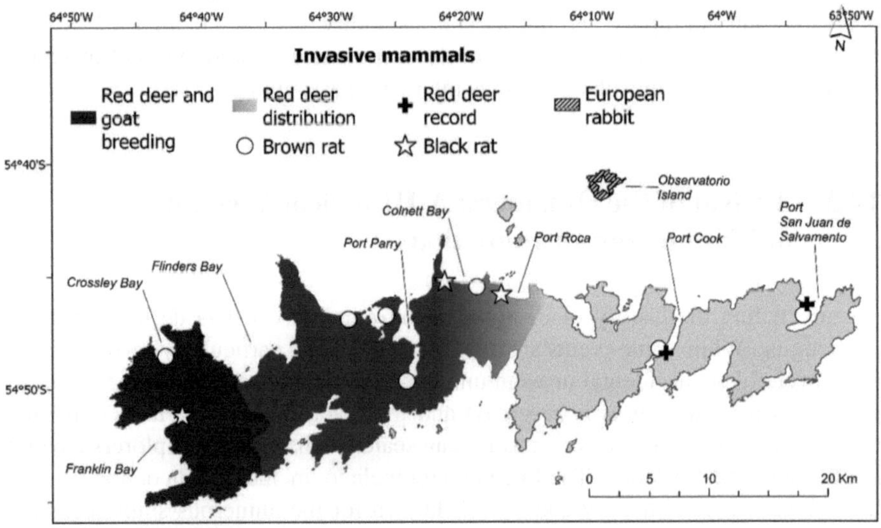

Fig. 14.1 Distribution of the four established invasive species present on Isla de los Estados and Observatorio Island. Red deer and goat breeding refers to the area where breeding of these species is confirmed. Red deer distribution refers to the area that holds established groups of red deer, based on sightings, presence of high-traffic animal trails and faeces; while the shading gradient represents the hypothesized decreasing deer density toward the east

releasing seven pairs of grey and black rabbits on Observatorio Island in 1911, and by 1913 the island "*was mined with rabbits*" (Vairo 1997).

Domestic cattle (*Bos taurus*) were unsuccessfully introduced to Port Cook between 1899 and 1902 (Massoia and Chebez 1993), and again around 1964 in Crossley Bay (Goodall 1978). By 1971, a small group was observed by Dudley and Crow (1983), who noted that "*the 'herd' of cattle numbered eight in 1971 and consisted entirely of steers, which do not appear to have caused much damage to the vegetation and certainly do not pose a threat in terms of a population explosion*." During the same expedition, conducted by the Research Vessel *Hero* cruise 71-5, feral goats were also recorded at multiple sites across the island, including Crossley and Flinders Bays, and Cape San Bartolomé in Lopez Peninsula (Dudley and Crow 1983). Cattle were last observed in Crossley Bay in 1982 (Massoia and Chebez 1993) and have presumably gone extinct since then. In contrast, goats remain established in the island's western region (Fig. 14.1).

Although by 1972 it was already recognized that non-native species posed a threat to the island's ecosystem, as noted by Imshaug (1972): "*It is hoped that every effort will be made to prevent further introductions, since Isla de los Estados still has many large areas of natural ecosystems which await detailed study*"; his warning went unheeded. In 1973, six red deer were released at Crossley Bay by the Argentine Navy, which had assumed administrative control of the island by 1936 (Valenzuela et al. 2023). Zooarchaeological evidence confirms the historical absence of guanacos (*Lama guanicoe*) on the island, a species widely consumed on the Isla Grande of Tierra del Fuego, highlighting the introduction of large herbivores as a novel event in the island's ecological history (Horwitz and Weissel 2011).

More recently, companion animals such as domestic cats (*Felis silvestris catus*) and dogs (*Canis lupus familiaris*) have been introduced. In 2011, five cats were brought to Observatorio Island to control rat populations. Four were later retrieved and returned to the Isla Grande of Tierra del Fuego, but one remained in the wild. This specimen was seen in a field campaign in 2016 (Raya Rey, personal communication) and its carcass was found in 2021 (Balza et al. 2022). At Port Parry, the island's only permanent settlement, between two and three domestic dogs are present year-round and were observed occasionally at Franklin Bay (~15 km away). In 2010, two dogs were observed in this area, and a massive number of Magellanic penguins (*Spheniscus magellanicus*) were found dead just outside their nests, exhibiting signs of canine predation (Fig. 14.2). That year one dog was captured and taken to the Isla Grande of Tierra del Fuego (Raya Rey, personal communication).

Four of the established invasive mammal species on Isla de los Estados are included in the IUCN list of the world's 100 worst invasive species due to their destructive ecological potential: the red deer, goat, the European rabbit and black rat (Lowe et al. 2000) (Fig. 14.3). Remarkably, Isla de los Estados is now the southernmost island known to host a feral goat population, as goats have been eradicated or have gone extinct on other sub-Antarctic islands (Brown et al. 2022). The same is true for red deer.

It is noteworthy that several invasive species well established in the Isla Grande of Tierra del Fuego, are absent from Isla de los Estados. These include the American

Fig. 14.2 Consequences of companionship species: (**a**) Dog (*Canis lupus familiaris*) sighting in Franklin Bay, approximately 15 km away from the only human settlement of the island, (**b**) Dead Magellanic penguins (*Spheniscus magellanicus*) found outside their nest nearby. (Credits: Andrea Raya Rey)

mink (*Neovison vison*), the North American beaver (*Castor canadensis*), the muskrat (*Ondatra zibethicus*) and the chilla fox (*Lycalopex griseus*) (Valenzuela et al. 2023). This highlights the importance of the Le Maire Strait as a barrier to animal movement between Isla Grande of Tierra del Fuego and Isla de los Estados, even for the semi-aquatic species.

Fig. 14.3 Current invasive mammals species present on Isla de los Estados: (**a**) rat (*Rattus* sp.), (**b**) feral goat (*Capra hircus*), (**c**) red deer (*Cervus elaphus*) and (**d**) European rabbit (*Oryctolagus cuniculus*). (Credits: Sergio Anselmino (**a**), Amira Salom (**b**, **c**), Emiliano Arona (**d**))

14.4 The Case of Isla de los Estados: What We Know So Far

Assessing the impact of an introduced species on the ecosystem it has invaded is a challenging task (Parker et al. 1999), particularly because baseline data from before the invasion are often lacking. This is the case for Isla de los Estados. Despite the considerable time elapsed since the introduction of deer, goats and rats on the island, our understanding of their ecological impacts remains limited. Only in the past decade have there been efforts to investigate some of the potential impacts of these invasive species (Fig. 14.4). Although many knowledge gaps persist, recent studies, including those on the role of exotic species in the trophic ecology of an endemic raptor and the impact of deer and goats on the island's vegetation, are helping to build a more comprehensive understanding of the ecological implications of these species on the island's ecosystems.

Prior to the 1990s, ungulates were restricted to the western half of the island. It was generally assumed that the steep terrain of the two mountains of about 550 m.a.s.l. that separate the deep fjord of Port Parry to the north and Torlaschi Bay

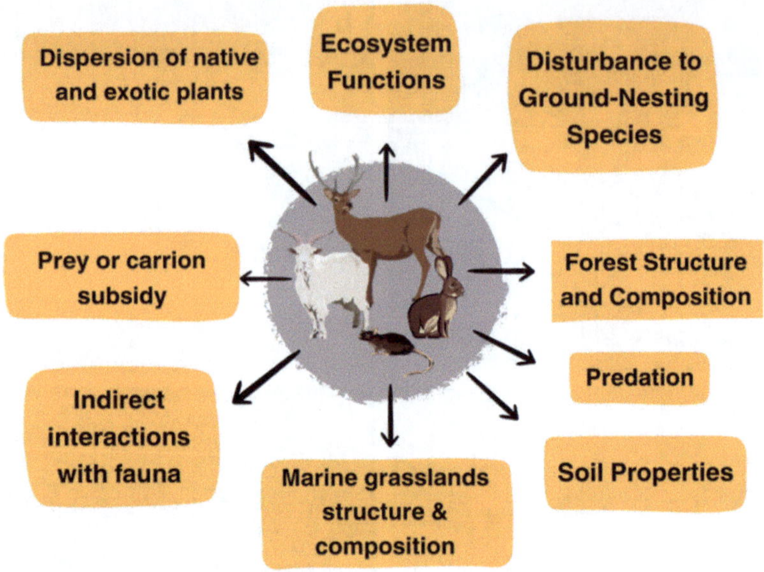

Fig. 14.4 Schematic conceptual model of the potential ecological impacts of introduced mammals (rats, rabbits, red deer, and goats) on Isla de los Estados

to the south, represented an insurmountable obstacle for these herbivores. Currently, red deer are more abundant and widely distributed than goats in this western sector, being recorded in all systematically surveyed sites, while goats appeared in less than half of them (Salom 2025). However, since the 1990s, opportunistic sightings of red deer groups and a pilot faecal pellet survey in Colnett Bay and Port Roca areas (Raya Rey, personal communication) indicate that red deer have expanded their range (Fig. 14.1). This is supported by camera trap data obtained during a Southern river otter survey in Colnett Bay (Claverie et al. 2021; Valenzuela, personal communication), where even a calf was recorded. We hypothesized that red deer breeding populations are currently expanding to areas adjacent to their historic western distribution. Nonetheless, deer density seems to decrease toward the east, as only scarce records of footprints and faeces were found in Port Cook and San Juan de Salvamento Bay (Raya Rey, personal communication) (Fig. 14.1).

To assess ungulates foraging behaviour and selection, vegetation availability was estimated through floristic surveys and diet was evaluated via microhistological examination of red deer and goat faeces (Salom et al. 2025a). Both species exhibited broad diets during spring and summer, consuming ~67% of the plant richness detected in the field. However, their diets focused on 5–6 key species, mainly shrubs and forbs, along with *Nothofagus* spp. (trees) and *Anthoxanthum* spp. (grasses), indicating intermediate, selective, and opportunistic foraging (Salom et al. 2025a). Notably, exotic forbs were detected in faeces but not in vegetation plots, suggesting they occur at low frequencies or are being consumed outside the study area (e.g. in Port Parry, a site with more human visitation). This raise concerns that ungulates could aid in spreading exotic plants. These findings also highlight the need for biosecurity measures, as growing ecotourism in the island may introduce new invasive species that ungulates could then help disperse (Barros et al. 2022).

Further diet analysis revealed overlap in selected items, with both species selecting trees, particularly *Drimys winteri* and *Nothofagus* spp., and grasses, particularly *Poa flabellata*, and *Anthoxanthum* spp., all species characteristic of subantarctic evergreen forests and maritime tussock grasslands, respectively (Salom et al. 2025a). The overlap in selection suggests that herbivory may significantly impact these habitats, especially if deer and goats exert additive or synergistic pressures on native flora. These findings are consistent with earlier reports of degraded forests in the island's northwest: "*...the forest is atypical of the generally encountered Magellanic evergreen forest and demonstrates an impoverished understory. This is largely due to the effect of drying winds and the ravages of goats... Although considerable litter occurs on the forest floor, there is a marked sparseness of bryophytes and filmy ferns. Similarly, the angiospermous understory contains as the predominant elements only depauperate Lebetanthus myrsinites and Berberis ilicifolia...*" (Dudley and Crow 1983, citing Goodall 1978).

To assess the herbivore effects on the early life stages of key forest woody species, we conducted a transplant exclosure experiment for 3 years (2019–2021). We monitored the survival and growth of seedlings of two dominant tree species (*Nothofagus betuloides* and *Drimys winteri*) and two commonly browsed understory shrubs (*Berberis ilicifolia* and *Chiliotrichum diffusum*) under two treatments: exclusion and exposure to herbivores. Results showed that exclosures presented higher survival and growth, which suggest that introduced herbivores may be limiting forest regeneration (Salom 2025). If that is the case, such impacts may lead to forest degradation and reduced landscape heterogeneity.

In parallel, in 2018 we established exclosures in the maritime tussock grasslands, another habitat heavily targeted by ungulates, to evaluate vegetation recovery in the absence of herbivory. These grasslands are of great conservation value, as they represent critical breeding sites for seabirds and marine mammals, including threatened species such as the Magellanic penguin (*Spheniscus magellanicus*), the Southern rockhopper penguin (*Eudyptes chrysocome*), the endemic striated caracara (*Phalcoboenus australis*), and the South American sea lion (*Otaria flavescens*) (Schiavini and Raya Rey 2001). On Isla de los Estados, roughly 70% of bird species are associated with these habitats (Chebez and Bertonatti 1994). Additionally, these ecosystems function as important carbon sinks (Payne et al. 2019). Therefore, selective grazing on dominant grassland species, such as *Poa flabellata*, may reduce vegetation cover and structural complexity, with cascading effects on carbon storage, nesting sites, and faunal refuge.

Vegetation recovery is a slow process, shaped by species' growth rates, seed availability and prior disturbance, but preliminary results are promising. After 4 years, exclosures showed greater plant cover and signs of recovery (Fig. 14.5), suggesting that herbivore exclusion facilitate restoration (Salom 2025). This aligns with previous studies suggesting that exotic herbivores may be impacting indirectly striated caracara by reducing available nesting sites in the tussock habitats (Balza et al.

Fig. 14.5 Difference observed in maritime tussock grassland cover between an herbivore exclosure and adjacent grazed area after 4 years of exclusion (2018–2022). The visual contrast highlights the impact of introduced herbivores on vegetation structure. (Credits: Ulises Balza)

2017). However, a direct link between ungulate presence and caracara low chick density remains unconfirmed (Balza et al. 2023).

A key concern is that forage scarcity during winter, due to snow cover, may intensify ungulate grazing on *P. flabellata*, as has been observed on other islands (Leader-Williams et al. 1981). Tussock grasslands tolerate seasonal disturbance from native fauna, as seabirds and marine mammals are present only during spring and summer. In contrast, introduced herbivores graze year-round and may heavily browse *P. flabellata* in autumn and winter, hindering the regrowth that usually takes place during these seasons when disturbance is minimal (Smith 1985; Leader-Williams et al. 1987; Strange et al. 1988). Mammalian herbivory has been reported to cause excessive damage, including tussock mortality (Leader-Williams et al. 1987).

In addition to vegetation impacts, introduced herbivores may directly disturb ground-nesting birds by trampling colonies or displacing adults (Carlberg et al. 2022). Although we have not formally assessed this, field observations commonly record red deer and goat herds feeding within active seabird colonies (Fig. 14.6). A southern giant petrel (*Macronectes giganteus*) colony at Lopez Peninsula has recently disappeared (Raya Rey, personal communication), raising concerns about possible herbivore-related disturbance and/or predation by rats, as these has been recognized as a main threat in the breeding season for albatrosses and large petrel species (Phillips et al. 2016). Similarly, camera traps installed at sooty shearwater (*Ardenna grisea*) burrows in a recently found colony (Raya Rey et al. 2024) and striated caracara nests recorded rats (*Rattus* spp.), though at present no predatory

Fig. 14.6 Introduced herbivores disturbance on seabird colonies: (**a**) Red deer grazing on *P. flabellata* near a Southern rockhopper penguin nest, (**b**) a herd of goats resting and foraging next to a Southern rockhopper penguin colony. (Credits: Ulises Balza (**a**), Amira Salom (**b**))

event was recorded (Raya Rey unpublished data). Additionally, camera traps placed to monitor the endangered Southern river otter have recorded goats and red deer at known burrows, where trampling may render them unsuitable for use. Rats (*Rattus* spp.), also observed in camera footage, could act as additional threats via disease transmission (Claverie et al. 2021). A dog was even detected at a burrow near the navy station, suggesting risks from free-ranging domestic animals.

On Observatorio island, although the ecological impacts of rabbits and rats remain poorly documented, field observations indicate that rabbits may be contributing to localized overgrazing and soil disturbance, particularly in tussock grasslands. For instance, camera trap footage has captured individuals feeding on medium-sized tussocks (Cano, personal communication). They may also compete with burrow-nesting seabirds, such as the Magellanic penguin for nesting space, as observed in other islands (Young 1981). Rabbits are widespread across the island and show high phenotypic variability, with coat colors ranging from albino and blonde to grey and black. This broad variation suggests weak or absent natural selection on coat color (Cano, personal communication). In parallel, rats are also suspected to pose threats to native bird and invertebrate communities, but their population status and ecological impacts remain largely unknown.

Dietary analyses of the striated caracara from Observatorio island indicate that both rabbits and rats serve as food for reproductive adults (Cano et al., in rev). However, it remains unclear whether these animals are hunted or consumed as carrion. Additionally, turkey vultures (*Cathartes aura*) and Chilean skuas (*Stercorarius chilensis*) have also been observed feeding on rabbit carcasses, suggesting that carrion of this species may represent a notable food source.

Conversely, seasonal diet analyses of striated caracaras breeding in Franklin Bay, Isla de los Estados, suggest that red deer and goats do not represent a significant food subsidy for this raptor (Balza et al. 2020). This emphasizes the species' reliance on marine-derived food sources, with seabirds dominating the diet during the breeding season. While terrestrial inputs increased to ~40% during the non-breeding season, the role of ungulates remained negligible. Rats may contribute occasionally, but due to overlapping isotopic signals with marine prey, their exact contribution remains uncertain.

Overall, the ecological impacts of invasive mammals span multiple habitats, species and processes, making a comprehensive assessment challenging. However, the evidence brings us closer to understanding the roles of these introduced species and highlights the urgent need for sustained research, monitoring and actions to guide conservation efforts across Isla de los Estados.

14.5 Filling the Gaps: Research Needs and Management Pathways

Effective conservation and management on Isla de los Estados require detailed, site-specific information. This includes a clear understanding of the ecological impacts of invasive mammals, the feasibility and risks of potential control or eradication measures, and the response of native ecosystems to interventions (Zavaleta et al. 2001; Simberloff et al. 2013).

A critical first step is to address current knowledge gaps regarding the population dynamics and ecological behaviors of introduced species. Key parameters such as abundance, distribution, habitat use, movement patterns, and reproductive phenology remain largely unknown. In the case of red deer, given recent sightings outside their previously assumed range, we highlight the need to expand surveys eastward and develop habitat selection models to predict distribution, incorporating fine-scale data on population and environment. The recent development of vegetation maps for Isla de los Estados will facilitate this process (Salom et al. 2025b).

Herbivore impacts on other habitats apart from forest and grasslands demands investigation, particularly in moorlands and *Marsippospermum grandiflorum* meadows. This widespread rush, largely avoided by herbivores (Salom et al. 2025a), may be benefiting from competitive release or increased niche availability as selected woody species are browsed (Fig. 14.7; Veblen et al. 1989; Mark et al. 1991; Bellingham et al. 2010).

Exclosure experiments in forest and grasslands offer a promising avenue for assessing the ecological outcomes of herbivores control. To enhance our understanding and predictive capacity, we also recommend establishing widespread vegetation monitoring plots to count with baseline data before any eradication or control efforts is applied. This will help evaluate the outcomes of any management action, determine whether active restoration is needed and refine models of ecological response. Long-term monitoring will be key for adaptive management and for advancing the broader science of invasive species control in remote island ecosystems like Isla de los Estados.

Similarly, understanding rat population dynamics is critical to identify high-risk areas and periods, especially for vulnerable and poorly known native species. Little is known, for example, about the recently rediscovered endemic Chuani-sin mouse or the island's bat populations (Massoia and Chebez 1993; Sánchez et al. 2020; Schiavini and Valenzuela, Chap. 13, this volume). On Isla de los Estados, rats are also suspected to pose a significant threat to birds and invertebrates, being seabird populations of particular concern. Species such as dolphin gulls (*Leucophaeus scoresbii*) and sooty shearwaters may be especially vulnerable to egg and chick predation. Camera traps have recorded brown rats inside the burrows of sooty shearwaters (Raya Rey, unpublished), which echoes earlier findings of rat predation at this species' nests (Lyver et al. 2000). Therefore, assessing rats' interactions with native species and with each other should be a priority in future studies.

Before implementing rodent or herbivore eradication programs, particularly in Observatorio island, it is important to assess the reliance of native scavengers or predators on invasive mammals for food during harsh seasons, a pattern seen in raptors worldwide (Speziale and Lambertucci 2013). Control or eradication of rodents could lead to unexpected consequences such as secondary poisoning, food shortages, or dietary shifts, as observed in Galapagos hawks after invasive species removal (Rivera-Parra et al. 2012; Jaramillo et al. 2016). Winter diets of raptors on Observatorio island should be studied before any intervention, especially to evaluate the importance of introduced mammals as seasonal subsidies (e.g., sheep as winter subsidy in the Malvinas Islands, Rexer-Huber and Bildstein (2013)).

While species like Southern rockhopper and Magellanic penguins are more aggressive and may deter predators, the actual impact of invasive mammals on their breeding success remains unclear. Burrow occupancy by penguins and other seabirds may also be affected by rabbits. Additionally, the role of invasive mammals in

Fig. 14.7 Vegetation in summer at Franklin Bay in: (**a**) December 1999 and (**b**) December 2017. Note the increase in *Marsippospermum grandiflorum* cover and a reduction in shrub cover. (Credits: Fernando Biganzoli (**a**), Amira Salom (**b**))

nutrient cycling and their influence on plant communities has not been studied, despite possible cascading effects on the flow of marine-derived nutrients via seabirds (Benkwitt et al. 2021; Carlberg et al. 2022).

There is also a pressing need for integrated management and cooperation with the island's military personnel. Measures should include (a) halting the introduction of pet animals, (b) restricting their free roaming, (c) improve waste management to reduce food availability for rats, and (d) implement biosecurity protocols to prevent the spread of invasive weeds. Similar protocols should be applied to ecotourism operators and research teams.

While many knowledge gaps remain regarding invasive species on Isla de los Estados, existing evidence clearly demonstrates their potential for significant ecological impacts. This highlights the urgency of timely intervention, particularly for species listed among the world's worst invasive species by the IUCN. In such cases, the lack of complete information should not delay efforts to mitigate their effects. We hope this chapter contributes to the development of effective management strategies that support the conservation of this unique and protected island ecosystem.

Competing Interests On behalf of all authors, the corresponding author declares no conflict of interest. This work was made possible thanks to the support of the Agencia Nacional de Promoción Científica y Tecnológica (PICT 2017 N° 897), the Antarctic Research Trust and the Wildlife Conservation Society Argentina. We thank Argentine Army, the General Direction of Protected Areas and Biodiversity of Tierra del Fuego Province Government and the National Park Administration for their institutional support. Special thanks to the crew of the ship ARA Granville and ARA Malvinas (Argentine Army), the Orion and Explorer cruises (National Geographic and Lindblad Expeditions), and to Henk Boersma, Darrel Day, Charlie Flesch, Guillaume, and Micky Porco Fisher from the Sarah Vorwek, Spirit of Sidney, Fernande, Flores, Puerto Beagle vessels, respectively, for safely transporting us to and from the island. We also thank Andrea Raya Rey, Fernando Biganzoli, Ulises Balza, Nicolás Lois, Francisco Zunino, Alvaro González Calderón, Rubén Rosales, Juan Bareiro, Fabio Ariza, Andrés Capdevielle, Carolina Bruzzesse, Gimena Pizzarello, Guillermo Retamar, Gustavo Retamar, Gonzalo Sabattini, Dante Francomano, Bruno Riviotti, Debora Anibaldi, Nahuel Borras, Augusto Di Camilis, Martin Cabo, Pablo Rizzo, Marcos Rolón, Gastón Pechin, Pablo Machado, Daniel Testoni, Constanza Lobo, Rodrigo Iturraspe, Diego Ramos, Patricio Blanco, Ignacio Domato, Romina Mansilla, Guillermo Gargiulo, Lucas Lanusse, Iñaqui Gómez, Maria Gowland, Fernando Encinas, Emanuel Mendieta and Luciana Mestre for their assistance with logistics and/or fieldwork. We are also grateful to Vanessa Lencinas, Laura Borrelli, Dave latham, Lucia Rodriguez Planes, Joaquín Cano, and Cecilia Latham for helping with plant identification, sample processing and data analysis, and to Alexandra Elbakyan for facilitating access to scientific literature.

References

Atkinson IAE (1985) The spread of commensal species of *Rattus* to oceanic islands and their effects on Island avifaunas. In: Moors PJ (ed) Conservation of Island birds, vol 3. ICBP Tech Publ, Cambridge, pp 35–81

Augustine DJ, McNaughton SJ (1998) Ungulate effects on the functional species composition of plant communities: herbivore selectivity and plant tolerance. J Wildl Manag 62(4):1165. https://doi.org/10.2307/3801981

Balza U, Lois NA, Raya Rey A (2017) Status and reproductive outcome of the breeding population of striated caracaras (*Phalcoboenus australis*) at Franklin Bay. Wilson J Ornithol 129(4):890–898

Balza U, Lois NA, Polito MJ, Pütz K, Salom A, Raya Rey A (2020) The dynamic trophic niche of an Island bird of prey. Ecol Evol 10(21):12264–12276

Balza U, Liljesthröm M, Pimper L, Franco Navarro I, Cañas Barrovecchio S, Zunino F, Domato I, Acardi S, Iturraspe R, Cano J, Raya Rey A (2022) Status of breeding birds at Observatorio and Goffré Islands, Argentina. Polar Biol 45:667–673. https://doi.org/10.1007/s00300-022-03019-2

Balza U, Lois NA, Pizzarello G, Lopez ME, Lois-Milevicich J, Raya-Rey A (2023) Variation in nest-site occupancy and breeding outcome in a threatened Subantarctic raptor. Ibis 166:504–517. https://doi.org/10.1111/ibi.13234

Barbar F, Lambertucci SA (2018) The roles of leporid species that have been translocated: a review of their ecosystem effects as native and exotic species. Mammal Rev 48(4):245–260

Barbar F, Lambertucci SA (2023) Invasive herbivores shape food web structure: European rabbit and hare acting as primary prey are conservation challenges. Biol Conserv 281:110016

Barrette M, Bélanger L, De Grandpré L, Royo AA (2017) Demographic disequilibrium caused by canopy gap expansion and recruitment failure triggers forest cover loss. For Ecol Manag 401:117–124. https://doi.org/10.1016/j.foreco.2017.07.012

Barrios-Garcia MN, Relva MA, Kitzberger T (2012) Patterns of use and damage by exotic deer on native plant communities in northwestern Patagonia. Eur J Wildl Res 58(1):137–146. https://doi.org/10.1007/s10344-011-0554-6

Barros A, Shackleton R, Rew LJ, Pizarro C, Pauchard A (eds) (2022) Tourism, recreation and biological invasions. CABI, Wallingford

Bellard C, Cassey P, Blackburn TM (2016) Alien species as a driver of recent extinctions. Biol Lett 12:20150623. https://doi.org/10.1098/rsbl.2015.0623

Bellingham PJ, Wiser SK, Wright AE, Cameron EK, Forester LJ (2010) Disperser communities and legacies of goat grazing determine forest succession on the remote Three Kings Islands. N Z Biol Conserv 143(4):926–938. https://doi.org/10.1016/j.biocon.2010.01.001

Benkwitt CE, Gunn RL, Le Corre M, Carr P, Graham NA (2021) Rat eradication restores nutrient subsidies from seabirds across terrestrial and marine ecosystems. Curr Biol 31(12):2704–2711

Boivin NL, Zeder MA, Fuller DQ, Crowther A, Larson G, Erlandson JM, Denham T, Petraglia MD (2016) Ecological consequences of human niche construction: examining long-term anthropogenic shaping of global species distributions. Proc Natl Acad Sci 113:201525200

Brown DA, Broome KG, Timpson KG (2022) The eradication of feral goats from Auckland Island. N Z J Ecol 46(3):3486. https://doi.org/10.20417/nzjecol.46.3486

Campbell DJ, Rudge MR (1984) Vegetation changes induced over ten years by goats and pigs at port Ross, Auckland Islands (Subantarctic). N Z J Ecol 7:103–118

Carlberg RA, Karris G, Verma M, Foufopoulos J (2022) Food versus disturbance: contradictory effects of human activities on an opportunistic seabird breeding in an oligotrophic marine system. Diversity 14(6):421

Chebez JC, Bertonatti CC (1994) La avifauna de la Isla de los Estados, Islas de Año Nuevo y mar circundante (Tierra del Fuego, Argentina). Monografía Especial LOLA 1:1–64

Chown SL, Rodrigues ASL, Gremmen NJM, Gaston KJ (2001) World heritage status and conservation of Southern Ocean islands. Conserv Biol 15(3):550–557. https://doi.org/10.1046/j.1523-1739.2001.015003550.x

Chynoweth MW, Litton CM, Lepczyk CA, Hess SC, Cordell S (2013) Biology and impacts of pacific Island invasive species. 9. *Capra hircus*, the feral goat (mammalia: Bovidae) 1. Pac Sci 67(2):141–156. https://doi.org/10.2984/67.2.1

Claverie A, Valenzuela A, Anderson C (2021) Isla De Los Huillines: Allá donde nace La cordillera de los Andes. La Lupa. Colección fueguina de divulgación científica 18:34–37

Clout MN, Russell JC (2008) The invasion ecology of mammals: a global perspective. Wildl Res 35(3):180–184

Coblentz BE (1978) The effects of feral goats (*Capra hircus*) on Island ecosystems. Biol Conserv 13(4):279–286. https://doi.org/10.1016/0006-3207(78)90038-1

Connan M, Jones CW, Risi MM, Smyth LK, Oppel S, Perold V, Steven KL, Daling R, Ryan PG (2024) First evidence of mouse predation killing adult great albatrosses. Biol Invasions 26(1):25–31

Courchamp F, Chapuis JL, Pascal M (2003) Mammal invaders on islands: impact, control and control impact. Biol Rev Camb Philos Soc 78(3):347–383. https://doi.org/10.1017/S1464793102006061

Dabbene R (1902) Fauna magallánica, mamíferos y aves de la Tierra del Fuego e islas adyacentes. Anales del Museo Nacional de Buenos Aires. Sfr3 I: 341–409

Dabbene R (1904) Viaje a la Tierra del Fuego e Isla de los Estados. Boletín del Instituto Geográfico Argentino 21:3–78

Davies D, Dilley BJ, Bond AL, Cuthbert RJ, Ryan PG (2015) Trends and tactics of mouse predation on Tristan Albatross *Diomedea dabbenena* chicks at Gough Island, South Atlantic Ocean. Avian Conservation & Ecology 10(1):5

Dudley GR, Crow GE (1983) A contribution to the flora and vegetation of Isla de los Estados (Staten Island), Tierra del Fuego, Argentina. Antarct Res Ser 37. American Geophysical Union, Washington, DC

Ebenhard T (1988) Introduced birds and mammals and their ecological effects. Swedish Wildl Res Viltrevy 13:1–107

Flux JE, Fullagar PJ (1992) World distribution of the rabbit *Oryctolagus cuniculus* on islands. Mammal Rev 22:151–205

Furness R (1988) Predation on ground-nesting seabirds by Island populations of red deer *Cervus elaphus* and sheep *Ovis*. J Zool 216:565–573

Gillham ME (1963) Some interactions of plants, rabbits and seabirds on south African islands. J Ecol 51:275–294

Goodall RNP (1978) Report on the small cetaceans stranded on the coasts of Tierra del Fuego. Sci Rep Whales Res Inst 30:197–230

Harper GA, Bunbury N (2015) Invasive rats on tropical islands: their population biology and impacts on native species. Glob Ecol Conserv 3:607–627

Harris DB, Macdonald DW (2007) Interference competition between introduced black rats and endemic Galapagos rice rats. Ecology 88:2330–2344

Horwitz VD, Weissel MN (2011) Arqueología de Isla de los Estados: la frontera de la abundancia. In: Zangrando AF, Tessone A, Vázquez M (eds) Los Cazadores-Recolectores Del Extremo Oriental Fueguino: Arqueología de Península Mitre e Isla de Los Estados. Sociedad Argentina de Antropología, Buenos Aires, p 312

Imshaug HA (1972) R/V Hero cruise 71-5 to Isla de los Estados. Antarct J US 7(2):42–44

Jaramillo M, Donaghy-Cannon M, Vargas FH, Parker PG (2016) The diet of the Galapagos hawk (*Buteo galapagoensis*) before and after goat eradication. J Raptor Res 50(1):33–44

Jones HP, Tershy BR, Zavaleta ES, Croll DA, Keitt BS, Finkelstein ME, Howald GR (2008) Severity of the effects of invasive rats on seabirds: a global review. Conserv Biol 22(1):16–26

Kier G, Kreft H, Lee TM, Jetz W, Ibisch P, Nowicki C, Mutke J, Barthlott W (2009) A global assessment of endemism and species richness across Island and mainland regions. Proc Natl Acad Sci 106:9322–9327

Kossoff A, Schueller S, Nossan H, Slack I, Avramidis P, Foufopoulos J (2024) Ecological impacts of introduced European rabbits (*Oryctolagus cuniculus*) on Island ecosystems in the Mediterranean. Diversity 16(4):244

Kühnemann O (1976) Observaciones ecológicas sobre la vegetación marina y terrestre de la Isla de los Estados (Tierra del Fuego, Argentina). Ecosur 3(6):121–248

Latorre L, Larrinaga AR, Santamaría L (2013) Rats and seabirds: effects of egg size on predation risk and the potential of conditioned taste aversion as a mitigation method. PLoS One 8(9):e76138

Leader-Williams N, Scott TA, Pratt RM (1981) Forage selection by introduced reindeer on South Georgia, and its consequences for the flora. J Appl Ecol 18(1):83–106

Leader-Williams N, Smith RL, Rothery P (1987) Influence of introduced reindeer on the vegetation of South Georgia: results from a long-term exclusion experiment. J Appl Ecol 24:801–822

Long J (2003) Introduced mammals of the world: their history, distribution and influence. CSIRO, Collingwood

Lowe S, Browne M, Boudjelas S, De Poorter M (2000) 100 of the world's worst invasive alien species: a selection from the global invasive species database. The Invasive Species Specialist Group (ISSG), Auckland

Lyver PO, Moller H, Robertson CJ (2000) Predation of sooty shearwater *Puffinus griseus* colonies on the New Zealand mainland: is there safety in numbers? Pac Conserv Biol 5(4):347–357

Mack RN, Simberloff D, Lonsdale WM, Evans H, Clout M, Bazzaz FA (2000) Biotic invasions: causes, epidemiology, global consequences, and control. Ecol Appl 10(3):689–710. https://doi.org/10.1890/1051-0761(2000)010[0689:BICEGC]2.0.CO;2

Marchant R, Kefford BJ, Wasley J, King CK, Doube J, Nugegoda D (2011) Response of stream invertebrate communities to vegetation damage from overgrazing by exotic rabbits on subantarctic Macquarie Island. Mar Freshw Res 62(4):404–413

Mark AF, Baylis GTS, Dickinson KJM (1991) Monitoring the impact of deer on vegetation condition of Secretary Island, Fiordland National Park, New Zealand: a clear case for deer control and ecological restoration. J R Soc N Z 21:43–54

Martin JL, Stockton SA, Allombert S, Gaston AJ (2010) Top-down and bottom-up consequences of unchecked ungulate browsing on plant and animal diversity in temperate forests: lessons from a deer introduction. Biol Invasions 12(2):353–371. https://doi.org/10.1007/s10530-009-9628-8

Massoia E, Chebez JC (1993) Mamíferos silvestres del archipiélago fueguino. L.O.L.A, Buenos Aires

McCreless EE, Huff DD, Croll DA, Tershy BR, Spatz DR, Holmes ND, Butchart SHM, Wilcox C (2016) Past and estimated future impact of invasive alien mammals on insular threatened vertebrate populations. Nat Commun 7:1–11. https://doi.org/10.1038/ncomms12488

Nuñez MA, Simberloff D (2005) Invasive species and the cultural keystone species concept. Ecol Soc 10:r4

Parker IM, Simberloff D, Lonsdale WM, Goodell K, Wonham M, Kareiva PM et al (1999) Impact: toward a framework for understanding the ecological effects of invaders. Biol Invasions 1:3–19

Payne RJ, Ring-Hrubesh F, Rush G, Sloan TJ, Evans CD, Mauquoy D (2019) Peatland initiation and carbon accumulation in The Falkland Islands. Quat Sci Rev 212:213–218. https://doi.org/10.1016/j.quascirev.2019.03.022

Payró RJ (1898) La Australia Argentina: excursión periodística a las costas patagónicas, Tierra del Fuego é Isla de los Estados; con una carta-prólogo del general Bartolomé Mitre. Centro Editor de America Latina, Buenos Aires

Phillips RA, Gales R, Baker GB et al (2016) The conservation status and priorities for albatrosses and large petrels. Biol Conserv 201:169–183. https://doi.org/10.1016/j.biocon.2016.06.017

Pine RH, Angle JP, Bridge D (1978) Mammals from the sea, mainland and islands at the southern tip of South America. Mammalia 42(1):105–114. https://doi.org/10.1515/mamm.1978.42.1.105

Raya Rey A, Balza U, Domato I, Salom A, Pütz K (2024) Breeding confirmation and habitat utilization of Sooty Shearwater (*Ardenna grisea*) on Isla de los Estados. Argentina Polar Biology 47(10):1159–1164

Reaser JK, Meyerson LA, Cronk Q, De Poorter M, Eldrege LG, Green E, Kairo M, Latasi P, Mack RN, Mauremootoo J, O'Dowd D, Orapa W, Sastroutomo S, Saunders A, Shine C, Thrainsson S, Vaiutu L (2007) Ecological and socioeconomic impacts of invasive alien species in Island ecosystems. Environ Conserv 34(2):98–111. https://doi.org/10.1017/S0376892907003815

Rexer-Huber K, Bildstein KL (2013) Winter diet of striated caracara *Phalcoboenus australis* (Aves, Polyborinae) at a farm settlement on The Falkland Islands. Polar Biol 36:437–443. https://doi.org/10.1007/s00300-012-1266-4

Ricciardi A, Hoopes MF, Marchetti MP, Lockwood JL (2013) Progress toward understanding the ecological impacts of nonnative species. Ecol Monogr 83(3):263–282

Rivera-Parra JL, Levenstein KM, Bednarz JC, Vargas FH, Carrion V, Parker PG (2012) Implications of goat eradication on the survivorship of the Galapagos hawk. J Wildl Manag 76(6):1197–1204

Rodriguez-Cabal MA, Barrios-Garcia MN, Amico GC, Aizen MA, Sanders NJ (2013) Node-by-node disassembly of a mutualistic interaction web driven by species introductions. Proc Natl Acad Sci 110(41):16503–16507. https://doi.org/10.1073/pnas.1300131110

Russell JC, Horn SR, Miskelly CM, Sagar RL, Taylor RH (2020) Introduced land mammals and their impacts on the birds of the subantarctic Auckland Islands. Notornis 67(1):247–268

Salom A (2025) Herbívoros introducidos en Isla de los Estados: estado poblacional e impactos producidos por el ciervo colorado (*Cervus elaphus*) y la cabra (*Capra hircus*). Dissertation, Universidad de Buenos Aires

Salom A, Borrelli L, Testoni D, Ramos DP, Raya Rey A, Biganzoli F (2025a) Forage selection overlap between co-occurring introduced ungulates: insights for conservation management of a sub-Antarctic Island. Mamm Biol 105:403–414

Salom A, Cano J, Testoni D, Mansilla R, Mestre L, Collado L, Biganzoli F (2025b) Land cover map of Isla de los Estados, Año Nuevo islands (Observatorio, Goffré, Zeballos and Elizalde Islands) and adjacent islets. Consejo Nacional de Investigaciones Científicas y Técnicas. Figshare. Figure https://doi.org/10.6084/m9.figshare.29391008.v1

Sánchez J, Teta PV, Poljak S, Lanusse L, Balza U, Salom A, Lois NA, Lizarralde MS (2020) Rediscovery of mouse of Chuanisin, *Akodon llanoi* Pine 1976 (Rodentia, Cricetidae), with comments on its conservation status. Notas sobre Mamíferos Sudamericanos 2:1–8

Sax DF, Gaines SD (2008) Species invasions and extinction: the future of native biodiversity on islands. Proc Natl Acad Sci 105(Suppl 1):11490–11497. https://doi.org/10.1073/PNAS.0802290105

Schiavini A, Raya Rey A (2001) Aves y mamíferos marinos en Tierra del Fuego: estado de situación, interacción con actividades humanas y recomendaciones para su manejo (ed. Natural FP). Puerto Madryn

Scott JJ, Kirkpatrick JB (2008) Rabbits, landslips and vegetation change on the coastal slopes of subantarctic Macquarie Island, 1980–2007: implications for management. Polar Biol 31:409–419. https://doi.org/10.1007/s00300-007-0367-y

Shaw J, Terauds A, Bergstrom D (2011) Rapid commencement of ecosystem recovery following aerial baiting on sub-Antarctic Macquarie Island. Ecol Manag Restor 12(3):241–244. https://doi.org/10.1111/j.1442-8903.2011.00611.x

Shiels AB (2011) Frugivory by introduced black rats (*Rattus rattus*) promotes dispersal of invasive plant seeds. Biol Invasions 13:781–792. https://doi.org/10.1007/s10530-010-9868-7

Simberloff D, Martin JL, Genovesi P, Maris V, Wardle DA, Aronson J, Courchamp F, Galil B, García-Berthou E, Pascal M, Pyšek P (2013) Impacts of biological invasions: what's what and the way forward. Trends Ecol Evol 28(1):58–66

Smith RL (1985) Growth and production of *Poa flabellata* in relation to soil nutrient status and exposure at South Georgia. In: Antarctic nutrient cycles and food webs. Springer, Berlin, Heidelberg, pp 221–228. https://doi.org/10.1007/978-3-642-82275-9_31

Spatz DR, Jones HP, Bonnaud E, Kappes P, Holmes ND, Guzmán YB (2023) Invasive species threats to seabirds. In: Young L, VanderWerf E (eds) Conservation of marine birds. Academic, London, pp 97–130

Spear D, Chown SL (2009) Non-indigenous ungulates as a threat to biodiversity. J Zool 279(1):1–17. https://doi.org/10.1111/j.1469-7998.2009.00604.x

Speziale KL, Lambertucci SA (2013) The effect of introduced species on raptors. J Raptor Res 47(2):133–144

Strange IJ, Parry CJ, Parry MC, Wood RW (1988) Tussac grass in the Falklands. Falkland Islands Foundation, Brighton

Tershy BR, Shen H, Newton KM, Holmes ND, Croll DA (2015) The importance of islands for the protection of biological and linguistic diversity. Bioscience 65(6):592–597. https://doi.org/10.1093/biosci/biv031

Towns DR, Atkinson IAE, Daugherty CH (2006) Have the harmful effects of introduced rats on islands been exaggerated? Biol Invasions 8(4):863–891. https://doi.org/10.1007/s10530-005-0421-z

Turbott EG (1948) Effects of goats on Great Island, three kings, with description of vegetation quadrats. Records of the Auckland Institute Museum 3:253–272

Vairo CP (1997) La Isla de los Estados y el Faro del Fin del Mundo. Zaguier y Urruty Publications, Ushuaia, p 208

Valenzuela AEJ, Anderson CB, Ballari SA, Ojeda RA (2023) Introduced invasive mammals of Argentina, 1st edn. Sociedad Argentina para Estudio de los Mamíferos SAREM, Mendoza

Vazquez MS, Gonzalez DV, Amico GC (2023) Herbivores but not vegans: deer as nest predators. Austral Ecol 48(7):1460–1465

Veblen TT, Mermoz M, Martin C, Ramilo E (1989) Effects of exotic deer on forest regeneration and composition in northern Patagonia. J Appl Ecol 26(2):711–724. https://doi.org/10.2307/2404095

Vitousek PM, D'antonio CM, Loope LL, Rejmanek M, Westbrooks R (1997) Introduced species: a significant component of human-caused global change. N Z J Ecol 21:1–16

Weimerskirch H, Zotier R, Jouventin P (1989) The avifauna of the Kerguelen Islands. Emu 89(1):15–29

Young C (1981) Rabbit eradication on islands of the Western Australian coast. Swans 11:13–16

Zavaleta ES, Hobbs RJ, Mooney HA (2001) Viewing invasive species removal in a whole-ecosystem context. Trends Ecol Evol 16(8):454–459

Chapter 15
The Path Ahead: Stewardship of Isla de los Estados in a Changing World

Andrea Raya Rey and Ulises Balza

Abstract This final chapter reflects on Isla de los Estados (Chuani-sin) as both a rediscovered wilderness and a shared responsibility. Drawing from the fragmented yet growing body of knowledge presented throughout the volume, it frames the island as a living laboratory, a place where resilience, biodiversity, and ecological recovery still persist despite historical disturbance. It emphasizes the urgent need for stewardship in the face of climate change, invasive species, and emerging tourism pressures. Through imagined future scenarios, it challenges readers to move from observation to action. The chapter concludes with a call to researchers, policymakers, and citizens alike: the future of Isla de los Estados is not yet written, it depends on the choices we make.

Keywords Subantarctic ecosystem · Socioecological system · Resilience · Management plan · Science based conservation

This book has carried us through windswept cliffs, beneath underwater forests, into hidden nesting burrows, and across fog-veiled landscapes. Each chapter has pulled back the curtain on a place long seen only through the haze, what Darwin once described as a mysterious land glimpsed "amidst the clouds". In rediscovering Isla de los Estados (Chuani-sin), we have moved beyond myth into memory, beyond

A. Raya Rey (✉)
Centro Austral de Investigaciones Científicas (CADIC-CONICET), Ushuaia, Tierra del Fuego, Argentina

Instituto de Ciencias Polares, Ambiente y Recursos Naturales, Universidad Nacional de Tierra del Fuego, Ushuaia, Tierra del Fuego, Argentina

Wildlife Conservation Society, Representación Argentina, Buenos Aires, Argentina
e-mail: arayarey@cadic-conicet.gob.ar

U. Balza
Centro Austral de Investigaciones Científicas (CADIC-CONICET), Ushuaia, Tierra del Fuego, Argentina

A. Raya Rey, U. Balza (eds.), *Biodiversity and Conservation of Isla de los Estados (Chuani-sin)*, Natural and Social Sciences of Patagonia,
https://doi.org/10.1007/978-3-032-13630-5_15

outline into detail, transforming *terra incognita* into a land known not only through science, but through history, presence, and care.

The knowledge gathered here does not arrive easily. As we've seen, much of it came by chance: a bird observed while undertaking military service, a new species revealed by a diver on an improvised expedition, a discovery made while seeking shelter. Yet, taken together, these fragments offer a portrait of a place few have truly known. The island reveals itself slowly, reluctantly, often only to those who endure its harshness with humility and patience. And when it does, the view is breathtaking.

But now, with knowledge comes responsibility.

15.1 Guardianship in a Changing World

Isla de los Estados occupies a strategic position within Argentina's geopolitical framework, especially in relation to the nation's sovereignty claims over the Malvinas Islands and Antarctica (Dicenta and Gerrard 2023; Gioffreda 2021).[1]

The forces reshaping our planet are not distant. Climate change alters ocean currents, shifts wind regimes, and redraws species' ranges, even here, at the end of the world (Risaro et al. 2022; Ponce and Fernández, Chap. 3, this volume). Invasive mammals, once inadvertently introduced by sealers and sailors, continue to disturb native plant communities and endanger nesting sites of sensitive species. Tourism, if unmanaged, could echo and upscale these disruptions in a modern form.

Yet amidst these pressures, the island has shown remarkable resilience. As Salom and Schiavini (Chap. 14, this volume) show, even modest exclusion experiments reveal how quickly native vegetation can recover once disturbances are removed. The recent attempt by the locally extinct King Penguin to breed again on the island (Balza et al., Chap. 12, this volume) stands as a powerful symbol of nature's potential return. But while such signs are encouraging, regeneration is not guaranteed. The recovery of native grasslands, for instance, depends on deliberate action—such as the management of invasive species and restoration efforts. The scars are real, but so is the promise of renewal. Realizing that promise, however, requires long-term political commitment and sustained financial and technical investment.

What fascinates and inspires us is that despite intermittent but active human presence throughout history (see Zangrando et al., Chap. 3, this volume), the island has retained much of its ecological integrity. It is as if the island forgets, softening the traces of extractive histories and offering a second chance. This resilience must not

[1] The Malvinas, the South Georgia and South Sanwdich Islands are non-Self-Governing Territories currently administered by the United Kingdom of Great Britain and Northern Ireland. A dispute exists between the Governments of Argentina and the United Kingdom concerning sovereignty over the Malvinas and the South Georgia and South Sandwich Islands. The Special Committee on decolonization of the United Nations recognizes Argentina's sovereignty claim and requests the Governments of Argentina and the United Kingdom to resume negotiations to find a peaceful solution to the sovereignty dispute.

be mistaken for invulnerability, but it gives us a window, perhaps a last one, to act meaningfully.

Imagine the year 2050. In one possible future, Isla de los Estados thrives, protected, understood, and accessible to all, especially to Fuegians, who live so close yet have long remained distant from it. We take no pride in the fact that, historically, access to the island has been a privilege reserved for a few. True conservation demands that more people, particularly those from nearby communities, could experience, value, and protect it firsthand.

Achieving this vision requires not only lasting political will, but also robust scientific knowledge, knowledge that is globally relevant, yet grounded in local realities. This book was largely written by Argentine scientists from the public science system, a system currently under serious threat (Carignano and Jaworski 2019; Debat 2025). The road to 2050 will undoubtedly pass through political and economic uncertainty. But with sustained scientific monitoring and community-led conservation, Isla de los Estados could become a model for protecting other fragile and remote places around the world.

These are not fantasy; they are plausible trajectories. The choices we make now will shape which path unfolds. That includes the choice of doing nothing.

Listen closely: the island speaks. In gull cries, in waves against stone, in the flight of a Sooty Shearwater overhead. In the quiet regeneration of a patch of moss where deer no longer tread. In the stories told by scientists, rangers, and sailors.

It speaks through the pages of this book, through every species profile, map, and observation log. It invites us to listen, not only as scholars or tourists, but as stewards.

15.2 From Knowledge to Action

This volume hopefully offers more than a catalog of facts, it offers a lens through which to see Isla de los Estados as both ancient and urgent, wild and wounded, resilient and vulnerable. It reminds us that knowing a place is never neutral. To observe is to engage. To document is to become responsible.

To researchers: may your curiosity be guided by respect, and your science by solidarity.

To policymakers: may you commit to long-term protection, grounded in local knowledge and informed by global science. Management plans and concrete actions are essential for lasting conservation.

To readers, especially the citizens of Tierra del Fuego: may you carry this island in your thoughts, and act in its name.

Because the island does not need saving, it needs guardians. It needs witnesses. It needs us.

At the edge of the known world, Isla de los Estados (Chuani-sin) reminds us of what still remains wild, and of what still can be restored. Its future does not depend on keeping it untouched, but on caring for it with intention. Thriving, here, means more than survival: it means ecosystems recovering through active restoration, and human presence guided by knowledge, respect, and sustainable use. With well-planned actions, from managing invasive species to enabling local communities to connect with and protect the island, Isla de los Estados can become a living example of how nature and people can flourish together.

References

Carignano HA, Jaworski JP (2019) Argentina's subpar investment in science. Science 363(6428):702–702. https://doi.org/10.1126/science.aaw3872

Debat H (2025) The collapse of Argentine science under Javier Milei: one year of unprecedented defunding. Info 3(1):25.007. https://doi.org/10.3145/infonomy.25.007

Dicenta M, Gerrard AC (2023) Ecotourism, infrastructures, and the drama of sovereignty on a border Island. J Lat Am Caribb Anth 28(4):298–309. https://doi.org/10.1111/jlca.12696

Gioffreda C (2021) Los espacios vitales del sur argentino: El Atlántico Sur y el futuro de la Antártida. URVIO 30:40–57. https://doi.org/10.17141/urvio.30.2021.4609

Risaro DB, Chidichimo MP, Piola AR (2022) Interannual variability and trends of sea surface temperature around southern South America. Front Mar Sci 9:829144. https://doi.org/10.3389/fmars.2022.829144

Correction to: Biodiversity and Conservation of Isla de los Estados (Chuani-sin)

Andrea Raya Rey and Ulises Balza

Correction to:
A. Raya Rey, U. Balza (eds.), *Biodiversity and Conservation of Isla de los Estados (Chuani-sin)*, Natural and Social Sciences of Patagonia, https://doi.org/10.1007/978-3-032-13630-5

The original versions of the Chapters 5, 9, and 12 were inadvertently published before the incorporation of the final corrections. These chapters have now been updated with the following important changes:

Chapter 5: On page 79, the inappropriate inclusion of an interrogation symbol "?" at the end of last line in the "Abstract" section has been removed.

Chapter 9: Table 9.1, reported for formatting issue spanning pages 161 to 174, has been corrected/revised and presented.

Chapter 12: The belated request to revise the existing text featured in Table 12.1 (page 229) from "Northern house wren" to "Southern house wren" has been implemented.

The updated version of this book can be found at https://doi.org/10.1007/978-3-032-13630-5

A. Raya Rey, U. Balza (eds.), *Biodiversity and Conservation of Isla de los Estados (Chuani-sin)*, Natural and Social Sciences of Patagonia,
https://doi.org/10.1007/978-3-032-13630-5_16

MIX
Papier aus verantwortungsvollen Quellen
Paper from responsible sources
FSC® C105338

If you have any concerns about our products,
you can contact us on
ProductSafety@springernature.com

In case Publisher is established outside the EU,
the EU authorized representative is:
Springer Nature Customer Service Center GmbH
Europaplatz 3, 69115 Heidelberg, Germany

Printed by Libri Plureos GmbH
in Hamburg, Germany